Introduction to Spatial Mapping of Biomolecules by Imaging Mass Spectrometry

Introduction to Spatial Mapping of Biomolecules by Imaging Mass Spectrometry

Bindesh Shrestha, Ph.D.

ELSEVIER

Elsevier
Radarweg 29, PO Box 211, 1000 AE Amsterdam, Netherlands
The Boulevard, Langford Lane, Kidlington, Oxford OX5 1GB, United Kingdom
50 Hampshire Street, 5th Floor, Cambridge, MA 02139, United States

British Library Cataloguing-in-Publication Data
A catalogue record for this book is available from the British Library

Library of Congress Cataloging-in-Publication Data
A catalog record for this book is available from the Library of Congress

ISBN: 978-0-12-818998-6

Publisher: Susan Dennis
Acquisitions Editor: Kathryn Eryilmaz
Editorial Project Manager: Lena Sparks
Production Project Manager: Bharatwaj Varatharajan
Cover Designer: Miles Hitchen

Typeset by Aptara, New Delhi, India

Contents

Preface

The idea about an introductory book on imaging mass spectrometry (MS) germinated when I first talked to Kathryn Eryilmaz from Elsevier at the annual American Society for Mass Spectrometry Conference. I mainly told her about what I wanted to see in such a book from the perspective of many scientists who are interested in bringing the imaging MS to their labs to benefit their research. Many of my suggestions were based on what I noticed in my job as a subject matter expert for imaging MS working with other scientists, from experts to newly curious. Most of them needed a simple guide that encompassed various aspects of imaging MS.

Later, when I was approached with writing or editing one, I took some time to think mainly because my last book on *Single Cell Metabolism* took more effort than anticipated and took quite a long time to finish. I wrote a book proposal on my initial thoughts on the book. The proposal was sent to several reviewers. Thank you for the constructive suggestions that prompted me to reframe the focus of my book. The book got a slow start, mostly because I kept revising the style and type of content. Additionally, as a new dad of twin boys, time was scarce. However, when not distracted, I got a lot done on the weekends and super early mornings before the world around me slowly woke up.

As COVID-19 hit and life around me changed, I had more time during the weekend. I was on temporary leave for a few months, which turned out to be the most productive time for writing this book. Most of the book was finished during that time. Now, I see the value of a sabbatical in an academic setting. One of the simple routines of meeting with the Elsevier project manager, Lena Sparks, periodically helped in pushing the book along. I made a Gantt chart with a progress bar for each chapter—a trick I learned when I was a Research Director for a DARPA project.

I am delighted with the final result. The aim of the book is to give a comprehensive introduction to many aspects of imaging MS for biomedical applications. The book is not an encyclopedia for everything on imaging MS or a compilation of detailed step-by-step protocols. I encourage readers to consult other reviews published each year on various aspects of imaging MS and, if more interested, dive deep into the original publication. The imaging MS community is still small, and most of us, when you ask for help, more than willing to exchange ideas and collaborate. I have aimed to include fundamental topics that are used for biomedical imaging applications and wrote in the aspect of a brief review and perspective on the topic.

One of my favorite outcomes of this book was that I got to learn a lot also while working on the book. Some topics, such as ionization, I was extremely familiar with, but many others I learned as I went along. It was a rewarding experience.

My parents, Govinda Ram and Sharda, and wife Lauren have been very supportive during the writing of the book. Thank you for your support and love. Also, thank you for your love, my daughter, Munchie, who is a happy pug, and my toddler twin boys, Kai and Neel.

Bindesh Shrestha
Swampscott, MA

Chapter 1

Fundamentals of imaging mass spectrometry

Chapter outline

Imaging mass spectrometry (MS) is an analytical technique that can visualize the distribution of molecules using a tool called a mass spectrometer. In the case of biological samples, the visualized molecules include any species that can be detected by the mass spectrometer, such as metabolites, drugs, lipids, peptides, and proteins.

A mass spectrometer is an analytical tool that can measure the mass-to-charge ratio (m/z) of ions of molecules detected in a sample. The molecular analysis of bulk biological samples is done after the sample is homogenized, extracted, or processed according to the molecule-of-interest. The processed aliquot is injected into liquid chromatography for separation. The separated molecules are ionized using an electrospray mechanism and detected by a mass spectrometer. The detected m/z ion intensities are assigned to molecules and examined for the presence of one or multiple molecules or their up/downregulation. This general bulk analysis workflow template is widely used in all MS-based omics analyses, such as metabolomics, proteomics, as well as in quantitative and qualitative molecular analyses. Such bulk tissue analysis can provide information on the identity and the concentration of molecule-of-interest in a bulk sample but does not provide insight into the location of a molecule within the sample. In the imaging MS workflow, a thin slice of tissue is sectioned instead of homogenization of the chunk of tissue. The sectioning is followed by sample preparation steps such as matrix application, and finally, the sample is analyzed by desorption/sampling and ionization using a mass spectrometer. A comparison of workflow between bulk analysis of tissue and imaging MS is illustrated in Fig. 1.1.

Principles of imaging mass spectrometry

In all imaging MS techniques, molecules are extracted from the sample surface, such as tissue sections. The extracted material is ionized by an "ionization

Introduction to Spatial Mapping of Biomolecules by Imaging Mass Spectrometry. DOI: 10.1016/C2018-0-03962-6

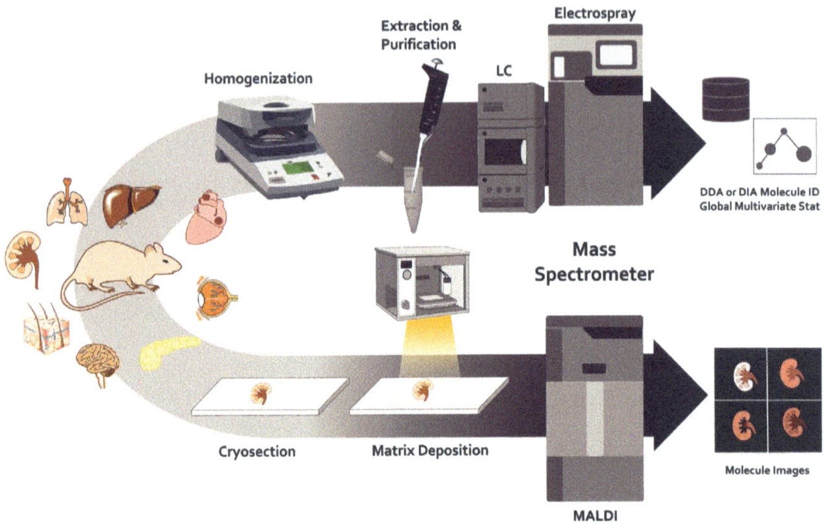

FIG. 1.1 A comparison between bulk tissue analysis workflow using electrospray liquid chromatography–mass spectrometer and tissue imaging using matrix-assisted laser desorption/ionization mass spectrometer.

source" as intact molecular species, or as their fragments, or tagged surrogates, and then detected by a "mass analyzer" component. The visual distributions of detected molecules are created by aligning the intensity of ions with the location of their ionization. Spatial imaging by mass spectrometers dates back to the 1960s and 1970s using secondary ions or laser as a desorption and ionization source.[35,36] There are several reviews on imaging MS, some of the selected reviews from the last 10 years are included in the references herein.[1–27]

In an overwhelming majority of the imaging MS techniques, spatial distribution is constructed by analyzing one pixel at a time by an ionization source

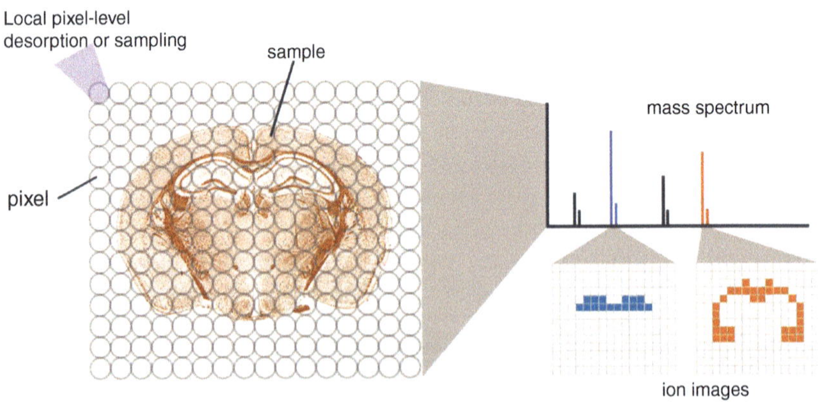

FIG. 1.2 General representation of imaging MS workflow, where a tissue section such as mouse brain section is interrogated pixel-by-pixel producing mass spectrum for each pixel, and the ion intensity distribution for each ion in each pixel is plotted as a false-color image.

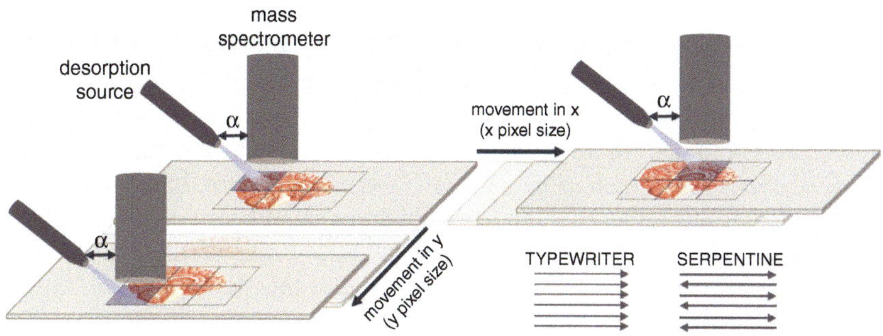

FIG. 1.3 During the imaging MS acquisition, the molecules on the tissue sections are sampled by a focused desorption mechanism. By moving the sample stage in a prescribed manner, such as in typewriter or serpentine modes, spatial information of molecules on each pixel can be obtained.

capable of regional extraction or desorption, and often, ionization of molecules from a sample. A pictorial representation of the imaging MS is given in Fig. 1.2. Pixel-by-pixel analysis is usually achieved by moving samples placed on a two-dimensional translational stage, as depicted in Fig. 1.3. The pixel-by-pixel movement can either be in a typewriter mode, where acquisition is made in one lateral direction, such as left-to-right or top-to-bottom or vice versa. Alternatively, acquisition can be made in a serpentine manner where acquisition switches between both lateral directions.

Each pixel corresponds to an *x–y* coordinate location on the sample and has a unique mass spectrum. A mass spectrum consists of values of detected ions, represented by a mass-to-charge ratio (*m/z*), and their corresponding intensities. As a simple illustration, the top portion of Fig. 1.4 has four mass spectra consisting of three same ions with *m/z* ratios of 123, 456, and 789. The heights of

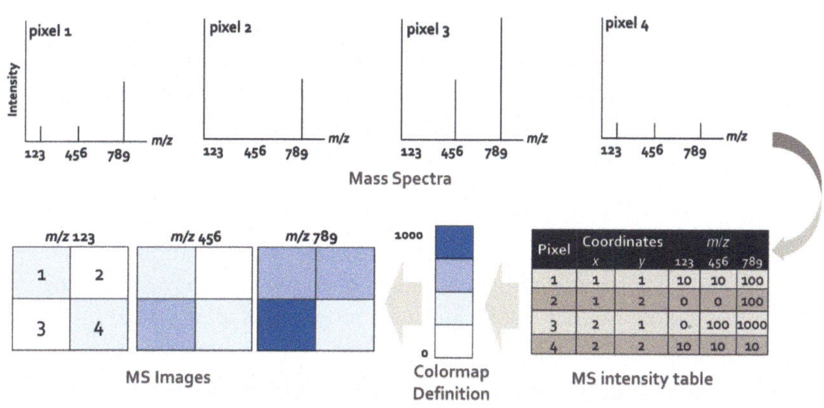

FIG. 1.4 A conceptual framework of imaging MS data is illustrated by a four-pixel image consisting of three ions. Mass spectra are generated for each pixel can be converted to ion intensities of all the ions at each pixel. Finally, MS images for all ions are obtained by correlating their ion intensities on each pixel with a colormap definition.

each ion in mass spectra plots denote their intensities. This plot can also be represented as an intensity table. MS images are created by assigning a false-color intensity for each ion at each pixel, as shown in the figure. The pixel-by-pixel imaging MS workflow is most commonly used and also sometimes referred to as microprobe mode. In an alternative instrumental setup, called microscope mode, the ions are imaged using a position-sensitive detector. In theory, microprobe imaging instruments should have higher spatial because the resolution is not dictated by the optical properties of the laser beam but the ion optics of the mass spectrometer with the ability to image below the diffraction limit.[28] Microscope mode can potentially have higher throughput but often lower mass resolving power.[29–30] Almost all of the imaging MS studies discussed in this book and the literature is done pixel-by-pixel or in microprobe mode.

In principle, each pixel of imaging MS data can be considered as an independent MS experiment. Like any MS experiments, we can either aim to detect all the ions in the acquired mass range, called here as profiling MS imaging, or aim a selected mass or fragment, named here as targeted MS imaging. In profiling MS imaging experiments, all the ions above a threshold detection limit in that biological microenvironment are detected. In this manner, the spatial distribution of hundreds to thousands of molecules can be obtained without a strict selection criterion, labeling, or a priori knowledge. It should be noted that there is a degree of inherent selection due to the choice of sample preparation workflow, experimental parameters or conditions, or type of instrumentation.

Advantages and limitation of imaging mass spectrometry

There are many advantages of imaging MS over other molecular imaging techniques. Most of the imaging MS workflow does not require any label, probe, or tracer. Unlabeled imaging allows for visualization of molecules that cannot be labeled, as well as makes it possible for the a priori discovery experiments where we do not know what molecules are present. Multiplex molecular imaging, i.e., imaging of multiple molecules in a single experiment, is not only possible, but it is a norm in the majority of imaging MS experiments. Fig. 1.5 shows a panel of desorption electrospray ionization (DESI) MS images of sagittal brain section of three ions simultaneously imaged in the same acquisition without any labeling.

Imaging MS is capable of providing a molecular visualization of a specific molecule instead of an indication for the presence of a class of molecules. For example, Fig. 1.6 shows the optical image of the whole-body mouse section (a), corresponding DESI MS/MS image of propranolol after 1 hour of dosing (b), optical image after 1 hour of dosing with [3H] propranolol (c), and corresponding whole-body autoradiography image (d). DESI showed the drug was detected in the brain, lung, and stomach regions. The autoradiogram also showed the detection but cannot differentiate between the drug or its radiolabeled metabolites.[31] Another example is an autoradiogram map of the sagittal

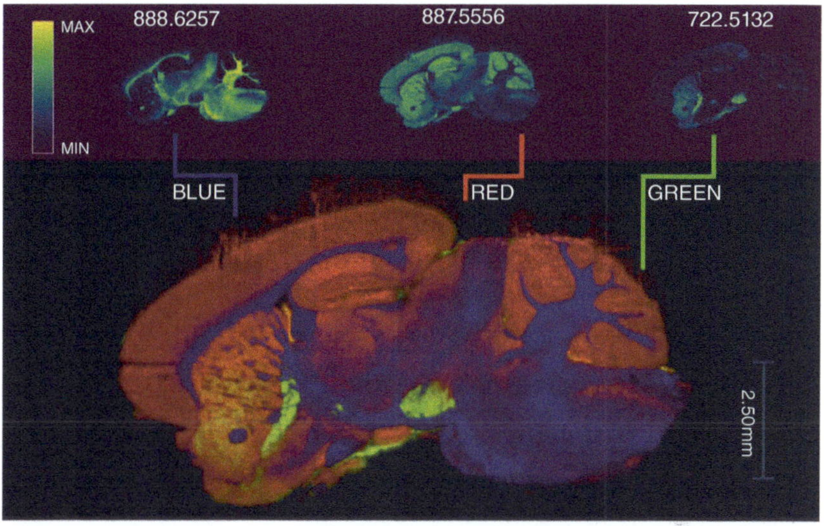

FIG. 1.5 Visualization of three molecular lipid ions using desorption electrospray ionization (DESI) imaging mass sagittal rodent brain section without any labeling is shown on top. The bottom shows an ion overlay image of the same three ions showing contextual distribution.

kidney section of mice after subcutaneous administration of carbon-14 isotope of a drug shows the distribution of the drug and two metabolites of drug that retained the radiolabel unchanged between two timepoints.[32] In contrast, matrix-assisted laser desorption/ionization (MALDI) images of the similar tissue show the drug was predominantly detected in the outer medulla and cortex after 30 minutes of dose. However, after 2 hours, the drug was detected only in the inner medulla, and one metabolite was predominantly detected in the cortex in both time points, but with higher abundance at the later timepoint. All MALDI MS images were acquired without any labeling. Imaging MS also has a high dynamic range spanning over several decades of concentration and is quantitative. Quantitative analysis of imaging MS is discussed more in detail in another chapter.

Some of the most common utility of imaging MS applications, until early 2020, were found by examining the keyword of imaging MS literature in the National Institutes of Health PubMed database.[33] Aside from the general imaging terms, the most popular author keywords in the order of the highest frequency are proteomics, metabolomics, lipids, biomarkers, lipidomics, cancer, phospholipids, brain, pharmacokinetics, Alzheimer's disease, atherosclerosis, cell imaging, and metabolism. If we count the most frequently appearing words in the abstract as well, they are humans, animals, male, female, mice, proteomics, rats, middle-aged, proteome, brain, adult, biomarkers. A pictorial cloud representation of this is given in Fig. 1.7. From this simple analysis, we can gauge that imaging MS has been mostly focused on biomolecular or pharmacokinetics analysis of tissue in diseases such as cancer.

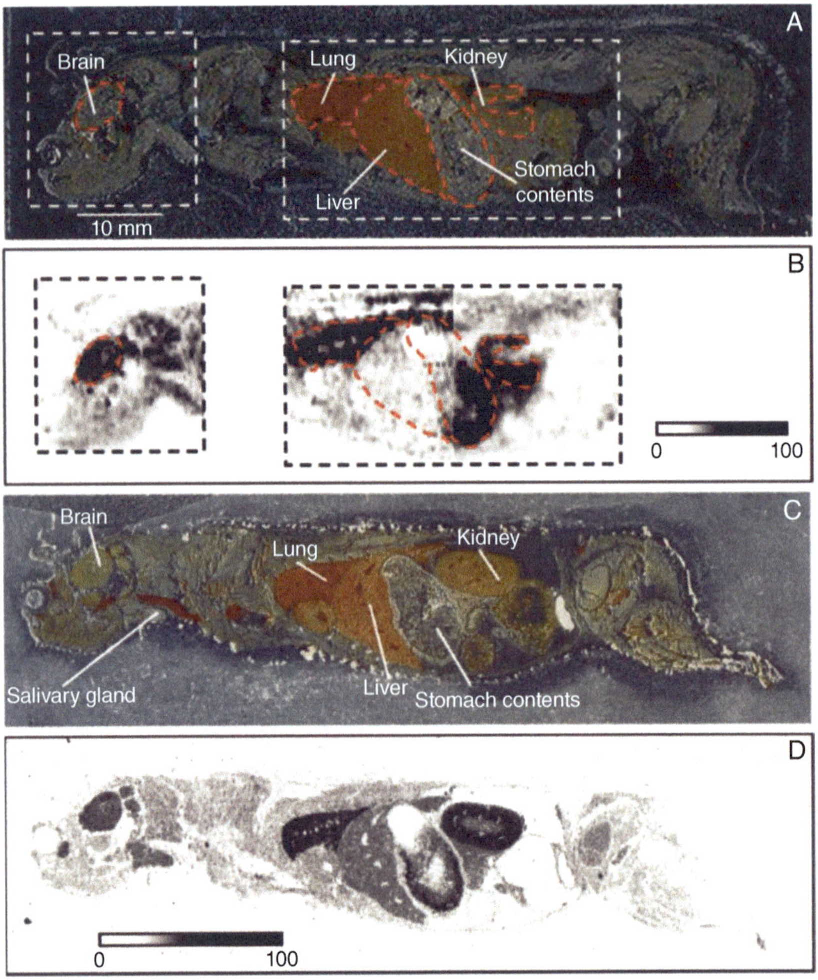

FIG. 1.6 A comparison of desorption electrospray ionization (DESI) mass spectrometry image (B) with whole-body autoradiography (WBA) image is shown with their respective optical image at (A) and (C). Adapted with permission from Kertesz V, Van Berkel GJ, Vavrek M, Koeplinger KA, Schneider BB, Covey TR. *Anal Chem*. 2008;80:5168–5177. Copyright 2007 American Chemical Society.

Most of the advantages and limitations of imaging MS stems from being a mass spectrometric technique. The major limitation is that the molecule needs to be ionized and detected at the physiological quantity present in the ex vivo biological samples. Depending on the goal of the imaging experiment, the requirement for MS imaging applications may differ a lot. Higher requirements usually mean more time or resources. Fig. 1.8 depicts an example of subjective requirements for two hypothetical-related applications, whole-body imaging of cancer drug and diagnosis of a cancerous tumor by imaging MS. Drug imaging would require a confident molecular identification of the drug and its metabolites, as well as, higher quantitative assay. In contrast, cancer diagnosis would require

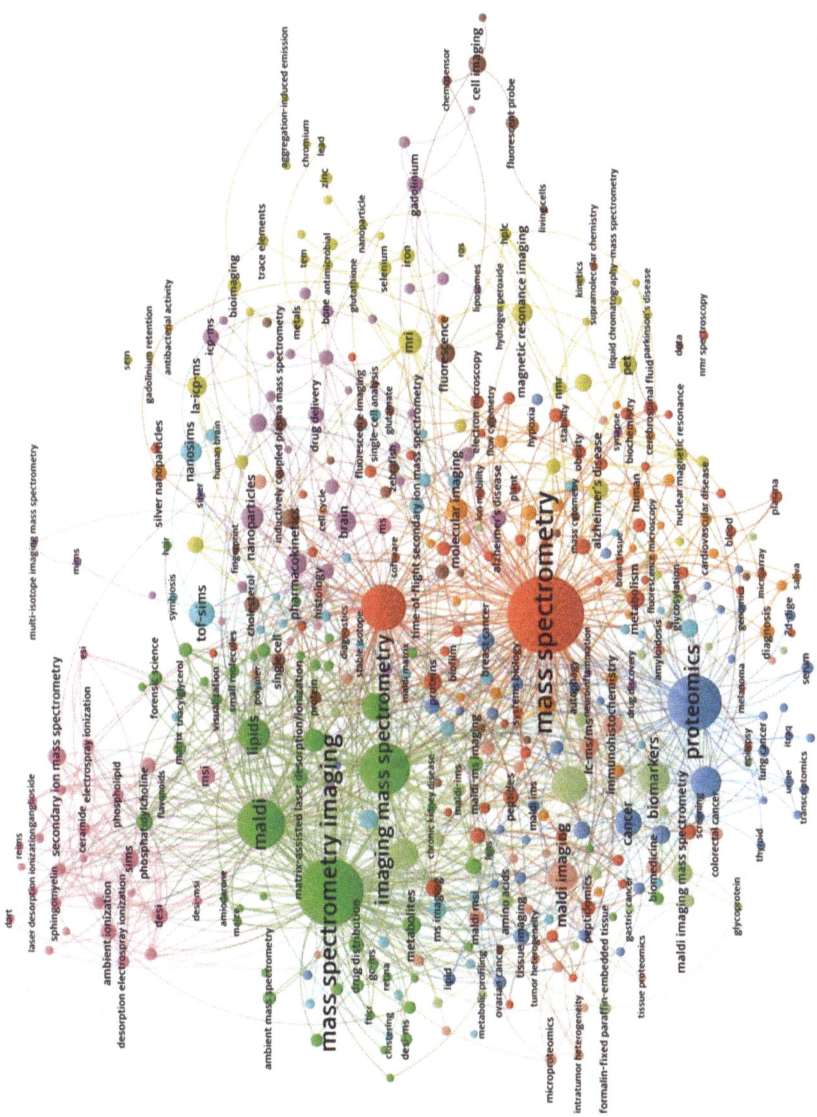

FIG. 1.7 Bibliometric map of the author's keyword of imaging mass spectrometry in National Institutes of Health PubMed data, created using VOSviewer software.[33–34]

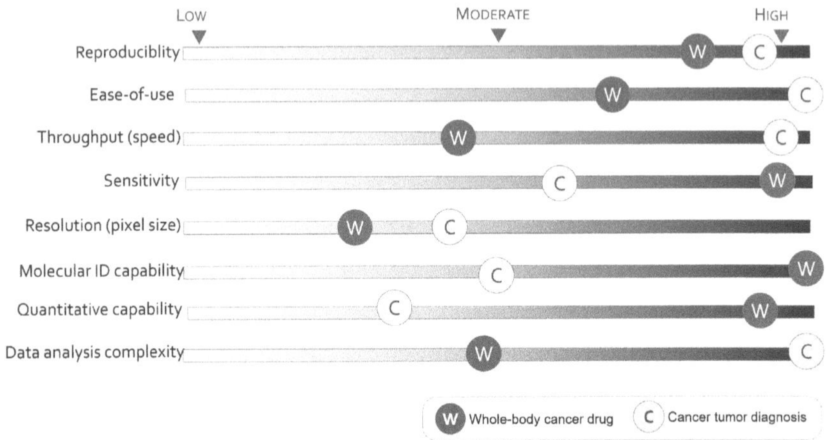

FIG. 1.8 Subjective requirements for two hypothetical biomedical imaging MS applications, whole-body cancer drug imaging, and diagnosis of a cancerous tumor by MS imaging are presented as examples. The requirements may drastically change depending on the specific application on hand.

higher throughput and must be very easy to use for nonexperts. Imaging MS has a wide variety of applications than these two hypothetical examples. However, with each set of unique applications, the required toolsets for imaging MS will change. Luckily, the research and development in imaging MS are proliferating and improving the availability of better protocols, software, and hardware. Imaging MS, as an analytical tool, has a unique ability to analyze thousands of molecules in biological systems without any a priori labeling. Some of those molecules cannot be directly mapped using any other tools. With broader adoption of imaging MS, we can anticipate improvements in reproducibility from the sample preparation to data acquisition to data analysis leading to the more widespread use of imaging MS in biomedical research followed by translational research and ultimately in routine clinical usage.

References

1. van Hove ERA, Smith DF, Heeren RM. A concise review of mass spectrometry imaging. *J Chromatogr A*. 2010;1217:3946–3954.
2. Schwamborn K, Caprioli RM. MALDI imaging mass spectrometry–painting molecular pictures. *Mol Oncol*. 2010;4:529–538.
3. Castellino S, Groseclose MR, Wagner D. MALDI imaging mass spectrometry: bridging biology and chemistry in drug development. *Bioanalysis*. 2011;3:2427–2441.
4. Seeley EH, Caprioli RM. MALDI imaging mass spectrometry of human tissue: method challenges and clinical perspectives. *Trends Biotechnol*. 2011;29:136–143.
5. Vickerman JC. Molecular imaging and depth profiling by mass spectrometry—SIMS, MALDI or DESI?. *Analyst*. 2011;136:2199–2217.
6. Goto-Inoue N, Hayasaka T, Zaima N, Setou M. Imaging mass spectrometry for lipidomics. *Biochimica et Biophysica Acta (BBA) - Molecular and Cell Biology of Lipids*. 2011;1811:961–969.

7. Norris JL, Caprioli RM. Analysis of tissue specimens by matrix-assisted laser desorption/ionization imaging mass spectrometry in biological and clinical research. *Chem Rev.* 2013;113:2309–2342.
8. Wu C, Dill AL, Eberlin LS, Cooks RG, Ifa DR. Mass spectrometry imaging under ambient conditions. *Mass Spectrom Rev.* 2013;32:218–243.
9. Römpp A, Spengler B. Mass spectrometry imaging with high resolution in mass and space. *Histochem Cell Biol.* 2013;139:759–783.
10. Ferguson CN, Fowler JW, Waxer JF, Gatti RA, Loo JA. Mass spectrometry-based tissue imaging of small molecules*Advancements of Mass Spectrometry in Biomedical Research*: Springer; 2014:283–299.
11. Bodzon-Kulakowska A, Suder P. Imaging mass spectrometry: instrumentation, applications, and combination with other visualization techniques. *Mass Spectrom Rev.* 2016;35:147–169.
12. Schwamborn K, Kriegsmann M, Weichert W. MALDI imaging mass spectrometry—from bench to bedside. *Biochim Biophy Acta Proteins Proteom.* 2017;1865:776–783.
13. Baker TC, Han J, Borchers CH. Recent advancements in matrix-assisted laser desorption/ionization mass spectrometry imaging. *Curr Opin Biotechnol.* 2017;43:62–69.
14. Siegel TP, Hamm G, Bunch J, Cappell J, Fletcher JS, Schwamborn K. Mass spectrometry imaging and integration with other imaging modalities for greater molecular understanding of biological tissues. *Mol Imaging Biol.* 2018;20:888–901.
15. Buchberger AR, DeLaney K, Johnson J, Li L. Mass spectrometry imaging: a review of emerging advancements and future insights. *Anal Chem.* 2018;90:240–265.
16. Xue J, Bai Y, Liu H. Recent advances in ambient mass spectrometry imaging. *TrAC Trends Anal Chem.* 2019:115659.
17. Gilmore IS, Heiles S, Pieterse CL. Metabolic imaging at the single-cell scale: recent advances in mass spectrometry imaging. *Annu Rev Anal Chem.* 2019;12:201–224.
18. Xiao Y, Deng J, Yao Y, Fang L, Yang Y, Luan T. Recent advances of ambient mass spectrometry imaging for biological tissues: a review. *Anal Chim Acta.* 2020;1117:74–88.
19. Alexandrov T. Spatial metabolomics and imaging mass spectrometry in the age of artificial intelligence. *Annu Rev Biomed Data Sci.* 2020;3.
20. Aichler M, Walch A. MALDI Imaging mass spectrometry: current frontiers and perspectives in pathology research and practice. *Lab Investig.* 2015;95:422–431.
21. Norris JL, Caprioli RM. Imaging mass spectrometry: a new tool for pathology in a molecular age. *Proteomics Clin Appl.* 2013;7:733–738.
22. Spengler B. Mass spectrometry imaging of biomolecular information. *Anal Chem.* 2015;87:64–82.
23. Shariatgorji M, Svenningsson P, Andrén PE. Mass spectrometry imaging, an emerging technology in neuropsychopharmacology. *Neuropsychopharmacology.* 2014;39:34–49.
24. Johnson Jr RW, Talaty N. Tissue imaging by mass spectrometry: a practical guide for the medicinal chemist. *ACS Med Chem Lett.* 2019;10:161–167.
25. Samarah LZ, Vertes A. Mass spectrometry imaging based on laser desorption ionization from inorganic and nanophotonic platforms*View*; (n/a):20200063 n/a.
26. Spraker JE, Luu GT, Sanchez LM. Imaging mass spectrometry for natural products discovery: a review of ionization methods. *Nat Prod Rep.* 2020;37:150–162.
27. Goodwin RJA, Takats Z, Bunch J. A critical and concise review of mass spectrometry applied to imaging in drug discovery. *Slas Discov Adv Sci Drug Discov.* 2020, doi:2472555220941843
28. Luxembourg SL, McDonnell LA, Mize TH, Heeren RM. Infrared mass spectrometric imaging below the diffraction limit. *J Proteome Res.* 2005;4:671–673.

29. Luxembourg SL, Mize TH, McDonnell LA, Heeren RM. High-spatial resolution mass spectrometric imaging of peptide and protein distributions on a surface. *Anal Chem.* 2004;76:5339–5344.
30. Spengler B. MALDI-mass spectrometry imaging. In: Peter-Katalinic FHJ, ed. *MALDI MS: A Practical Guide to Instrumentation, Methods, and Applications.* 2nd ed.: Wiley-VCH Verlag GmbH & Co. KGaA; 2014:133–167.
31. Kertesz V, Van Berkel GJ, Vavrek M, Koeplinger KA, Schneider BB, Covey TR. Comparison of drug distribution images from whole-body thin tissue sections obtained using desorption electrospray ionization tandem mass spectrometry and autoradiography. *Anal Chem.* 2008;80:5168–5177.
32. Goodwin RJA, Nilsson A, Mackay CL, Swales JG, Johansson MK, Billger M, Andrén PE, Iverson SL. Exemplifying the screening power of mass spectrometry imaging over label-based technologies for simultaneous monitoring of drug and metabolite distributions in tissue sections. *J Biomol Screen.* 2015;21:187–193.
33. Data from PubMed. Accessed on July 2020.
34. Van Eck N, Waltman L. Software survey: VOSviewer, a computer program for bibliometric mapping. *Scientometrics.* 2010;84:523–538.
35. Liebl H. Ion microprobe mass analyzer. *Journal of Applied Physics.* 1967;38(13):5277–5283.2.
36. Wechsung R, Hillenkamp F, Kaufmann R, Nitsche R, Unsöld E, Vogt H. LAMMA - a new laser - microprobe - mass - analyzer. *Microsc Acta Suppl.* 1978 (2):281–296.

Chapter 2

Ionization sources for imaging mass spectrometry

Chapter Outline

Imaging mass spectrometry (MS) is a surface analytical technique that provides a two-dimensional (2D) spatial plot of ions. In imaging MS, chemicals on samples are extracted, ionized, and detected by a mass spectrometer. Ion source for imaging MS system needs to extract and ionize chemicals from a localized area of the sample. The localized area is referred to as a pixel. The extraction and ionization step can happen synchronously or disjointedly. In order to create an image, data are acquired at each pixel using the ion source, most often sequentially. Spatial distribution for an ion is constructed by plotting its intensity at each of the acquired pixels in alignment with extraction coordinates.

In addition to its ability to perform a localized sampling/ionization, for many applications, it is preferred that the ion source can soft ionize intact molecular ions. MS analysis of biomolecule leapfrogged after the adoption of the two soft ionization techniques, electrospray ionization and matrix-assisted laser desorption/ionization (MALDI), in the 1990s. In 2002, Nobel Prize in Chemistry was awarded to John Bennett Fenn and Koichi Tanaka for their work in electrospray ionization and soft laser desorption, respectively, in recognition of the breakthrough soft ionization *"for their development of soft desorption ionization methods for mass spectrometric analyses of biological macromolecules."*[1] Today, electrospray ionization has become the default ionization source in the majority of the MS assay. Electrospray enables the analysis of large biomolecules through the production of multiply charged ions by solubilizing the sample in a liquid phase. In contrast, MALDI analyzes biomolecules in a solid-phase sample.[2] MALDI is extensively used in high

Introduction to Spatial Mapping of Biomolecules by Imaging Mass Spectrometry. DOI: 10.1016/C2018-0-03962-6

throughput clinical setting for microbial analysis.[3] Most importantly, MALDI is well suited for local ionization due to the use of a focused laser beam as a desorption and ionization source.

Generally, only a single imaging ion source is available with a mass spectrometer due to technical, logistics, or commercial issues. Sometimes more than one imaging ion source is commercially integrated with a single mass spectrometer. For example, desorption electrospray ionization (DESI) and MALDI are available with a single quadrupole time-of-flight mass spectrometer with ion mobility separation.[4] Often, the existing ion source infrastructure of a mass spectrometer is modified to add another imaging modality. For example, picosecond infrared (IR) laser imaging technique was added to existing commercial DESI imaging ion source,[5] atmospheric pressure (AP) MALDI, and laser ablation electrospray ionization (LAESI) was installed on the same quadrupole time-of-flight mass spectrometer.[6]

There are several imaging MS ion sources, and all of them are not discussed here. If we review the scientific literature (until 2019) for imaging MS and ion source, with no surprise, MALDI has the highest number of publications, followed by secondary ion MS (SIMS). There is an increasing trend for the new breed of ion sources, known as ambient ionization sources (e.g., desorption electrospray ionization or DESI). The trends in the citations are shown in Fig. 2.1.

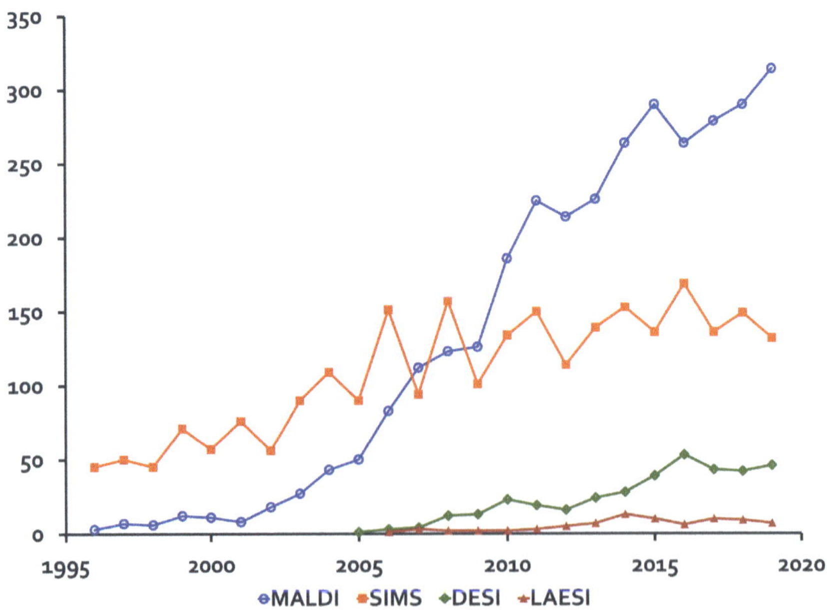

FIG. 2.1 Citation trends for ion source and imaging mass spectrometry show the domination of MALDI ion source for imaging.

Matrix-assisted laser desorption/ionization

MALDI is the most widely used imaging MS ion source. MALDI was first described by Franz Hillenkamp and colleagues in the mid-1980s, and imaging was introduced in the late 1990s by Richard Caprioli.[7–8] There has been quite an effort in the understanding mechanism of ion formation in MALDI, which is reviewed here.[9–10] In brief, primary ionization occurs during or shortly after the irradiance of an ultraviolet laser pulse (λ = 337 nm, 355 nm) followed by secondary reactions in the desorbed expanding plume. A simple schematic representation of the MALDI ion source is given in Fig. 2.2A. All commercial

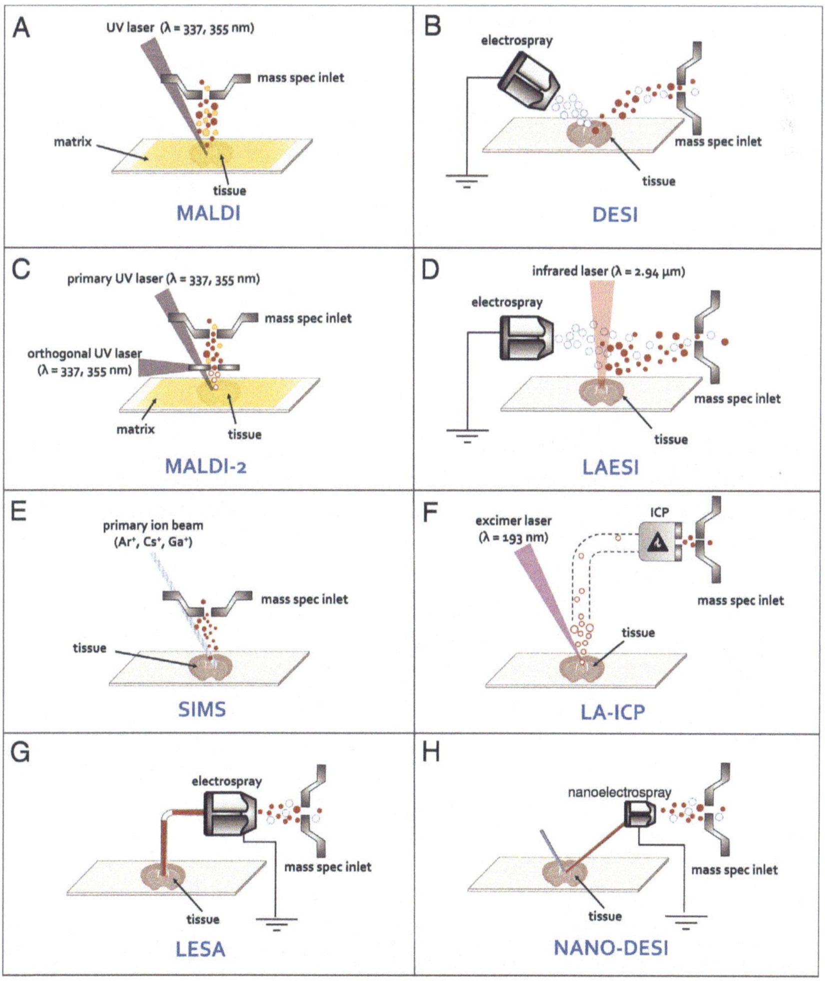

FIG. 2.2 Conceptual representation of ion sources used for imaging MS.

MALDI ion source utilizes ultraviolet irradiance, but some homebuilt systems have used other wavelength, such as IR laser pulse.[11–12]

In an overwhelming majority, laser desorption in MALDI is performed by focusing the beam on top of the sample, also referred to as reflection mode. In transmission mode, the beam can be focused via the back of the sample through the tissue slide to achieve a smaller laser spot size.[13] A detailed discussion of this is provided in another chapter on spatial resolution in imaging MS.

MALDI requires the application of an external compound called matrix onto the sample in order to ionize molecules. There are many types of matrices from graphene,[14] glycerol,[15] nanostructures,[16] (but most common are organic molecules such as 2,5-dihydroxybenzoic acid, -α-cyano-4-hydroxycinnamic acid, 1,5-diaminonapthalene, 9-aminoacridine. The application of matrix to sample surface extracts analyte from the sample surface and forms cocrystallized analyte and matrix crystals. The laser energy is absorbed by the matrix leading to the soft ionization of analytes. Many matrix compounds have been developed for a specific class of molecules. For instance, the matrix 5-chloro-2-mercaptobenzothiazole matrix has been found as an excellent matrix for the detection of gangliosides species in the brain tissue.[17] In addition to serving as ionization aid, some matrices react and derivatize analytes making them easier to be detected. For example, a fluoromethylpyridinium-based reactive matrix can covalently react to molecules containing phenolic hydroxyl or amine groups, such as neurotransmitters leading to imaging of low-abundance neurotransmitters in a brain tissue sample.[18] Matrix is applied to a sample is generally two ways; either matrix is sublimated onto the sample under vacuum or matrix sprayed onto the sample after dissolving in an organic-aqueous solvent.

MALDI started with protein and peptide imaging but quickly demonstrated the capability of lipid imaging.[19] In the early stage of its development, MALDI was not used for the analysis of small molecules, such as metabolites, often due to interferences from the applied organic matrix ions. Due to improvements in mass resolution of mass spectrometers and the development of new matrices, MALDI is now routinely used for imaging metabolites and other small molecules, such as drugs.[20] Depending on instrumentation, sample, and analyte, the spatial resolution or pixel size of MALDI hovers around tens of microns, making it just shy of mammalian single-cell imaging resolution.

Several matrix-free laser desorption ionization platforms that use MALDI sources, such as nanostructure-initiator mass spectrometry[21] and uniform silicon nanopost arrays,[16] have been developed for imaging MS. Instead of a coating sample with a matrix compound, the tissue section is placed on a specialized surface, such as nanostructures, to induce nanophotonic soft ionization. Some of these novel platforms have demonstrated enhanced ionization efficiency for specific class molecules. For instance, neutral lipids, which are hard to detect in MALDI, have been easily detected in tissues using nanopost arrays providing complementary molecular analysis.[22]

Desorption electrospray ionization

DESI is an ambient ionization source capable of performing direct surface analysis.[23] All ambient ionization source operates under ambient conditions, such as AP, requires little-to-no sample preparation, and able to ionize sample directly. Ions are extracted and ionized in concert from a sample surface in DESI when electrospray plume pneumatic assisted by a gas desorbs analyte from samples into the gas phase. The resulting ions analytes are carried into a mass spectrometer inlet capillary and transferred for analysis. Unlike MALDI, ambient ionization techniques, such as DESI, do not require matrix application and need to be introduced to a vacuum chamber for analysis. Most of the ambient ionization techniques like DESI can use standard AP mass spectrometer inlet developed for electrospray.

A conceptual schematic representation of DESI is given in Fig. 2.2B. The proposed DESI ionization mechanism involves the "droplet pickup" of the analyte, followed by ion formation by solvent evaporation. Initially, the sample surface, along with the analyte, is thinly coated by the solvent, followed by the collision of second sets of droplets that pickup and desorb the wetted analyte, and finally ion evaporation solvated droplets to gas-phase ions.[24–25]

DESI imaging MS is performed by obtaining a pixel-by-pixel mass spectra of a tissue by impinging the focused electrospray droplets on a confined point, such as a pixel. Thus, the spatial resolution (pixel size) of DESI imaging is primarily defined by the area of the electrospray impact on the sample surface. A stable and highly focused electrospray beam is capable of producing high spatial resolution DESI imaging experiments under 20 μm in diameter. Besides, these imaging sprayers need to generate a robust MS signal from the first pixel to the last millionth pixel across a few centimeter-square area. DESI has been utilized in two-dimensional molecular imaging of lipids in brain tissues,[26–27] breast cancer tissues.[28] DESI is a minimally destructive technique often allowing complementary analyses of the same sample by other imaging modalities postimaging.[29]

In terms of ionization, DESI is similar to electrospray. The soft ionization characteristics of DESI measured by internal energy distributions of its ions using the survival yield of thermometer ions showed similar internal energy distributions as electrospray—suggesting a similar ionization process.[30] A comparison of salt tolerance of DESI with that of electrospray ionization showed a better tolerance for DESI. The lesser degree of ion suppression is useful during the direct analysis of samples with high salt content such as tissue.[31] A post photoionization (PI) assembly using a portable krypton lamp was coupled with for imaging both polar and nonpolar compounds in tissue sections. DESI/PI had higher signal intensities for nonpolar molecules.[32] Compared to DESI, neutral biomolecules (e.g., creatine, cholesterol, and GalCer lipids) had higher signal in DESI/PI in the positive ion mode, as well as, glutamine, lipids (e.g., HexCer, PE, and PE-O) in negative ion mode. Similar to DESI, desorption electro-flow focusing ionization utilizes electro-flow focusing with DESI. In desorption electro-flow focusing ionization, a concentric low pressure laminar gas flow focuses a solvent through a recessed electrospray emitter. The electro-flow focusing produces a steady liquid jet smaller than the orifice capable of a higher spatial resolution imaging.[33–34]

MALDI-2

MALDI combined with laser-induced postionization (MALDI-2) is a recently introduced variant of MALDI.[35] MALDI-2 increases the sensitivity of suppressed molecules by postionization using a second orthogonal UV laser beam, which intersects the expanding MALDI plume at a short distance (0.5 mm) above the sample surface. In the proposed mechanism, the second postionization laser produces additional matrix ions while the expanding MALDI plume is decelerated, leading to a gas-phase reaction environment.[36] A newly charged matrix transfer is its charge to neutral molecules yielding up to two–threefold higher ionization for glycolipids, metabolites, pharmaceuticals, oligosaccharides, etc.[37–38] By increasing overall ion yield and sensitivity, MALDI-2 can map the chemical classes typically inaccessible by MALDI.[39] A conceptual schematic of MALDI-2 is given in Fig. 2.2C. Similar to MALDI, the MALDI-2 has also coupled with a transmission geometry optical configuration.[40] In transmission mode, the laser beam is focused through behind the sample to achieve reduced focal spot size for smaller pixel size. MALDI-2 helps improve sensitivity loss due to a decrease in spot size, enabling higher spatial resolution images.

Laser ablation electrospray ionization

LAESI analyzes samples by producing neutrals by the mid-IR laser ablation (LA) followed by ionization by an electrospray. There are other techniques similar to LAESI, such as IR matrix-assisted laser desorption electrospray ionization (MALDESI)[41] and electrospray laser desorption ionization (ELDI).[42] For our discussions here, all of the techniques are categorized under the term "LAESI." A schematic representation of the LAESI ion source is shown in Fig. 2.2D. LAESI has been utilized for the analysis of metabolites from various plant organs, human bodily fluids, and animal tissue.[6,43] LAESI coupled with fiber has been extensively used in the single-cell analysis of plant tissue [44–45] and has introduced the concept of cell-by-cell molecular imaging, where each cell was treated as a natural voxel or three-dimensional pixel for imaging MS.[46]

In LAESI, the neutrals are produced by a 2.94 μm wavelength IR laser pulse because of the strong absorption of the water molecules due to OH vibrations at that wavelength. This effectively makes native water content of any tissue as a matrix to couple the laser energy. The neutral particulate ejected by the recoil pressure in the ablation plume is intercepted by the electrospray and ionized. The ions are sampled inside an AP mass spectrometer inlet designed for electrospray. The ion yield in LAESI strongly depends on the properties of the electrospray, LA, and intercept geometry of ablation electrospray.

Laser ablation atmospheric pressure photoionization (LAAPPI) is another ambient ion source similar to LAESI, where samples are ablated using an IR laser running at 2.94 μm wavelength. In LAAPPI, the ablation plume is intercepted with a hot solvent jet (e.g., toluene or anisole) and ionized in the gas phase by AP PI with an ultraviolet krypton discharge lamp.[47] LAAPPI provided complementary molecular detection to electrospray-based ionization of LAESI. For example, LAAPPI imaging was able to detect larger nonpolar compounds, such as sesquiterpenes and triterpenoid derivatives, from the sage (*Salvia officinalis*) leaves.[48] Using novel IR beam focusing strategies, LAAPPI and LAESI are able to perform sub-100 μm spatial resolution imaging MS of the rodent brain.[49]

Secondary ion MS

SIMS produces ions by directing the primary ion beam at the sample surface and ejecting secondary ions for analysis. In SIMS, high energy primary ions made from ion clusters (e.g., C_{60}^{+},[50] water clusters[51]) or ion clusters (e.g., Ar^{+}, Ga^{+}) are bombarded toward the sample surface, penetrating the sample surface and inducing a collision cascade at the surface resulting in the release of secondary ions. The secondary ions are released when the kinetic energy is increased above the binding energy to a few Armstrong of substrate depth.[52] There are two types of strategies for the bombardment of the primary ion beam, static and dynamic. In static SIMS has lower ion doses than dynamic SIMS, making dynamic SIMS more destructive but also useful for depth profiling.[53]

Imaging is performed by rastering the primary ion beam across the tissue section. SIMS can produce a highly focused analytical beam and thus have a higher spatial resolution than MALDI or other ionization based on LA or fluid desorption. The primary ions of SIMS are not limited by the diffraction limit of laser light or constrained set by optical components. A conceptual representation of the SIMS ion source is provided in Fig. 2.2E. As an ion source, SIMS is suitable for imaging small molecules with *m/z* value less than a thousand due to extensive surface fragmentation. However, recently gas cluster ion beam SIMS has reported low chemical damage enabling the detection of intact molecular ions of metabolites, such as one associated with purine biosynthetic pathways, in single cells.[54]

Recently, SIMS ion source has been used in multiplexed ion beam imaging (MIBI), where it has been used to analyze metal-tagged antibodies that serve as surrogates for larger biomolecules, such as proteins like FOXP3, CD3 etc.[55] In MIBI, rare earth metals are conjugated with antibodies in a similar manner as fluorophores or chromogens. During analysis, the metals mass tags are liberated, ionized, and analyzed by mass spectrometer as a reporter for antibody-bound protein molecules. Rare earth metals, such as lanthanides, are employed to label antibodies because they do not naturally occur in typical biological specimens. The analysis of metals by mass spectrometer instead of detection of emission of fluorophore has many advantages—(a) larger multiplex capability due to a larger number of potential metal tags, (b) less spectral overlap of signal because of discrete metal ion signal in a mass spectrometer, (c) lack of autofluorescence background, and (d) larger dynamic range.[56] MIBI imaging has been able to generate a molecular image at cellular resolution in formalin-fixed, paraffin-embedded clinical tissue sections. Sample preparation includes staining the tissue with a mixture of antibodies with individual elemental reporters such as indium (In), neodymium (Nd), gadolinium (Gd), etc. During imaging, the primary ion beam ionizes multiple elemental reporters representing dozens of proteins within the sample.

Laser ablation inductively coupled plasma

In LA inductively coupled plasma (ICP), a portion of the material is removed from the sample and introduced to an ICP for elemental ionization. Similar to MIBI, LA-ICP has been used for imaging metal tags in antibodies. Several rare earth metals can be conjugated with antibodies and serve as surrogate reporters for proteins. The spatial map of the elemental composition of the sample is acquired by rastering the laser beam on a sample like MALDI. Several types of lasers have been employed for ablation, from Ruby lasers to excimer or Nd:YAG lasers.[57] LA-ICP have been used to map the distribution of trace elements (e.g., phosphorus, sulfur, iron, copper, and zinc) in brain tissue section.[58] This technique is often referred to as imaging mass cytometry. In both SIMS-based MIBI and ICP-based mass cytometry, the metal labels conjugated to the antibodies found to proteins on tissue are analyzed by a mass spectrometer. In SIMS, metal ions are produced using a primary ion beam, while in LA-ICP, metal tags are first ablated and then ionized by ICP. A conceptual representation of LA-ICP or mass cytometry ion source is shown in Fig. 2.2F Recently, atmospheric pressure (AP) MALDI imaging MS using in-line plasma induced post-ionization showed a significant ionization enhancement for a range of compounds.[62]

Liquid extraction surface analysis

Liquid extraction surface analysis (LESA) is an ionization source that is able to directly extract analytes from thin tissue sections using a liquid extraction and subsequent ionization using electrospray. Liquid-microjunction electrospray is useful for tissue profiling or very coarse imaging of tissue. Various configurations are commercially available as SepQuant dropletProbe or LESA. A schematic rendering of LESA is shown in Fig. 2.2G.

Surface sampling using liquid microjunction followed by an optional liquid chromatography separation and ionization by electrospray have been used for various organs (brain, lung, kidney, and liver) and drugs from whole-body tissue sections, natural product analysis, or food analysis.[59–60] The average pixel size of

LESA is much bigger than MALDI or DESI due to the larger size of the droplet. However, due to the efficient nature of extraction yields higher sensitivity. Also, when coupled with chromatographic separation, it provides better selectivity and sensitivity. LESA can be an excellent tool for whole-body imaging and to confirm the molecular identity by liquid chromatography tandem MS.

Nanospray desorption electrospray ionization

Nanospray desorption electrospray Ionization (nano-DESI) is one of the ambient surface ionization techniques that allow analysis without much sample preparation. Nano-DESI utilizes a liquid junction as a solvent droplet bridge between the primary capillary that supplies solvent for extraction of analyte from tissue and a secondary capillary that ionizes the extracted solvent.[61] Like DESI, molecular images are acquired by moving the sample stage under the nano-DESI probe, which is constantly in contact with the sample surface. Nano-DESI has been employed for imaging of several types of a small molecule within the tissue sections. A schematic representation of nano-DESI is shown in Fig. 2.2H.[15]

Conclusions

It would be remissive not to state that only selected ion sources used for imaging MS are discussed here. Many other ion sources have been used for imaging MS and many more being developed in academia and industry. Each ion source has its advantages and limitations depending on the application workflow, hardware robustness, software development, and commercial availability. In addition to the development of the new ion source, it is notable to state the advancement in the hardware and workflow of the existing ion sources. Gas cluster ion beam SIMS and MALDI combined with laser-induced postionization (MALDI-2) are exemplary examples of hardware advancement on established tools. While MIBI and Cytof have elevated the utilization of the established ion source by just changing the sample preparation workflows. They have introduced a new breed of users to assess the imaging of many different kinds of molecules unimaginable before.

References

1. Advanced information on the Nobel Prize in chemistry 2002. Available at: www.nobel.se/chemistry/laureates/2002/chemadv.pdf. Accessed on 23 January, 2021.
2. Hillenkamp F, Jaskolla TW, Karas M. The MALDI process and method. *MALDI MS. A practical guide to instrumentation*. In: Hillenkamp F, Peter-Katalinic J, eds. *Methods, and Applications*. 2nd ed. Weinheim, Germany: Wiley Blackwell; 2014.
3. Carbonnelle E, Mesquita C, Bille E, et al. MALDI-TOF mass spectrometry tools for bacterial identification in clinical microbiology laboratory. *Clin Biochem*. 2011;44:104–109.
4. Škrášková K, Claude E, Jones EA, Towers M, Ellis SR, Heeren RM. Enhanced capabilities for imaging gangliosides in murine brain with matrix-assisted laser desorption/ionization and desorption electrospray ionization mass spectrometry coupled to ion mobility separation. *Methods*. 2016;104:69–78.

5. Katz L, Woolman M, Talbot F, et al. Dual laser and desorption electrospray ionization mass spectrometry imaging using the same interface. *Anal Chem.* 2020;92:6349–6357.
6. Shrestha B, Nemes P, Nazarian J, Hathout Y, Hoffman EP, Vertes A. Direct analysis of lipids and small metabolites in mouse brain tissue by AP IR-MALDI and reactive LAESI mass spectrometry. *Analyst.* 2010;135:751–758.
7. Karas M, Hillenkamp F. Laser desorption ionization of proteins with molecular masses exceeding 10,000 daltons. *Anal Chem.* 1988;60:2299–2301.
8. Caprioli RM, Farmer TB, Gile J. Molecular imaging of biological samples: localization of peptides and proteins using MALDI-TOF MS. *Anal Chem.* 1997;69:4751–4760.
9. Knochenmuss R. Ion formation mechanisms in UV-MALDI. *Analyst.* 2006;131:966–986.
10. Dreisewerd K. The desorption process in MALDI. *Chem Rev.* 2003;103:395–426.
11. Berkenkamp S, Kirpekar F, Hillenkamp F. Infrared MALDI mass spectrometry of large nucleic acids. *Science.* 1998;281:260–262.
12. Li Y, Shrestha B, Vertes A. Atmospheric pressure molecular imaging by infrared MALDI mass spectrometry. *Anal Chem.* 2007;79:523–532.
13. Zavalin A, Todd EM, Rawhouser PD, Yang J, Norris JL, Caprioli RM. Direct imaging of single cells and tissue at sub-cellular spatial resolution using transmission geometry MALDI MS. *J Mass Spectrom.* 2012;47:1473–1481.
14. Dong X, Cheng J, Li J, Wang Y. Graphene as a novel matrix for the analysis of small molecules by MALDI-TOF MS. *Anal Chem.* 2010;82:6208–6214.
15. Bergman H-M, Lundin E, Andersson M, Lanekoff I. Quantitative mass spectrometry imaging of small-molecule neurotransmitters in rat brain tissue sections using nanospray desorption electrospray ionization. *Analyst.* 2016;141:3686–3695.
16. Stopka SA, Rong C, Korte AR, et al. Molecular imaging of biological samples on nanophotonic laser desorption ionization platforms. *Angew Chem Int Ed.* 2016;55:4482–4486.
17. Whitehead SN, Chan KHN, Gangaraju S, Slinn J, Li J, Hou ST. Imaging mass spectrometry detection of gangliosides species in the mouse brain following transient focal cerebral ischemia and long-term recovery. *PloS One.* 2011;6:e20808 -e20808.
18. Shariatgorji M, Nilsson A, Fridjonsdottir E, et al. Comprehensive mapping of neurotransmitter networks by MALDI–MS imaging. *Nat Methods.* 2019;16:1021–1028.
19. Chaurand P, Schwartz SA, Caprioli RM. *Peer Reviewed: Profiling and Imaging Proteins in Tissue Sections by MS*: ACS Publications; 2004.
20. Svatoš A. Mass spectrometric imaging of small molecules. *Trends Biotechnol.* 2010; 28:425–434.
21. Patti GJ, Shriver LP, Wassif CA, et al. Nanostructure-initiator mass spectrometry (NIMS) imaging of brain cholesterol metabolites in Smith-Lemli-Opitz syndrome. *Neuroscience.* 2010;170:858–864.
22. Fincher JA, Jones DR, Korte AR, et al. Mass spectrometry imaging of lipids in human skin disease model hidradenitis suppurativa by laser desorption ionization from silicon nanopost arrays. *Sci Rep.* 2019;9:17508 17508.
23. Takáts Z, Wiseman JM, Gologan B, Cooks RG. Mass spectrometry sampling under ambient conditions with desorption electrospray ionization. *Science.* 2004;306:471–473.
24. Costa AB, Cooks RG. Simulation of atmospheric transport and droplet–thin film collisions in desorption electrospray ionization. *Chem Commun.* 2007:3915–3917.
25. Costa AB, Cooks RG. Simulated splashes: elucidating the mechanism of desorption electrospray ionization mass spectrometry. *Chem Phys Lett.* 2008;464:1–8.
26. Wiseman JM, Ifa DR, Song Q, Cooks RG. Tissue imaging at atmospheric pressure using desorption electrospray ionization (DESI) mass spectrometry. *Angew Chem Int Ed.* 2006;45:7188–7192.

27. Eberlin LS, Ferreira CR, Dill AL, Ifa DR, Cooks RG. Desorption electrospray ionization mass spectrometry for lipid characterization and biological tissue imaging. *Biochim Biophys Acta Mol Cell Biol Lipids*. 2011;1811:946–960.
28. Calligaris D, Caragacianu D, Liu X, et al. Application of desorption electrospray ionization mass spectrometry imaging in breast cancer margin analysis. *Proc Natl Acad Sci*. 2014;111:15184–15189.
29. Tata A, Woolman M, Ventura M, et al. Rapid detection of necrosis in breast cancer with desorption electrospray ionization mass spectrometry. *Sci Rep*. 2016;6:35374.
30. Nefliu M, Smith JN, Venter A, Cooks RG. Internal energy distributions in desorption electrospray ionization (DESI). *J Am Soc Mass Spectrom*. 2008;19:420–427.
31. Jackson AU, Talaty N, Cooks RG, Van Berkel GJ. Salt tolerance of desorption electrospray ionization (DESI). *J Am Soc Mass Spectrom*. 2007;18:2218–2225.
32. Liu C, Qi K, Yao L, et al. Imaging of polar and nonpolar species using compact desorption electrospray ionization/postphotoionization mass spectrometry. *Anal Chem*. 2019; 91:6616–6623.
33. Forbes TP, Sisco E. Chemical imaging of artificial fingerprints by desorption electro-flow focusing ionization mass spectrometry. *Analyst*. 2014;139:2982–2985.
34. Forbes TP, Sisco E. Mass spectrometry detection and imaging of inorganic and organic explosive device signatures using desorption electro-flow focusing ionization. *Anal Chem*. 2014;86:7788–7797.
35. Soltwisch J, Kettling H, Vens-Cappell S, Wiegelmann M, Müthing J, Dreisewerd K. Mass spectrometry imaging with laser-induced postionization. *Science*. 2015;348:211–215.
36. Soltwisch J, Heijs B, Koch A, Vens-Cappell S, Höhndorf J, Dreisewerd K. MALDI-2 on a trapped ion mobility quadrupole time-of-flight instrument for rapid mass spectrometry imaging and ion mobility separation of complex lipid profiles. *Anal Chem*. 2020.
37. Ellis SR, Soltwisch J, Paine MRL, Dreisewerd K, Heeren RMA. Laser post-ionisation combined with a high resolving power orbitrap mass spectrometer for enhanced MALDI-MS imaging of lipids. *Chem Commun (Camb)*. 2017;53:7246–7249.
38. Barré FPY, Paine MRL, Flinders B, et al. Enhanced sensitivity using MALDI imaging coupled with laser postionization (MALDI-2) for pharmaceutical research. *Anal Chem*. 2019;91:10840–10848.
39. Bowman AP, Bogie JFJ, Hendriks JJA, et al. Evaluation of lipid coverage and high spatial resolution MALDI-imaging capabilities of oversampling combined with laser post-ionisation. *Anal Bioanal Chem*. 2020;412:2277–2289.
40. Spivey EC, McMillen JC, Ryan DJ, Spraggins JM, Caprioli RM. Combining MALDI-2 and transmission geometry laser optics to achieve high sensitivity for ultra-high spatial resolution surface analysis. *J Mass Spectrom*. 2019;54:366–370.
41. Robichaud G, Barry JA, Garrard KP, Muddiman DC. Infrared matrix-assisted laser desorption electrospray ionization (IR-MALDESI) imaging source coupled to a FT-ICR mass spectrometer. *J Am Soc Mass Spectrom*. 2013;24:92–100.
42. Huang M-Z, Jhang S-S, Shiea J. Electrospray laser desorption ionization (ELDI) mass spectrometry for molecular imaging of small molecules on tissues*Mass Spectrometry Imaging of Small Molecules*: Springer; 2015:107–116.
43. Shrestha B, Sripadi P, Reschke BR, et al. Subcellular metabolite and lipid analysis of *Xenopus laevis* eggs by LAESI mass spectrometry. *PLoS One*. 2014:9.
44. Shrestha B, Vertes A. In situ metabolic profiling of single cells by laser ablation electrospray ionization mass spectrometry. *Anal Chem*. 2009;81:8265–8271.

45. Stopka SA, Khattar R, Agtuca BJ, et al. Metabolic noise and distinct subpopulations observed by single cell LAESI mass spectrometry of plant cells in situ. *Front Plant Sci*. 2018;9:1646.
46. Shrestha B, Patt JM, Vertes A. In situ cell-by-cell imaging and analysis of small cell populations by mass spectrometry. *Anal Chem*. 2011;83:2947–2955.
47. Vaikkinen A, Shrestha B, Kauppila TJ, Vertes A, Kostiainen R. Infrared laser ablation atmospheric pressure photoionization mass spectrometry. *Anal Chem*. 2012;84:1630–1636.
48. Vaikkinen A, Shrestha B, Koivisto J, Kostiainen R, Vertes A, Kauppila TJ. Laser ablation atmospheric pressure photoionization mass spectrometry imaging of phytochemicals from sage leaves. *Rapid Commun Mass Spectrom*. 2014;28:2490–2496.
49. Hieta J-P, Kopra J, Räikkönen H, Kauppila TJ, Kostiainen R. Sub-100 μm spatial resolution ambient mass spectrometry imaging of rodent brain with laser ablation atmospheric pressure photoionization (LAAPPI) and laser ablation electrospray ionization (LAESI). *Anal Chem*. 2020.
50. Fletcher JS, Lockyer NP, Vaidyanathan S, Vickerman JC. TOF-SIMS 3D biomolecular imaging of *Xenopus laevis* oocytes using buckminsterfullerene (C60) primary ions. *Anal Chem*. 2007;79:2199–2206.
51. Sheraz née Rabbani S, Barber A, Fletcher JS, Lockyer NP, Vickerman JC. Enhancing secondary ion yields in time of flight-secondary ion mass spectrometry using water cluster primary beams. *Anal Chem*. 2013;85:5654–5658.
52. van Hove ERA, Smith DF, Heeren RM. A concise review of mass spectrometry imaging. *J Chromatogr A*. 2010;1217:3946–3954.
53. Massonnet P, Heeren RMA. A concise tutorial review of TOF-SIMS based molecular and cellular imaging. *J Anal Atom Spectrom*. 2019;34:2217–2228.
54. Pareek V, Tian H, Winograd N, Benkovic SJ. Metabolomics and mass spectrometry imaging reveal channeled de novo purine synthesis in cells. *Science*. 2020;368:283–290.
55. Angelo M, Bendall SC, Finck R, et al. Multiplexed ion beam imaging of human breast tumors. *Nat Med*. 2014;20:436–442.
56. Levenson RM, Borowsky AD, Angelo M. Immunohistochemistry and mass spectrometry for highly multiplexed cellular molecular imaging. *Lab Invest*. 2015;95:397–405.
57. Russo RE, Mao X, Liu H, Gonzalez J, Mao SS. Laser ablation in analytical chemistry—a review. *Talanta*. 2002;57:425–451.
58. Becker JS, Zoriy MV, Pickhardt C, Palomero-Gallagher N, Zilles K. Imaging of copper, zinc, and other elements in thin section of human brain samples (hippocampus) by laser ablation inductively coupled plasma mass spectrometry. *Anal Chem*. 2005;77:3208–3216.
59. Kertesz V, Van Berkel GJ. Liquid microjunction surface sampling coupled with high-pressure liquid chromatography–electrospray ionization-mass spectrometry for analysis of drugs and metabolites in whole-body thin tissue sections. *Anal Chem*. 2010;82:5917–5921.
60. Eikel D, Henion J. Liquid extraction surface analysis (LESA) of food surfaces employing chip-based nano-electrospray mass spectrometry. *Rapid Commun Mass Spectrom*. 2011;25:2345–2354.
61. Roach PJ, Laskin J, Laskin A. Nanospray desorption electrospray ionization: an ambient method for liquid-extraction surface sampling in mass spectrometry. *Analyst*. 2010;135:2233–2236.
62. Elia, E.A., et al., Atmospheric Pressure MALDI Mass Spectrometry Imaging Using In-Line Plasma Induced Postionization. *Anal Chem.*, 2020.

Chapter 3

Sample preparation for imaging mass spectrometry

Chapter Outline

Imaging mass spectrometry (MS) is an analytical technique that can plot chemical or molecular species in a wide range of samples, most commonly in tissue sections. Imaging MS creates molecular distribution patterns by sequential pixel-by-pixel analysis of samples using direct ionization tools, such as matrix-assisted laser desorption/ionization (MALDI). Direct ion source has fewer sample preparation steps than other ion sources that analyze aliquots, such as electrospray ionization. Sample preparation is reduced even more for ambient ionization sources, designed to analyze samples in its native form with little or no sample preparation.[1] Several new ambient ionization sources have revolutionized the imaging MS,[2] such as desorption electrospray ionization (DESI) used for imaging metabolites without any sample preparation,[3] laser-ablation electrospray ionization (LAESI) used for cell-to-cell imaging.[4] Nevertheless, any imaging MS requires some sample preparation. The sample preparation steps can include simple mounting of tissue on a slide or more elaborate such as antibody tagging.

Careful and consistent sample handling and preparation are critical for reproducible MS or any laboratory analysis. Proper sample preparation is more critical for volume-limited analytical techniques such as imaging MS, where even minor variation in sample preparation can have a profound consequence on ionization and detection of types of molecules, signal intensity, and spatial

Introduction to Spatial Mapping of Biomolecules by Imaging Mass Spectrometry. DOI: 10.1016/C2018-0-03962-6

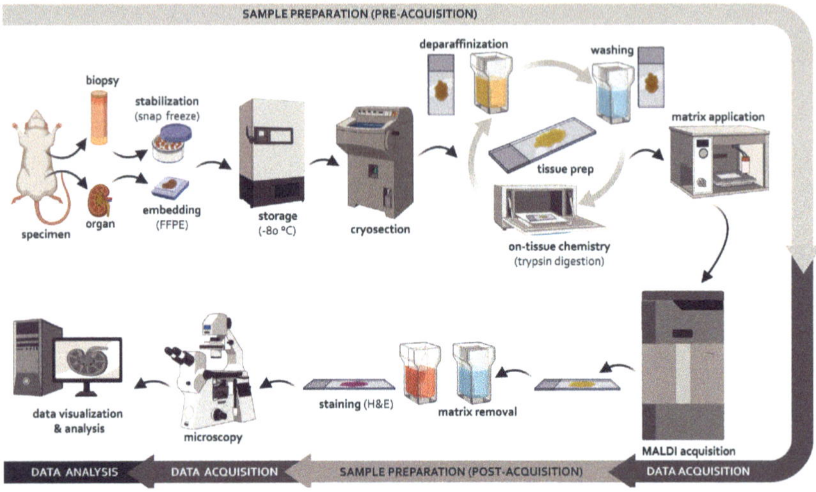

FIG. 3.1 A typical workflow for MALDI imaging mass spectrometry experiment is shown. Sample preparation plays important role before and after the analysis depending on the sample type and the goal of the imaging experiment.

distribution. For example, tissue incubation time or temperature with trypsin may significantly affect downstream imaging results. Alongside the new development in imaging MS instrumentation, many new sample preparation strategies have been reported, such as reactive matrix for derivatizing and ionizing hard-to-detect small metabolites within the tissue,[5] or streamline the sample preparation procedures for highest quality reproducible data.[6]

The sample preparation for imaging MS starts early on from collecting samples from sacrificed animal or removed tissues such as biopsy, stabilizing or embedding excised tissue, storing the tissue, sectioning, and thaw mounting to thin sections onto a microscope slide. The tissue section can further go through sample preparation depending on the type of analyte analyzed, such as washing, on-tissue chemical derivatization, or enzymatic digestion. For MALDI, the tissue section needs to be coated with a matrix for extraction and ionization. The analyzed tissue section may also go through more sample preparation after the data acquisition for staining is intended for microscopy. The potential sample preparation journey is conceptually shown in Fig. 3.1.

Sample preparation for imaging MS, particularly for MALDI, has been reviewed and discussed extensively in the referenced publications.[7–11] In general, the sample preparation step can be divided into four major stages: sample collection, sample processing, postsectioning treatments, and procedures done to improve ion yield—as shown in Fig. 3.2. Most of these reviews are focused on general sample preparation workflows for imaging mammalian tissue section, but some are focused on imaging of plant tissues.[11] A minor error or modification in the sample preparation step can lead to significant changes in MS imaging results. In this chapter, various aspects of sample preparation steps will be briefly discussed. Some sample preparation steps that need more details, such as sectioning and matrix applications, are discussed in separate chapters dedicated to them. For detailed protocols on the specific sample preparation, refer to the respective scientific literature on that procedure.

Sectioning and storage

The preparation of the tissue section is required for any imaging MS applications. Bulk tissue specimens are often collected and sectioned by scientists who do not analyze samples in the mass spectrometer. Typically, imaging MS uses

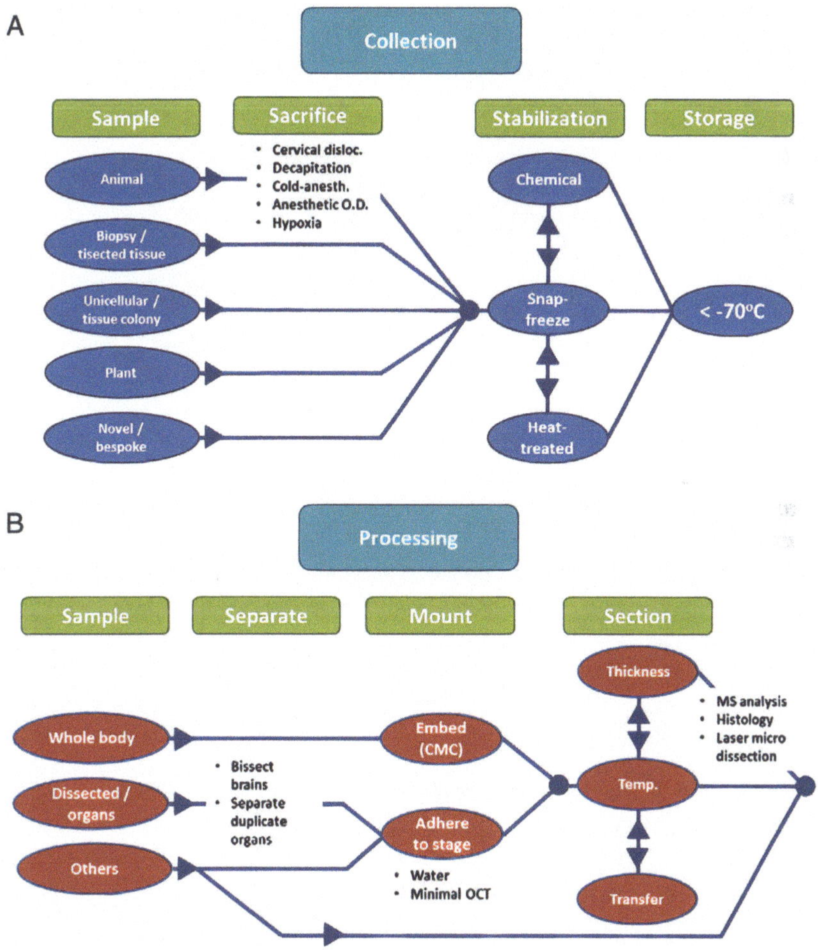

FIG. 3.2 Schematic of the major stages in sample preparation for imaging mass spectrometry broken down into the four major steps encompassing (A) collection, (B) processing, (C) postsectioning, and (D) ionization-aiding treatments. Adapted with the permission from Goodwin RJA. Sample preparation for mass spectrometry imaging: small mistakes can lead to big consequences. *J Proteomics*. 2012;75:4893–4911.[10]

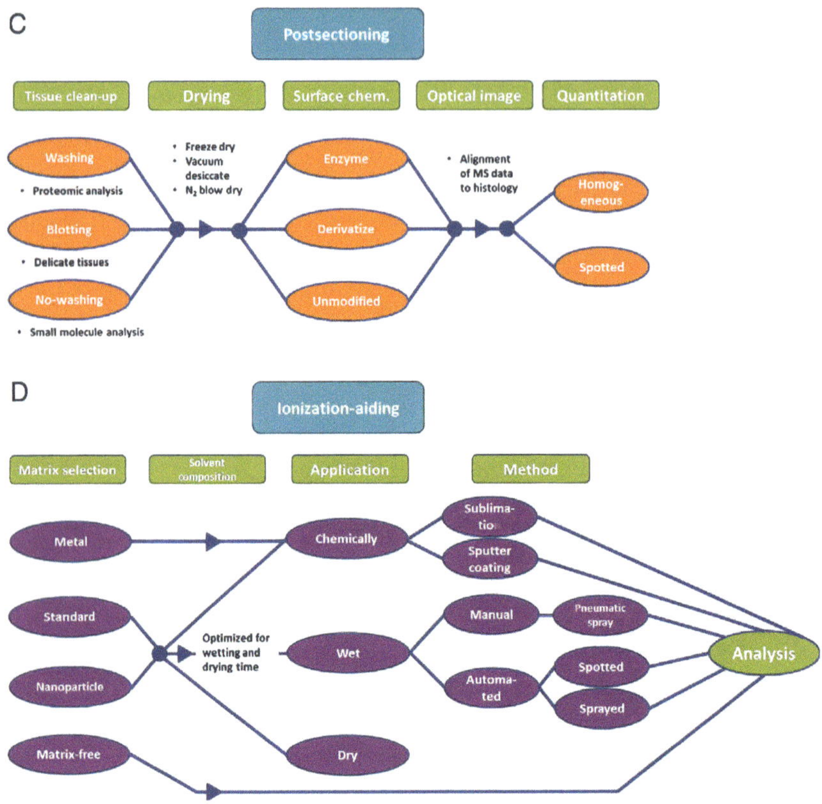

FIG. 3.2 Continued

thaw-mounted fresh-frozen sections cut around 10–20 microns thick without any embedding media such as optimal cutting temperature. While sectioning tissue, alignment, and orientation must be correctly noted. More details on tissue sectioning are discussed in another chapter.

Mounting

Most imaging MS applications use thaw mounting to affix a frozen section of tissue onto the slide. Briefly, a thin section of tissue is adhered to the slide by applying heat from the finger. For some larger and delicate sections, such as whole-body sections, a tape transfer method maintains tissue integrity during mounting. In this method, a double-sided adhesive tape is used to mount the tissue section on a slide. Some material cannot be easily sectioned due to their mailable or brittle property, such as drug implant, or are small, size such as hair. In such cases, the sample is mounted using adhesive tape. For example, the spatial distribution of drugs in polymeric implants before and after exposure to accelerated in vitro release media was imaged by mounting the implant using double-sided tape on a glass slide by DESI.[12] In other examples, metabolites in leaves and petals were analyzed by atmospheric pressure MALDI by affixing them using double-sided tape.[13] MALDI has imaged recreational drugs incorporated in human hair after attaching the hair to a glass slide using conductive carbon tape.[14]

Imprinting and stamping

Biological samples can be imprinted on a substrate by pressing the sample against the substrate. The imprint on the substrate is imaged like any other specimen by imaging MS tools. Biological samples imprinted on a silver surface using pressure have been imaged by secondary ion mass spectrometry (SIMS) imaging. The SIMS imaging of imprint was successful at mapping the spatial distribution of cholesterol and phosphatidylcholine in blood cells.[15] Fig. 3.3 shows scanning electron microscopy images cells imprinted on a substrate at various pressure and with or without moistening. Blotted imprints of tissue section coated with matrix have been used to image molecules over 50 kDa by MALDI.[16]

Imprints of plant seeds and mouse brain tissue on paper and thin-layer chromatography plates were successful at imaging metabolites and lipids by DESI.[17] DESI has also been utilized to image glycoalkaloids on a tape imprint of potatoes infected by the phytopathogen,[18] and metabolites in petals and leaves imprinted on thin-layer chromatography substrate.[19]

Imprinting is useful for imaging uneven soft surfaces, such as microbial cultures grown on agar media, that are difficult to section and are challenging to

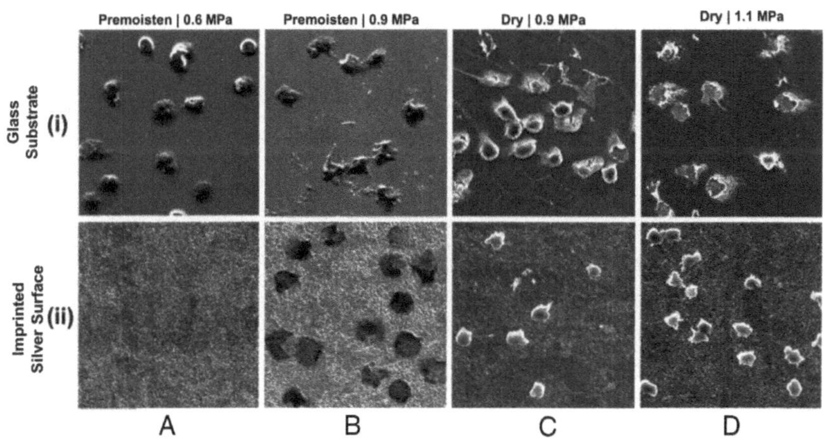

FIG. 3.3 Scanning electron microscopy (SEM) images of polymorphonuclear leukocytes on glass cell substrate (i) and imprinted silver surface (ii) obtained after imprinting at (A) 0.6 MPa and (B) 0.9 MPa after moistening of the cells immediately before imprinting over hot water and (C) 0.9 MPa and (D) 1.1 MPa after drying. Field of view, 100 × 100 μm^2. Adapted with permission from Sjövall P, Lausmaa J, Nygren H, Carlsson L, Malmberg P. Imaging of membrane lipids in single cells by imprint-imaging time-of-flight secondary ion mass spectrometry. *Anal Chem.* 2003;75:3429–3434.[15]

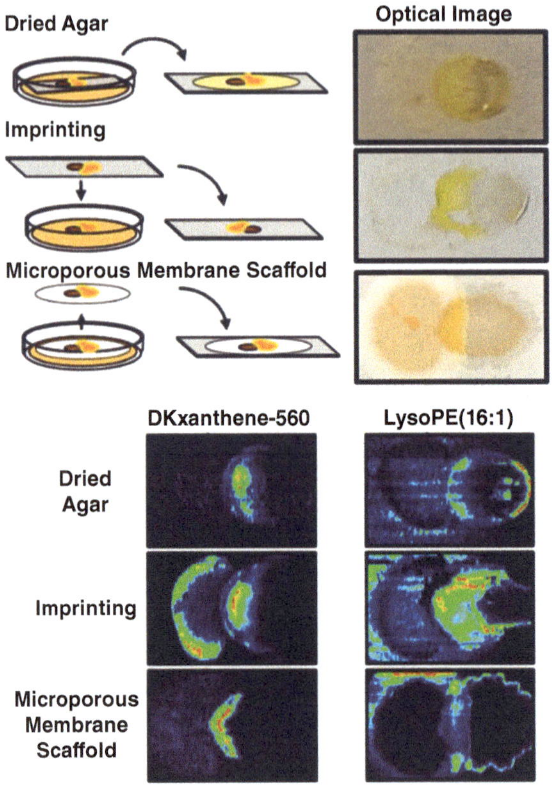

FIG. 3.4 Schematic representations of three sample preparation methods for imaging microbial colony by DESI, e.g., drying, imprinting, and microporous membrane scaffold, with optical images of microbial colonies on top. DESI image hf DKxanthene-560 and lysoPE 16:1 using each of the method. Adapted with permission from Ellis BM, Fischer CN, Martin LB, Bachmann BO, McLean JA. Spatiochemically profiling microbial interactions with membrane scaffolded desorption electrospray ionization-ion mobility-imaging mass spectrometry and unsupervised segmentation. *Anal Chem*. 2019;91:13703–13711. Copyright 2019 American Chemical Society.[22]

image directly. DESI imprint of agar culture imaging has been used to monitor secondary metabolites during antagonistic interaction of fungi.[20] Alternatively, culture on a thin solid agar after dehydration has also been imaged by DESI.[21] Three sampling methods for imaging microbial metabolites using DESI imaging, i.e., drying, imprinting, and microporous membrane scaffold was evaluated as shown in Fig. 3.4.[22] The dried agar samples were prepared by culturing microbial colonies on a removable glass slide placed inside the Petri dish with agar medium. The samples were imaged after drying and removing samples from the Petri dish. The imprinting of colonies was done by applying uniform pressure on a glass slide placed on the microbial colonies in agar for 10–20 seconds. The colony imprint was dried before imaging. The microporous membrane scaffold method involves stamping the colonies onto a sterile nylon membrane and placing the membrane on the agar medium. After culture, the membrane was removed, dried, and mounted to a glass slide for imaging MS by DESI.

Tissue storage

The sectioned tissues or bulk tissue can be stored inside a sealed container in −80°C freezer for later use. The specific storage and handling protocols vary a lot between labs. There is no comprehensive data on very long-term storage, i.e., months to years, for many types of tissues. The extent of molecular degradation will depend not only on storage conditions but also on the type of tissue and type of molecules. Some other studies that investigated aliquots also showed a low number of metabolomic and lipidomic changes for high species in unfractionated serum and pooled human plasma sample stored at −80°C.[23–24]

There are some imaging MS studies have shown changes in metabolite and lipid contents in kidney sections within a week.[25] The tissue section stored under nitrogen and −80°C showed the least degradation while one stored at room temperature showed the most. For endogenous metabolites, an optimal stabilization included embedding frozen tissue in 10% w/v gelatin solution, sectioning, immediate desiccation over dry nitrogen gas, vacuum-sealed packing, and storage at −80°C.[26]

Tissue drying

Tissue sections are dried after they are removed from storage in the freezer and before other sample preparation or acquisition. Drying tissue is a fixation step that minimizes any sample instability or potential delocalization due to the formation of ice or droplet coating due to room humidity. For imaging MS, tissue sections are usually dried by a home-built vacuum desiccation system for 10–15 minutes. Other drying methods used include freeze-drying for SIMS and MALDI,[27–28] dehydration by ethanol wash for MALDI,[29] drying under a gentle stream of oxygen-free nitrogen gas for matrix-free laser desorption ionization imaging.[30]

The tissue sections can undergo several sample processing steps before matrix application or imaging data acquisition by the mass spectrometer. The goals of sample preparation steps may be to enhance the detection of molecule-of-interest by removing other ion suppressing species or by chemically modifying it into a structure that can be easily detected by a mass spectrometer, improve quantitation or selection, or processing tissue for other complementary assays.

Tissue rinsing and incubation

Tissue washing

Tissue washing procedure can be used for improving the ionization and detection of molecule-of-interest by removing the abundant ion suppressing molecules, such as soluble endogenous metabolites and lipids, away from the tissue. The

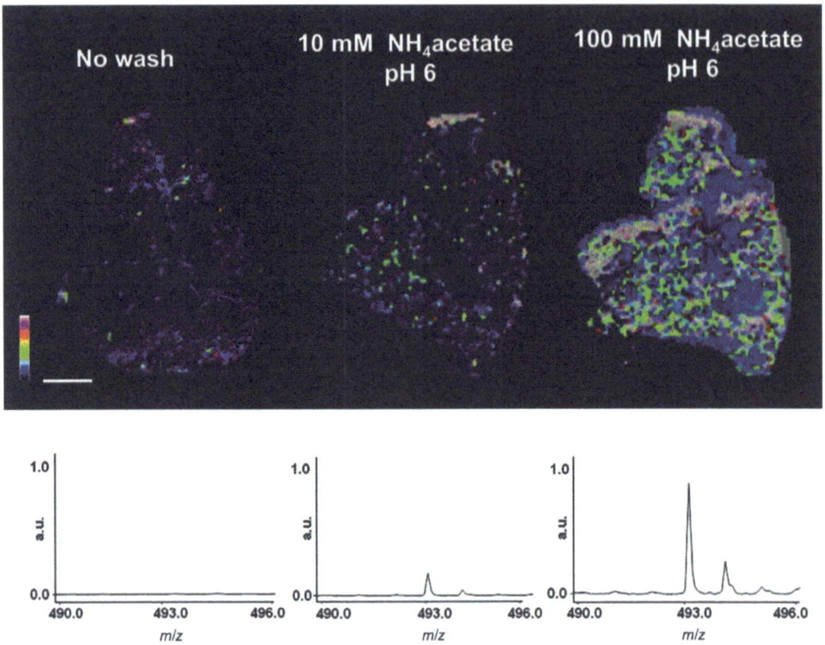

FIG. 3.5 The detection of compound C (*m/z* 493.2) administered by in vivo inhalation in serial rat lung tissue in tissue that was not washed, washed in 10 mM ammonium acetate at pH 6, and washed in 100 mM ammonium acetate solution at pH 6. The high concentration wash yielded in most intense signal. Adapted with permission from Shariatgorji M, Källback P, Gustavsson L, et al. Controlled-pH tissue cleanup protocol for signal enhancement of small molecule drugs analyzed by MALDI-MS imaging. *Anal Chem.* 2012;84:4603–4607. Copyright 2012 American Chemical Society.[31]

composition of tissue washing solvent depends on the solubility of molecules that are indented for removal and those that need to be avoided. For example, the detection of some low mass compounds, such as drugs, can be improved by washing tissue with a buffer solution at a set pH and concentration, where the molecule-of-interest had the lower solubility. An example of such washing for a drug at *m/z* 492 is shown in Fig. 3.5.[31] Without carefully crafted protocols, washing of tissue can dramatically change the composition of the tissue, lead to unintended delocalization, cross-contamination, or streaking.

The tissue washing step is also part of many other procedures such as deparaffinization, staining, etc. The most employed washing protocol involves removing lipids by a series of organic or alcohol washes for glycan or protein analysis. For example, washing tissue in cold baths of acetone for 30 seconds followed by 95% ethanol for 30 seconds, and finally chloroform for 60 seconds helps enable detection of large proteins at *m/z* values 44,600, 46,800, and 66,000 from a mouse brain section.[32] The effect on the sensitivity of lipid and protein for MALDI imaging after washing of serial section of tissue by several solvents is

Acetone	2×100% acetone
No wash	control
Toluene	2×100% toluene
Water	2×100% water
***t*-BME**	2×100% *tert*-butyl methyl ether
Hexane	2×100% hexane
Ethanol	70% ethanol | 95% ethanol
Chloroform	2×100% chloroform
Xylene	2×100% xylene
Methanol	70% methanol | 95% methanol
Acetic acid	70% ethanol | 90% ethanol/9% acetic acid
Isopropanol	isopropanol | 95% isopropanol

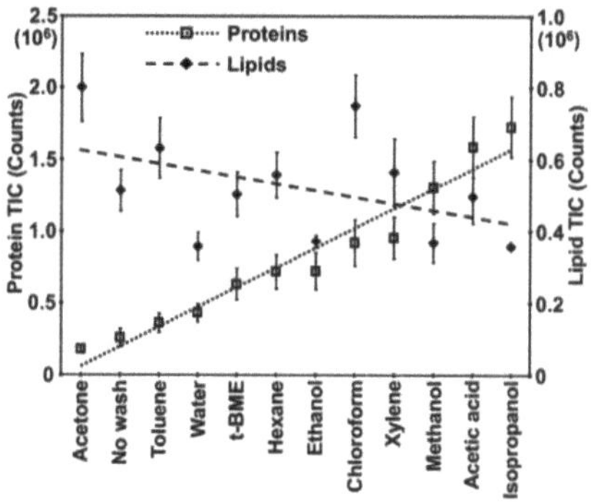

FIG. 3.6 Solvent compositions used for washing serial mouse liver section with total ion currents (TIC) of mass spectra profile of lipid profile (*m/z* 500–1100) and protein profile (m/z 2000–25,000). Adapted with permission from Seeley EH, Oppenheimer SR, Mi D, Chaurand P, Caprioli RM. Enhancement of protein sensitivity for MALDI imaging mass spectrometry after chemical treatment of tissue sections. J Am Soc Mass Spectrom. 2008;19:1069–1077. Copyright 2008 American Chemical Society.[33]

shown in Fig. 3.6[33]. Tissue sections were washed twice by immersing it for 30 seconds and promptly dried in a vacuum desiccator. Tissue washing can also be used for enhancing signals for some lipid species as well. Volatile buffers, such as ammonium formate or ammonium acetate at a specific pH, have been found to have improved MS signal of lipids in negative ion mode. This type of aqueous washes resulted in up to a fivefold increase was observed for various lipids, such as glycerophosphoinositols, glycerophosphates, glycerophosphoethanolamines, glycerophosphoserines, sulfatides, and gangliosides, across various tissue types.[34] In another study, three rinse protocols were evaluated for removing lipids for enhancing protein imaging by MALDI—a standard protocol used in classical histology of rinsing with 70% ethanol followed by 100% ethanol; a four-step rinse protocol of water 70% ethanol, 100% ethanol, Carnoy's fluid (ethanol:chloroform:acetic acid in 6:3:1 ratio); and a six-step rinse with 70% ethanol, 100% ethanol, Carnoy's fluid, 100% ethanol, H2O, 100% ethanol.[35] The six-step rinse provided the best result for imaging protein over 25 kDa. Rinsing alcohol fixes the tissue, which was followed by Carnoy's fluid rinse that removed most of the lipids, ethanol removed the chloroform from Carnoy's fluid, water rinsed off the salts.

A slide with a mounted tissue section can be rinsed or washed in several ways by alcohol or detergent washing. Tissue sections are washed to remove lipids and salts or can be a part of the immunohistochemistry (IHC) protocol. Tissue washing is also used to dehydrate tissue section by immersing the tissue section in increasing concentrations of alcohols with a gradual change in hydrophobicity for minimizing cell damage. Tissue washing is also used as part of many staining protocols. In addition to removing unwanted ions, tissue washing may be required to analyze compounds just beneath the surface of the tissue. For example, many flavonoids compounds in plant tissue, such as leaves, are intracellular metabolites underneath an epicuticular wax coat made of long-chain fatty acids. By dipping the leaf in chloroform, removes the epicuticular surface waxes soluble in chloroform, enabling the analysis of flavonoids. In Fig. 3.7, kaempferol is detected only after washing off the top layer by chloroform, while long-chain fatty acids are detected in unwashed tissue.[36]

Deparaffinization

Formalin-fixed paraffin-embedded (FFPE) slides are deparaffinized and rehydrated before antigen retrieval and IHC steps. There are many protocols for paraffin removal, but generally, paraffin is washed by successive xylene washes for a few minutes each, followed by a decreasing concentration of ethanol was for a few minutes each. Finally, the slide is rinsed by cold, where it is kept until antigen retrieval is performed to avoid any nonspecific antibody binding due to drying.

Pre-extraction

Tissue washing can be combined with some other pre-extraction step to improve the sensitivity of molecule-of-interest. For example, two drugs, cobimetinib and clozapine, were extracted and detected in dosed rodent brain sections after a combination of a buffer wash with ammonium citrate or ammonium formate and precoating for extraction with cyclohexane before the matrix application. This combined washing and pre-extraction approach gained an 8–20-fold in sensitivity for analyzing those xenobiotics on tissue as seen in Fig. 3.8.[37]

Other tissue incubation procedures

Besides rinsing tissue, tissue can be incubated or fixed for a specific application. Also, intact tissue can undergo washing or processing to improve the detection of certain molecules. For example, extracellular matrix (ECM) made of large, insoluble, and cross-linked proteins, are difficult to detect and imaged by imaging MS. The detection of ECM can be improved by decellularization of intact tissue by incubation in sodium dodecyl sulfate. The incubation removes the highly abundant molecules that dominate MALDI, but still preserved the ECM proteins.[38]

On-tissue chemistry

A molecule-of-interest within the tissue can undergo chemical changes before samples in several ways, such organic reaction leading to a derivatized compound, enzymatic digestion, breaking of bonds, etc. The goals of these chemical reactions are to improve or enable the imaging or perform them more quantitatively or selectively. Increasing metabolic coverage for imaging MS using on-tissue chemical modifications using three methods—coniferyl aldehyde for primary amines,

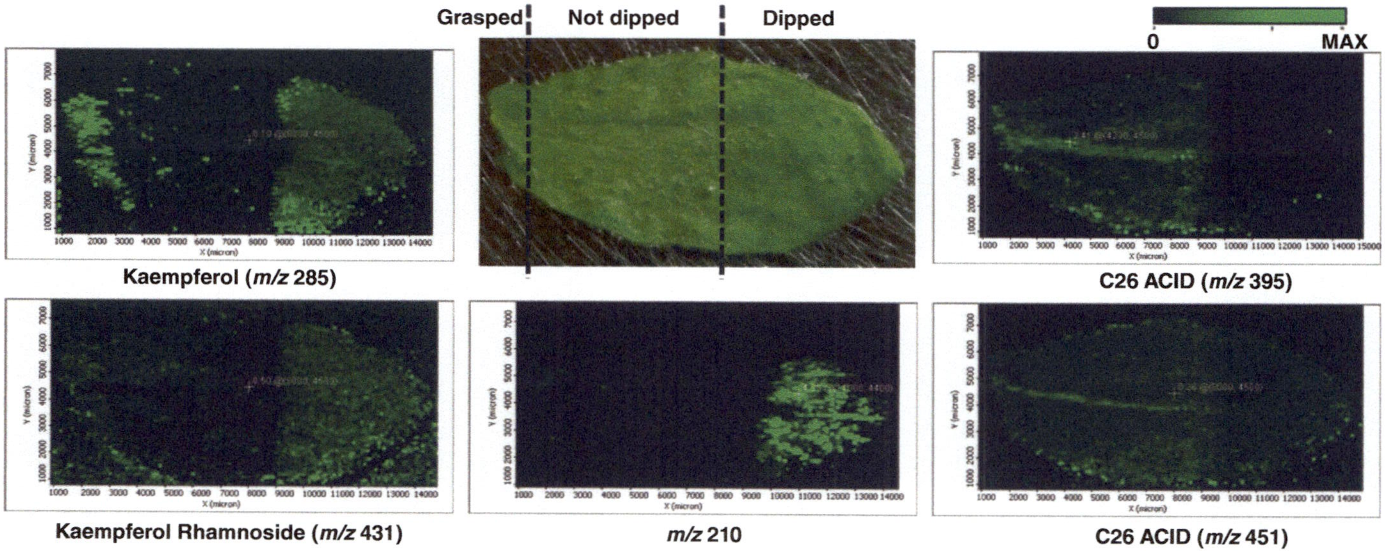

FIG. 3.7 An optical image of *Arabidopsis* leaf after dipping in chloroform is shown in the top middle. C26 and C30 fatty acids showed high abundance on the undipped portion of the tissue is shown on right, while kaempferol, its monoglycoside, and unidentified ion at m/z 210 showed high abundance in area washed by chloroform. Adapted with permission from Cha S, Zhang H, Ilarslan HI, et al. Direct profiling and imaging of plant metabolites in intact tissues by using colloidal graphite-assisted laser desorption ionization mass spectrometry. *Plant J*. 2008;55:348–360.[36]

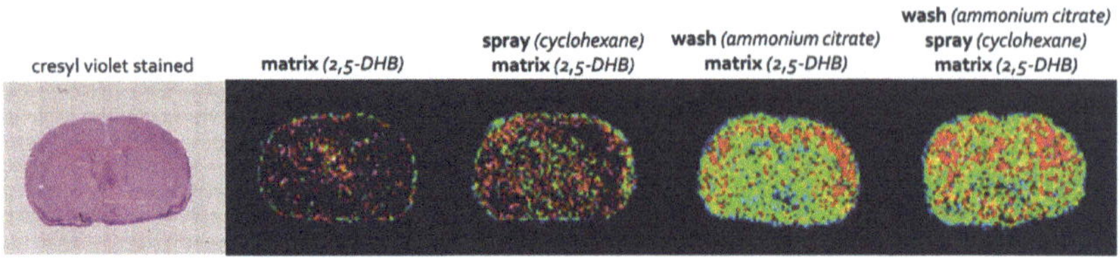

FIG. 3.8 The utility of buffer washing tissue and pre-extraction is shown for image of drug cobimetinib on coronal section of mouse brain is shown. The drug was detected confidently after washing with ammonium citrate buffer before matrix application. Further increase in ion intensity was noticed after spraying tissue for pre-extraction. Adapted with permission from Quiason CM, Shahidi-Latham SK. Imaging MALDI MS of dosed brain tissues utilizing an alternative analyte pre-extraction approach. *J Am Soc Mass Spectrom.* 2015;26:967–973. Copyright 2015 American Chemical Society.[37]

Girard's reagent T for carbonyl groups, and 2-picolylamine for carboxylic acids—have been discussed in the referenced publication.[39] The detection of molecule-of-interest is not improved if it cannot undergo ionization easily, regardless of other competing ion suppressing molecules in the tissue. In ambient ionization methods, such as DESI and LAESI, reagents can be added to the spray solvents and used to rapid derivatized desorbed targeted molecule species.[88,89]

On-tissue chemical derivatization

Chemical derivatization of tissue prior to imaging MS has been used for the detection of molecules-of-interest that are difficult to ionize or unstable during ionization.[91] On-tissue chemical derivatization has also applied for improving the selectivity of molecules-of-interest that are masked by other molecules with similar *m/z* values, as well as used to obtain more descriptive structural fragmentation for identification. A reactive compound can be added to the tissue to precisely modify the molecule-of-interest or class of molecules in the sample into surrogate molecular species that readily undergo analysis by a mass spectrometer.

Chemical derivatization has been used to measure drugs that cannot be easily detected due to background interferences from endogenous molecules. For example, isoniazid, an antituberculosis drug, in tissue was detected by MALDI MS/MS after reacting with *trans*-cinnamaldehyde.[40] Using a similar protocol, the sensitivity and specificity for detecting endogenous amine metabolites (e.g., amino acids, neurotransmitters) was boosted by on-tissue chemical derivatization with 4-hydroxy-3-methoxycinnamaldehyde.[41] The sensitivity of detection of a small organic molecule, 3-methoxysalicylamine, was improved by chemical derivatization with 1,1′-thiocarbonyldiimidazole into an easily detected oxo-thiazolidine derivative.[42] Steroids such as testosterone are challenging to image due to poor ionization and ion suppression. On-tissue chemical derivatization by Girard's reagent T increased the ionization efficiency of testosterone by adding a polar group enabling imaging of testosterone in testis tissues of mice treated with human chorionic gonadotropin.[43–44] MALDI image of two steroids,

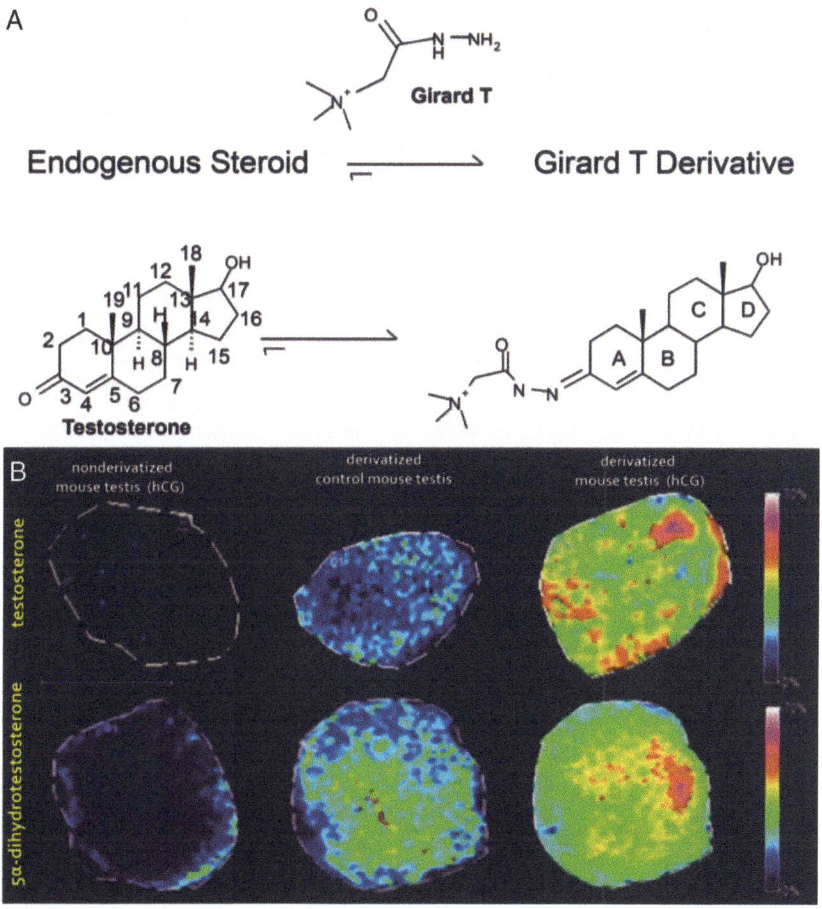

FIG. 3.9 (A) Derivatization reaction of testosterone with Girard T derivative, (B) testosterone and 5α-dihydrotestosterone could not be detected before the derivatization in mouse testes after stimulation with human chorionic gonadotrophin (hCG), but can be readily detected and imaged in both control and hCG stimulated mouse testes after the derivatization by Girard T. Adapted with permission from Cobice DF, Livingstone DEW, Mackay CL, et al. Spatial localization and quantitation of androgens in mouse testis by mass spectrometry imaging. *Anal Chem.* 2016;88:10362–10367. Copyright 2016 American Chemical Society.[43] Creative common.

testosterone, and 5α-dihydrotestosterone, is shown with and without derivatization by Girard's reagent T reagent in Fig. 3.9.[43] Girard's reagent T has also been used to find the distribution of administered triamcinolone acetonide drug to study osteoarthritis in cartilage.[45] On-tissue derivatization of lipopolysaccharide has been used for MALDI imaging of endotoxin, lipid A, in pathogens using high throughput lipid A hydrolysis method with citrate solution wash.[46]

The sialic acids of sialylated glycans were modified by methylamine and (7-azabenzotriazol-1-yloxy) trispyrrolidinophosphonium hexafluorophosphate enabling its analysis by MALDI can be analyzed by MALDI-MS without loss of the sialic acid moiety.[47] The ionization efficiency and detection sensitivity of glycans were improved by polycyclic aromatic hydrocarbon ethidium bromide derivatization.[48] In another chemical derivatization strategy, the sensitivity of N-glycans detection was dramatically improved by adding a permanent charge at the reducing end of the N-linked glycans by rapid deglycosylation with N-succinimidyloxycarbonylmethyl tris (2,4,6-trimethoxyphenyl) phosphonium bromide.[49]

The application of 6-aminoquinolyl-*N*-hydroxysuccinimidyl carbamate, also used for labeling for fluorescence detection in liquid chromatography, lead to improved detection of glycopeptides by MALDI.[50] Chemical derivatization of tissue by sulfonation agents added a negative charge at the N-terminus of a tryptic digested peptide for obtaining y fragment series leading to improved identification of proteins.[51] The ionization phosphopeptide was improved by subsiding negatively charged phosphate group of phosphoserine and phosphothreonine by a positively charged S-ethylpyridyl.[52]

In general, chemical derivatization is performed by either immersing the slide in the chemical reagent or spraying reagent with an automated sprayer or manual airbrush. In one example, the electrospray deposition of 2-picolylamine was able derivatize endogenous fatty acids. This derivatization improved the detection by threefolds compared to airbrush spray and also decreased analyte delocalization.[53] A fluoromethylpyridinium-based reactive matrix that acted as derivatization agent and MALDI matrix enabled imaging of low-abundance neurotransmitters at a high spatial resolution.[5]

Antigen retrieval

Most formalin-fixed tissue will need an antigen retrieval before analysis because of the formation of methylene cross-link bridge in proteins that mask antigenic sites. Antigen retrieval can be done by one of these methods—heat-mediated or heat-induced epitope retrieval, by enzymatic activity or protease-induced epitope retrieval, and reversal of cross-links using acids or room temperature epitope retrieval. All antigen retrieval methods break the methylene bridges and expose the antigenic for binding. The type of antigen retrieval method may depends on the type of antigens. A high concentration of enzymatic activity or acid or high heat can sometimes damage the morphology of the sections. Heat-induced epitope retrieval methods can be done in a pressure cooker, microwave, or vegetable steamer designed or modified for scientific application. Enzymatic antigen retrieval or chemical antigen retrieval can be done by pipetting enzyme on a slide by immersing a large batch of slides in the solution. Consult the manufacturer or vendor for the appropriate protocol. FFPE tissue is not particularly appropriate for lipid imaging since many lipids are removed during tissue processing. Some solvent-resistant lipids remain in tissue, but their extraction is stopped by the cross-link between proteins. The antigen retrieval step was used

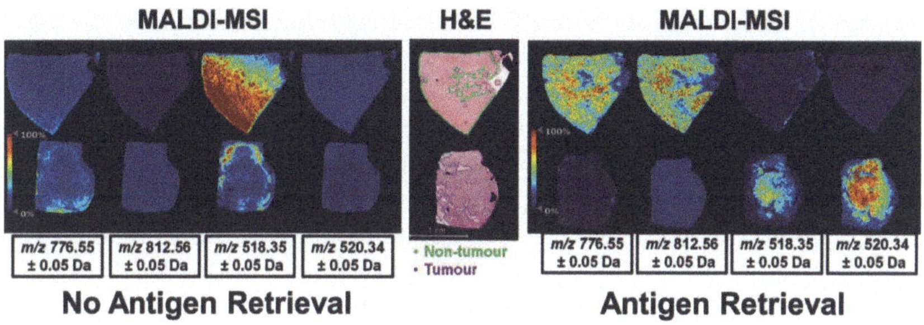

FIG. 3.10 MALDI images of serial sections analyzed without antigen retrieval at left compared with one performed after antigen retrieval is shown. Lipids at *m/z* 518.35 (LPC18:3) and at *m/z* 520.34 (LPC 18:2) had a higher intensity in the tumor region and at *m/z* 776.55 (PC(34:4) + Na) and at *m/z* 812.56 (PS 36:1) had a higher intensity in the nontumor control. Adapted with permission from Denti V, Piga I, Guarnerio S, et al. Antigen retrieval and its effect on the MALDI-MSI of lipids in formalin-fixed paraffin-embedded tissue. *J Am Soc Mass Spectrom.* 2020;31:1619–1624. Copyright 2020 American Chemical Society.[54]

for the extraction of such lipids. The effect of lipid imaging for heat-induced antigen retrieval performed in a citric acid buffer on FFPE human renal cancer tissue is shown in Fig. 3.10.[54] The result showed that tentatively identified lysophosphatidylcholines had a higher intensity in the tumor region, while phosphatidylcholines and phosphatidylserine had a higher intensity in the nontumor control after antigen retrieval.

Immunohistochemistry

Tissue may go through conventional IHC sample preparation steps before imaging MS data acquisition. Antibodies labeled with a unique elemental mass tags are used instead of typical fluorophores or chromogens. The mass tags are isotope of elements that do not naturally occur in biological tissue, such as lanthanides isotopes. These metal tags are liberated after the irradiance of laser or ion beams and detected by MSIHC techniques.[55] A highly multiplexed IHC imaging panel is possible by choosing metal labels that do not overlap. More than two dozen antibodies have been imaged simultaneously since *m/z* detection of mass tags do not suffer from spectral overlap seen with fluorophores. In scanning or imaging mass cytometry, laser-ablation inductively coupled plasma uses a high-intensity laser to liberate these mass tags for plasma ionization, and in multiplexed ion-beam imaging the mass tags are ionized by SIMS.[56–57]

MALDI's laser irradiance has been used for photocleaving and ionizing mass tags for detecting proteins in targeted multiplex mass spectrometry imaging or relatively quantifying the proteins in stable isotope label-based mass spectrometric imaging.[58–59] The discrete masses are released from their respective antibodies attached to target proteins. The labeled chromogens are deposited on stained to histological tissue sections enzymatically using the common IHC protocols. For example, in stable isotope label-based mass spectrometric imaging, the primary antibody that is specific to the biomarker and secondary antibody, conjugated with alkaline phosphatase and specific to the primary antibody, is incubated. Naphthol phosphate is cleaved by alkaline phosphatase to release naphthol that reacts with the chromogen to form a colored azo dye precipitates that fragments by MALDI laser to produce signature reporter ions.

On-tissue in situ enzymatic digestion

The detection of singly charged intact proteins from the tissue is challenging. Molecular coverage of intact protein and peptides is limited to the more abundant proteins at the lower mass end. The proteins within a tissue can be enzymatically cleaved into peptides and attached structures. The smaller molecular structure or peptides can be ionized and detected more readily by a mass spectrometer. For example, peptides from larger proteins, such as histone 2A and hemoglobin, can be imaged elucidating spatial map of larger proteins.[60] Minimizing analyte diffusion and accurately controlling enzymatic digestion process during the incubation is challenging, but often the most difficult aspect is analyzing and separating the complex mixture of peptides and other molecules resulting from the digestion process.

Trypsin, a serine protease that cleaves at the carboxylic side of lysine and arginine residues, can be homogeneously coated on a tissue by spraying for imaging MS of peptides. Before the enzyme application, the tissue is washed with an increasing concentration of alcohol and strong organic solvents to remove any other interfering molecule and dried to halt any potential enzymatic activity. The tissue is rehydrated in a buffer, such as ammonium bicarbonate at set alkaline pH, and washed by detergent such as, octyl-α/β-glucoside. The in situ digestion begins after the enzyme is applied to the tissue and incubated in a humidity chamber. A uniform application of enzymes across the tissue is needed to avoid any localized variation. Typically, the enzyme is coated on the tissue by spraying similar to the matrix application. A discrete spot analysis can be done by spotting small volumes of the enzyme on the tissue to carry out in situ protease digestion.[61] After the digestion step is complete, the matrix is applied to the tissue.

Imaging N-glycans from FFPE tissue sections also needs similar in situ digestion steps.[62–63] N-glycan is released from the tissue after deparaffinization, antigen retrieval, and enzyme digestion step. Peptide:N-glycosidase F or PNGase F, an amidase, perform deglycosylation by cleaving *N*-acetylglucosamine and asparagine residues of oligosaccharides from N-linked glycoproteins and glycopeptides. PNGase is applied to tissue similar to trypsin and matrix by sprayer and incubated in a humidity chamber at 37°C for a few hours. The freed the glycan residue that can be imaged by imaging MS, such as MALDI, after the application of the matrix.

Heat stabilization of tissue

Uncontrolled endogenous enzymatic activity may lead to rapid and uneven degradation of tissue leading to irreproducible results during long sample preparation or acquisition. Heat-stabilization method involving fast heating of tissue can inhibit postmortem enzymatic degradation of proteins, peptides, and metabolites. For example, microwave irradiation was used to halt proteomic degradation by inhibiting enzymatic activity, leading to an increase in the sensitivity of peptides and the detection of additional peptides.[64] The effectiveness of heating was assessed in a study where one hemisphere of mouse brains was heat treated, and another half was directly snap frozen after in liquid nitrogen while the other snap frozen without any heat treatment is shown in Fig. 3.11.[65] After thaw mounting of tissue sections, the slides were processed, stored immediately, or kept at room temperature for a defined period up to 5 minutes for snap frozen and up to 20 minutes for heat treated. MALDI Imaging showed that the intensity

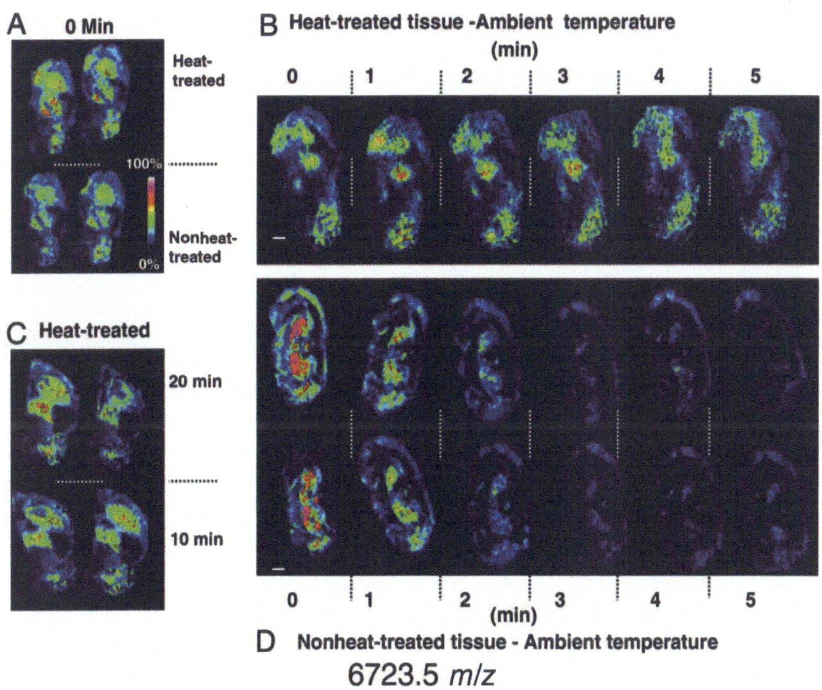

FIG. 3.11 A comparison of heat treatment for halting enzymatic activity is shown by comparing heat-treated and snap-frozen mouse brain tissue section. The sections were brought to room temperature for set amount of time before processing. MALDI images of a marker at *m/z* 6723.5 (A) heat-treated and not treated mouse brain sections at 0 minute shows similar intensity, (B) Heat-treated sections show no change, but untreated show significant decrease in intensity after 2 minutes, (C) heat treatment shows no change in intensities up to 20 minutes. Adapted with permission from Goodwin RJA, Lang AM, Allingham H, Borén M, Pitt AR. Stopping the clock on proteomic degradation by heat treatment at the point of tissue excision. *Proteomics*. 2010;10:1751–1761.[65]

of a marker at *m/z* 6723.5 was the same for all tissue sections at a 0-minute point and did not change in heat-treated mouse brain sections over 20 minutes. However, in a sample that was just snap frozen, it slowly began to reduce after 2 minutes. Heat stabilization of tissue has been reported to improve the detection of proteins in the kidney and testes tissue by liquid extraction surface analysis by a factor of two to four.[66] Adenine nucleotides ATP, ADP, and AMP in mouse brain sections were visualized after overcoming rapid postmortem degradation of nucleotides by heat stabilization.[67]

Instead of postmortem heat-stabilization methods, in situ funnel freezing used during the sacrifice of the mouse with fast thaw mounting of the brain sections was found to improve the stability of small metabolites such as glutamate, malate, taurine, inosine, etc.[68] In another study, a focused microwave irradiation method was used to reduce postmortem decomposition of metabolites, such as enzymatic biotransformation of glucose to lactate. Here the focused microwave irradiation method was found more efficient at halting enzymatic activities changing metabolite composition than compared to in situ freezing or posteuthanized freezing.[69]

In-plume microdroplet reactions

In plume, microdroplet reactions undergo at a significantly faster rate than bulk reaction, perhaps due to droplet evaporation and droplet confinement of reagents. For imaging MS, sample preparation steps for in-plume microdroplet reaction may involve including any technique that is utilized, adding reagent in an analytical sprayer or flooding vapor in a chamber during the ionization process.

In reactive DESI and reactive LAESI experiments, reagents are dissolved to the electrospray solvent. The solution seeded with dissolved reagent undergoes in situ reaction with the analyte during the imaging acquisition. In reactive DESI, the reagent ions react with the analytes on surfaces to form a reactant.[70] In reactive LAESI, the electrosprayed droplets containing reactants coalesce and react with the analyte present in the laser-ablation plume.[71] Reactive DESI has been used for improving detection of natural products of a marine alga by adding anions such Cl^-, Br^-, and CF_3COO^- in the DESI spray[72] perform imaging in of rat brain and zebrafish tissues with enhanced detection of lipids[73]. In-plume reactions of reactive LAESI have been used to obtain structure-specific fragmentation of lipid from the tissue without adding any reactant to the biological sample.[71] Reactive LAESI have also been used to study click-chemistry reactions, high yielding reactions commonly used in bioconjugation, in charged microdroplets and to perform imaging MS.[74]

Vapor exposure in gas cluster ion beams SIMS has been shown to enhance its molecular coverage. Trifluoroacetic acid vapor exposure during the imaging of lipids in mouse brain sections led to uncovering a more extensive range of lipid species in the white matter regions of the tissue.[75]

Adding chemicals to tissue

Chemical doping

Chemical doping includes any sample preparation steps that add chemical species without any reaction. Doping or spiking can take place before or after the tissue is dissected from an animal or can be injected to the animal before issue excision. For example, $^{13}C_6$-glucose was intraperitoneally injected with for visualization of cerebral metabolic fluxes of glucose in mice.[69] Chemicals can be also added to the tissue and used as internal mass calibrants. For example, internal calibrants deposited on tissue was used for recalibrating all MALDI spectra for obtaining more accurate *m/z* values.[76] Often a standard, such as leucine enkephalins (~200 pg/μL) is dissolved in DESI sprayer solvent to be used as a lock-mass to correct any potential mass drift of the quadrupole time-of-flight mass spectrometer.[3]

Internal standards are added onto adjacent control samples to quantitate concentrations of the analyte. Alternatively, an internal standard can be added to spray or matrix solution depending ionization source used. The examples include; quantitative MALDI imaging of cocaine from brain tissue by spotting its deuterated internal standard,[77] internal standard and matrix deposited using an inkjet desktop printer to correct different ionization efficiencies of lipids analyzed in the kidney by MALDI,[78] isotopically labeled internal standard deposited using an acoustic robotic spotter for absolute quantitative MALDI imaging of rifampicin on the liver,[79] adding internal standard to tissue homogenate to an analysis of erlotinib and its metabolites in rat tissue sections by MALDI, micropipetting internal standard on top of the tissues to quantify clozapine by DESI, adding internal standards to the solvent used for mapping distributions of nicotine in rat brain tissue by using nano-DESI,[80] quantifying endogenous lipids in single cells by adding phosphatidylcholine internal standard into the nano-DESI solvent.[81] In another example, internal standards are added to control tissue make mimetic tissue models for quantitative imaging. The mimetic models are made by serially freezing tissue homogenates with spiked internal standards in a mold. This results in a stepped concentration gradient of the spiked internal standards in tissue homogenate.[82]

Matrix for MALDI

An application of small molecules or nanoparticle as a matrix is required for ionization for MALDI. The sensitivity of ionization and spatial resolution is affected by the type of matrix and method of its application. When the matrix is applied to the tissue section, the analyte is cocrystallized with matrix crystals that are irradiated and ionized by a laser pulse. An ideal matrix application would be a uniform coating on the sample with small crystal size and produce efficient analyte extraction and ionization without any delocalization. There are many types of matrix molecules used in imaging, such as 2,5-dihydroxybenzoic acid, -α-cyano-4-hydroxycinnamic acid, 9-aminoacridine. Besides small organic molecules, many nanostructures have been used as matrices, such as carbon nanotubes,[83] gold nanoparticles,[84] colloidal graphite,[85] quantum dots,[86] silicon nanopost array,[87] etc. Application of these novel matrices can enable an analysis of hard to access drugs and molecules within the tissue. Some analytes are highly dependent on the type of MALDI used. Similarly, there are many types of equipment for automated matrix deposition, from nebulizer to inkjet printers. More details on the matrix application are discussed in another chapter.

Conclusions

Proper sample preparation is needed for accurate, reproducible, and sensitive imaging analysis.[90] Without sample preparation, molecular coverage of any imaging MS technique is limited to a handful of abundant and easily ionizable compounds, such as phospholipids in the tissue. Chemical modification strategies, such as enzymatic digestion and chemical derivatization, are expected to become more routine with an increase in demand for imaging a more diverse group of compounds. A decent sample preparation strategy is needed to avoid any delocalization or decomposition of the molecules, representing the accurate biological spatial distribution to nature as possible. Small variations in sample preparation steps or protocol can have a significant butterfly effect in the abundance and distribution of a molecule-of-interest sample. Well-documented and well-followed sample preparation is a prerequisite for any reproducible and repeatable imaging results. In the future, we can anticipate more elaborate and automated, precise sample preparation steps for imaging MS increasing its versatility and applicability.

References

1. Huang M-Z, Yuan C-H, Cheng S-C, Cho Y-T, Shiea J. Ambient ionization mass spectrometry. *Annu Rev Aanal Chem*. 2010;3:43–65.
2. Perez CJ, Bagga AK, Prova SS, Yousefi Taemeh M, Ifa DR. Review and perspectives on the applications of mass spectrometry imaging under ambient conditions. *Rapid Commun Mass Spectrom*. 2019;33:27–53.
3. Midey A, Olivos H, Shrestha B. Spatial mapping of cellular metabolites using desi ion mobility mass spectrometry*Single Cell Metabolism*: Springer; 2020:181–190.
4. Shrestha B, Patt JM, Vertes A. In situ cell-by-cell imaging and analysis of small cell populations by mass spectrometry. *Anal Chem*. 2011;83:2947–2955.
5. Shariatgorji M, Nilsson A, Fridjonsdottir E, et al. Comprehensive mapping of neurotransmitter networks by MALDI–MS imaging. *Nat Methods*. 2019;16:1021–1028.
6. Ly A, Longuespée R, Casadonte R, et al. Site-to-site reproducibility and spatial resolution in MALDI–MSI of peptides from formalin-fixed paraffin-embedded samples. *Proteomics Clin Appl*. 2019;13:1800029.
7. Schwartz SA, Reyzer ML, Caprioli RM. Direct tissue analysis using matrix-assisted laser desorption/ionization mass spectrometry: practical aspects of sample preparation. *J Mass Spectrom*. 2003;38:699–708.

8. Kaletaş BK, van der Wiel IM, Stauber J, et al. Sample preparation issues for tissue imaging by imaging MS. *Proteomics*. 2009;9:2622–2633.
9. Setou M. *Imaging Mass Spectrometry: Protocols for Mass Microscopy*: Springer Science & Business Media; 2010:11–91.
10. Goodwin RJA. Sample preparation for mass spectrometry imaging: small mistakes can lead to big consequences. *J Proteomics*. 2012;75:4893–4911.
11. Dong Y, Li B, Malitsky S, et al. Sample preparation for mass spectrometry imaging of plant tissues: a review. *Front Plant Sci*. 2016;7:60.
12. Pierson EE, Midey AJ, Forrest WP, et al. Direct drug analysis in polymeric implants using desorption electrospray ionization – mass spectrometry imaging (DESI-MSI). *Pharm Res*. 2020;37:107.
13. Li Y, Shrestha B, Vertes A. Atmospheric pressure infrared MALDI imaging mass spectrometry for plant metabolomics. *Anal Chem*. 2008;80:407–420.
14. Miki A, Katagi M, Kamata T, et al. MALDI-TOF and MALDI-FTICR imaging mass spectrometry of methamphetamine incorporated into hair. *J Mass Spectrom*. 2011;46:411–416.
15. Sjövall P, Lausmaa J, Nygren H, Carlsson L, Malmberg P. Imaging of membrane lipids in single cells by imprint-imaging time-of-flight secondary ion mass spectrometry. *Anal Chem*. 2003;75:3429–3434.
16. Caprioli RM, Farmer TB, Gile J. Molecular imaging of biological samples: localization of peptides and proteins using MALDI-TOF MS. *Anal Chem*. 1997;69:4751–4760.
17. Ifa DR, Srimany A, Eberlin LS, et al. Tissue imprint imaging by desorption electrospray ionization mass spectrometry. *Anal Methods*. 2011;3:1910–1912.
18. Tata A, Perez CJ, Hamid TS, Bayfield MA, Ifa DR. Analysis of metabolic changes in plant pathosystems by imprint imaging DESI-MS. *J Am Soc Mass Spectrom*. 2014;26:641–648.
19. Hemalatha R, Pradeep T. Understanding the molecular signatures in leaves and flowers by desorption electrospray ionization mass spectrometry (DESI MS) imaging. *J Agric Food Chem*. 2013;61:7477–7487.
20. Tata A, Perez C, Campos ML, Bayfield MA, Eberlin MN, Ifa DR. Imprint desorption electrospray ionization mass spectrometry imaging for monitoring secondary metabolites production during antagonistic interaction of fungi. *Anal Chem*. 2015;87:12298–12305.
21. Angolini CFF, Vendramini PH, Araújo FDS, et al. Direct protocol for ambient mass spectrometry imaging on agar culture. *Anal Chem*. 2015;87:6925–6930.
22. Ellis BM, Fischer CN, Martin LB, Bachmann BO, McLean JA. Spatiochemically profiling microbial interactions with membrane scaffolded desorption electrospray ionization-ion mobility-imaging mass spectrometry and unsupervised segmentation. *Anal Chem*. 2019;91:13703–13711.
23. Zivkovic AM, Wiest MM, Nguyen UT, Davis R, Watkins SM, German JB. Effects of sample handling and storage on quantitative lipid analysis in human serum. *Metabolomics*. 2009;5:507.
24. Haid M, Muschet C, Wahl S, et al. Long-term stability of human plasma metabolites during storage at– 80 C. *J Proteome Res*. 2018;17:203–211.
25. Lukowski J, Pamreddy A, Velickovic D, et al. Storage conditions of human kidney tissue sections affect spatial lipidomics analysis reproducibility. *J Am Soc Mass Spectrom*. 2020.
26. Swales JG, Dexter A, Hamm G, et al. Quantitation of endogenous metabolites in mouse tumors using mass-spectrometry imaging. *Anal Chem*. 2018;90:6051–6058.
27. Lanekoff I, Phan NT, Van Bell CT, Winograd N, Sjövall P, Ewing AG. Mass spectrometry imaging of freeze-dried membrane phospholipids of dividing *Tetrahymena pyriformis*. *Surf Interface Anal*. 2013;45:211–214.

28. Goodwin RJA, Scullion P, MacIntyre L, Watson DG, Pitt AR. Use of a solvent-free dry matrix coating for quantitative matrix-assisted laser desorption ionization imaging of 4-bromophenyl-1,4-diazabicyclo(3.2.2)nonane-4-carboxylate in rat brain and quantitative analysis of the drug from laser microdissected tissue regions. *Anal Chem*. 2010;82:3868–3873.
29. Goodwin RJ, Dungworth JC, Cobb SR, Pitt AR. Time-dependent evolution of tissue markers by MALDI-MS imaging. *Proteomics*. 2008;8:3801–3808.
30. Goodwin RJA, Pitt AR, Harrison D, et al. Matrix-free mass spectrometric imaging using laser desorption ionisation Fourier transform ion cyclotron resonance mass spectrometry. *Rapid Commun Mass Spectrom*. 2011;25:969–972.
31. Shariatgorji M, Källback P, Gustavsson L, et al. Controlled-pH tissue cleanup protocol for signal enhancement of small molecule drugs analyzed by MALDI-MS imaging. *Anal Chem*. 2012;84:4603–4607.
32. van Remoortere A, van Zeijl RJM, van den Oever N, et al. MALDI imaging and profiling MS of higher mass proteins from tissue. *J Am Soc Mass Spectrom*. 2010;21:1922–1929.
33. Seeley EH, Oppenheimer SR, Mi D, Chaurand P, Caprioli RM. Enhancement of protein sensitivity for MALDI imaging mass spectrometry after chemical treatment of tissue sections. *J Am Soc Mass Spectrom*. 2008;19:1069–1077.
34. Angel PM, Spraggins JM, Baldwin HS, Caprioli R. Enhanced sensitivity for high spatial resolution lipid analysis by negative ion mode matrix assisted laser desorption ionization imaging mass spectrometry. *Anal Chem*. 2012;84:1557–1564.
35. Yang J, Caprioli RM. Matrix sublimation/recrystallization for imaging proteins by mass spectrometry at high spatial resolution. *Anal Chem*. 2011;83:5728–5734.
36. Cha S, Zhang H, Ilarslan HI, et al. Direct profiling and imaging of plant metabolites in intact tissues by using colloidal graphite-assisted laser desorption ionization mass spectrometry. *Plant J*. 2008;55:348–360.
37. Quiason CM, Shahidi-Latham SK. Imaging MALDI MS of dosed brain tissues utilizing an alternative analyte pre-extraction approach. *J Am Soc Mass Spectrom*. 2015;26:967–973.
38. Gessel M, Spraggins JM, Voziyan P, Hudson BG, Caprioli RM. Decellularization of intact tissue enables MALDI imaging mass spectrometry analysis of the extracellular matrix. *J Mass Spectrom*. 2015;50:1288–1293.
39. Dueñas ME, Larson EA, Lee YJ. Toward mass spectrometry imaging in the metabolomics scale: increasing metabolic coverage through multiple on-tissue chemical modifications. *Front Plant Sci*. 2019;10:860 860.
40. Manier ML, Reyzer ML, Goh A, Dartois V, Via LE, Barry 3rd CE, Caprioli RM. Reagent precoated targets for rapid in-tissue derivatization of the anti-tuberculosis drug isoniazid followed by MALDI imaging mass spectrometry. *J Am Soc Mass Spectrom*. 2011;22:1409–1419.
41. Manier ML, Spraggins JM, Reyzer ML, Norris JL, Caprioli RM. A derivatization and validation strategy for determining the spatial localization of endogenous amine metabolites in tissues using MALDI imaging mass spectrometry. *J Mass Spectrom*. 2014;49:665–673.
42. Chacon A, Zagol-Ikapitte I, Amarnath V, et al. On-tissue chemical derivatization of 3-methoxysalicylamine for MALDI-imaging mass spectrometry. *J Mass Spectrom*. 2011;46:840–846.
43. Cobice DF, Livingstone DEW, Mackay CL, et al. Spatial localization and quantitation of androgens in mouse testis by mass spectrometry imaging. *Anal Chem*. 2016;88:10362–10367.
44. Shimma S, Kumada H-O, Taniguchi H, et al. Microscopic visualization of testosterone in mouse testis by use of imaging mass spectrometry. *Anal Bioanal Chem*. 2016;408:7607–7615.

45. Barré FPY, Flinders B, Garcia JP, et al. Derivatization strategies for the detection of triamcinolone acetonide in cartilage by using matrix-assisted laser desorption/ionization mass spectrometry imaging. *Anal Chem*. 2016;88:12051–12059.
46. Yang H, Chandler CE, Jackson SN, et al. On-tissue derivatization of lipopolysaccharide for detection of lipid A using MALDI-MSI. *Anal Chem*. 2020;92:13667–13671.
47. Liu X, Qiu H, Lee RK, Chen W, Li J. methylamidation for sialoglycomics by MALDI-MS: a facile derivatization strategy for both α2,3- and α2,6-linked sialic acids. *Anal Chem*. 2010;82:8300–8306.
48. Tong W, Han H, Song Z, et al. Chemical derivatization with a polycyclic aromatic hydrocarbon for highly sensitive detection of N-linked glycans using MALDI-TOF MS. *Anal Methods*. 2012;4:3531–3535.
49. Gao W, Li H, Liu Y, et al. Rapid and sensitive analysis of N-glycans by MALDI-MS using permanent charge derivatization and methylamidation. *Talanta*. 2016;161:554–559.
50. Ullmer R, Plematl A, Rizzi A. Derivatization by 6-aminoquinolyl-N-hydroxysuccinimidyl carbamate for enhancing the ionization yield of small peptides and glycopeptides in matrix-assisted laser desorption/ionization and electrospray ionization mass spectrometry. *Rapid Commun Mass Spectrom*. 2006;20:1469–1479.
51. Franck J, El Ayed M, Wisztorski M, Salzet M, Fournier I. On-tissue N-terminal peptide derivatizations for enhancing protein identification in MALDI mass spectrometric imaging strategies. *Anal Chem*. 2009;81:8305–8317.
52. Arrigoni G, Resjö S, Levander F, et al. Chemical derivatization of phosphoserine and phosphothreonine containing peptides to increase sensitivity for MALDI-based analysis and for selectivity of MS/MS analysis. *Proteomics*. 2006;6:757–766.
53. Wu Q, Comi TJ, Li B, Rubakhin SS, Sweedler JV. On-tissue derivatization via electrospray deposition for matrix-assisted laser desorption/ionization mass spectrometry imaging of endogenous fatty acids in rat brain tissues. *Anal Chem*. 2016;88:5988–5995.
54. Denti V, Piga I, Guarnerio S, et al. Antigen retrieval and its effect on the MALDI-MSI of lipids in formalin-fixed paraffin-embedded tissue. *J Am Soc Mass Spectrom*. 2020;31:1619–1624.
55. Levenson RM, Borowsky AD, Angelo M. Immunohistochemistry and mass spectrometry for highly multiplexed cellular molecular imaging. *Lab Invest*. 2015;95:397–405.
56. Giesen C, Wang HA, Schapiro D, et al. Highly multiplexed imaging of tumor tissues with subcellular resolution by mass cytometry. *Nat Methods*. 2014;11:417–422.
57. Angelo M, Bendall SC, Finck R, et al. Multiplexed ion beam imaging of human breast tumors. *Nat Med*. 2014;20:436–442.
58. Thiery G, Shchepinov MS, Southern EM, et al. Multiplex target protein imaging in tissue sections by mass spectrometry–TAMSIM. *Rapid Commun Mass Spectrom*. 2007;21:823–829.
59. Wang H, DeGnore JP, Kelly BD, True J, Garsha K, Bieniarz C. A technique for relative quantitation of cancer biomarkers in formalin-fixed, paraffin-embedded (FFPE) tissue using stable-isotope-label based mass spectrometry imaging (SILMSI). *J Mass Spectrom*. 2015;50:1088–1095.
60. Cole L, Djidja M, Bluff J, et al. Investigation of protein induction in tumour vascular targeted strategies by MALDI MSI. *Methods*. 2011;54:442–453.
61. Groseclose MR, Andersson M, Hardesty WM, Caprioli RM. Identification of proteins directly from tissue: in situ tryptic digestions coupled with imaging mass spectrometry. *J Mass Spectrom*. 2007;42:254–262.
62. Powers TW, Neely BA, Shao Y, et al. MALDI imaging mass spectrometry profiling of N-glycans in formalin-fixed paraffin embedded clinical tissue blocks and tissue microarrays. *PLoS One*. 2014;9:e106255.

63. Everest-Dass AV, Briggs MT, Kaur G, Oehler MK, Hoffmann P, Packer NH. N-glycan MALDI imaging mass spectrometry on formalin-fixed paraffin-embedded tissue enables the delineation of ovarian cancer tissues. *Mol Cell Proteomics*. 2016;15:3003–3016.
64. Che F-Y, Lim J, Pan H, Biswas R, Fricker LD. Quantitative neuropeptidomics of microwave-irradiated mouse brain and pituitary. *Mol Cell Proteomics*. 2005;4:1391–1405.
65. Goodwin RJA, Lang AM, Allingham H, Borén M, Pitt AR. Stopping the clock on proteomic degradation by heat treatment at the point of tissue excision. *Proteomics*. 2010;10: 1751–1761.
66. Griffiths RL, Simmonds AL, Swales JG, Goodwin RJA, Cooper HJ. LESA MS imaging of heat-preserved and frozen tissue: benefits of multistep static FAIMS. *Anal Chem*. 2018;90:13306–13314.
67. Blatherwick EQ, Svensson CI, Frenguelli BG, Scrivens JH. Localisation of adenine nucleotides in heat-stabilised mouse brains using ion mobility enabled MALDI imaging. *Int J Mass Spectrom*. 2013:19–27.
68. Mulder IA, Esteve C, Wermer MJ, et al. Funnel-freezing versus heat-stabilization for the visualization of metabolites by mass spectrometry imaging in a mouse stroke model. *Proteomics*. 2016;16:1652–1659.
69. Sugiura Y, Honda K, Kajimura M, Suematsu M. Visualization and quantification of cerebral metabolic fluxes of glucose in awake mice. *Proteomics*. 2014;14:829–838.
70. Chen H, Cotte-Rodríguez I, Cooks RG. cis-Diol functional group recognition by reactive desorption electrospray ionization (DESI). *Chem Commun*. 2006:597–599.
71. Shrestha B, Nemes P, Nazarian J, Hathout Y, Hoffman EP, Vertes A. Direct analysis of lipids and small metabolites in mouse brain tissue by AP IR-MALDI and reactive LAESI mass spectrometry. *Analyst*. 2010;135:751–758.
72. Nyadong L, Hohenstein EG, Galhena A, et al. Reactive desorption electrospray ionization mass spectrometry (DESI-MS) of natural products of a marine alga. *Anal Bioanal Chem*. 2009;394:245–254.
73. Lostun D, Perez CJ, Licence P, Barrett DA, Ifa DR. Reactive DESI-MS imaging of biological tissues with dicationic ion-pairing compounds. *Anal Chem*. 2015;87:3286–3293.
74. van Geenen FAMG, Franssen MCR, Zuilhof H, Nielen MWF. Reactive laser ablation electrospray ionization time-resolved mass spectrometry of click reactions. *Anal Chem*. 2018;90:10409–10416.
75. Angerer TB, Dowlatshahi Pour M, Malmberg P, Fletcher JS. Improved molecular imaging in rodent brain with time-of-flight-secondary ion mass spectrometry using gas cluster ion beams and reactive vapor exposure. *Anal Chem*. 2015;87:4305–4313.
76. Gustafsson JOR, Eddes JS, Meding S, et al. Internal calibrants allow high accuracy peptide matching between MALDI imaging MS and LC-MS/MS. *J Proteomics*. 2012;75:5093–5105.
77. Pirman DA, Reich RF, Kiss A, Heeren RMA, Yost RA. Quantitative MALDI tandem mass spectrometric imaging of cocaine from brain tissue with a deuterated internal standard. *Anal Chem*. 2013;85:1081–1089.
78. Aboulmagd S, Esteban-Fernández D, Moreno-Gordaliza E, et al. Dual internal standards with metals and molecules for MALDI imaging of kidney lipids. *Anal Chem*. 2017;89:12727–12734.
79. Prentice BM, Chumbley CW, Caprioli RM. Absolute quantification of rifampicin by MALDI imaging mass spectrometry using multiple TOF/TOF events in a single laser shot. *J Am Soc Mass Spectrom*. 2016;28:136–144.
80. Lanekoff I, Thomas M, Carson JP, Smith JN, Timchalk C, Laskin J. Imaging nicotine in rat brain tissue by use of nanospray desorption electrospray ionization mass spectrometry. *Anal Chem*. 2013;85:882–889.

81. Bergman H-M, Lanekoff I. Profiling and quantifying endogenous molecules in single cells using nano-DESI MS. *Analyst*. 2017;142:3639–3647.
82. Barry JA, Groseclose MR, Fraser DD, Castellino S, Revised preparation of a mimetic tissue model for quantitative imaging mass spectrometry. 2018.
83. Ren S-f, Zhang L, Cheng Z-h, Guo Y-l. Immobilized carbon nanotubes as matrix for MALDI-TOF-MS analysis: applications to neutral small carbohydrates. *J Am Soc Mass Spectrom*. 2005;16:333–339.
84. Su C-L, Tseng W-L. Gold nanoparticles as assisted matrix for determining neutral small carbohydrates through laser desorption/ionization time-of-flight mass spectrometry. *Anal Chem*. 2007;79:1626–1633.
85. Zhang H, Cha S, Yeung ES. Colloidal graphite-assisted laser desorption/ionization MS and MS n of small molecules. 2. Direct profiling and MS imaging of small metabolites from fruits. *Anal Chem*. 2007;79:6575–6584.
86. Shrivas K, Kailasa SK, Wu H-F. Quantum dots laser desorption/ionization MS: multifunctional CdSe quantum dots as the matrix, concentrating probes and acceleration for microwave enzymatic digestion for peptide analysis and high resolution detection of proteins in a linear MALDI-TOF MS. *Proteomics*. 2009;9:2656–2667.
87. Stopka SA, Rong C, Korte AR, et al. Molecular imaging of biological samples on nanophotonic laser desorption ionization platforms. *Angew Chem*. 2016;128:4558–4562.
88. Espy, R. D.; Wleklinski, M.; Yan, X.; Cooks, R. G., Beyond the flask: reactions on the fly in ambient mass spectrometry. *TrAC Trends in Analytical Chemistry* 2014, 57, 135–146.2.
89. Shrestha, B.; Nemes, P.; Nazarian, J.; Hathout, Y.; Hoffman, E. P.; Vertes, A., Direct analysis of lipids and small metabolites in mouse brain tissue by AP IR-MALDI and reactive LAESI mass spectrometry. *Analyst* 2010, 135 (4), 751–758.
90. Balluff, B.; Hopf, C.; Siegel, T.P.; Grabsch, H.I.; Heeren, R.M.A., Batch Effects in MALDI Mass Spectrometry Imaging. *J. Am. Soc. Mass Spectrom.* 2021, doi:10.1021/jasms.0c00393.
91. Harkin, C.; Smith, K.W.; Cruickshank, F.L.; Mackay, C.L.; Flinders, B.; Heeren, R.M.A.; Moore, T.; Brockbank, S.; Cobice, D.F., On-tissue chemical derivatization in mass spectrometry imaging. *Mass Spectrom. Rev.* 2021, doi:10.1002/mas.21680.

Chapter 4

Tissue sectioning for imaging mass spectrometry

Chapter Outline

Imaging mass spectrometry (MS) aims to correlate biochemical localization in tissue with its morphology. In order to do such correlation, a thin tissue section, typically between 5 and 15 microns thick, is mounted on a target plate such as a microscope glass slide. Tissue sections are prepared similarly as a frozen histological sections, where a knife attached in a microtome cuts a thin part from the desired location of the tissue and thaw-mounted.

Tissue sectioning for imaging MS should be carefully performed in order to avoid delocalization or degradation of molecules within the tissue. Suboptimal sample preparation, mishandling pre or post sample preparation, or improperly storing tissue may cause delocalization and degradation of the analytes within the tissue. Some excellent recommendations on methods for collection, handling, storage, retrieval, and distribution of biospecimen can be found at recommendation published by International Society for Biological and Environmental Repositories,[1] Biorepositories and Biospecimen Research Branch of National Cancer Institute,[2] National Heart, Lung, and Blood Institute Biorepository.[3] National Cancer Institute also maintains Biospecimen Research Database that contains a free and publicly accessible database with peer-reviewed and user-generated standard operating protocols for biospecimen collection curated by a Ph.D. scientist.[4]

Tissue collection

Extensive data on the biospecimen should be recorded during sample collection. At a minimum, organ, animal breed, age, gender, any disease status, or drug treatment must be noted. Also, details on any applicable biospecimen acquisition steps such as surgical or dissection procedure, surgical duration,

Introduction to Spatial Mapping of Biomolecules by Imaging Mass Spectrometry. DOI: 10.1016/C2018-0-03962-6

anesthesia, warm ischemia time, postmortem interval time should also be logged if possible. McDonnell et al.,[5] describe minimum reporting guidelines for tissue samples used for imaging MS experiments, which includes (a) tissue origin (institution, ethical approval); (b) specimen (species, age, sex, and organ); (c) sampling method (postmortem time, stabilization, resection status, and freezing method); (d) fixing or embedding method if any; and (e) morphological classification. The article also describes minimum reporting for sample preparation, data acquisition, mass spec preprocessing, imaging MS visualization, compound identification strategy, data analysis requirements to describe imaging MS experiments accurately.

Snap freezing

Tissue should be frozen or fixed as promptly as possible after cessation of circulation to avoid morphological distortions and damage due to tissue drying artifact, autolysis due to enzymatic activities, putrefaction, or decomposition by microorganisms. Freezing sample slowly cause morphological distortion of tissue because it leads to ice crystal formation that replaces the tissue architecture. In snap freezing, samples are frozen very rapidly so that water does not have time to form ice crystals. Snap freezing of fresh tissue is fast, compatible with molecular imaging techniques such as immunohistochemistry (IF), or imaging MS, and no cross-linking negating antigen retrieval step. The most critical factor that influences the rate of cooling is the size of the sample. Hence freezing small pieces of sample to minimize ice crystal artifact is recommended.[6]

In a snap-freezing method proposed for imaging MS by Schwartz et al., the tissue is loosely wrapped in aluminum foil and gently lowered in liquid nitrogen (LN2) for a minute or less. Avoiding rapid plunging of the tissue into the LN2 often leads to cracking.[7] LN2 is often used for the rapid freezing of tissue.[8] However, LN2 has a low specific heat constant, which means when it contacts warm tissues, it boils and forms a vapor barrier. The barrier acts as an insulator resulting in slow and uneven freezing with cracks, especially of large tissues. Instead of LN2, isopentane (2-methyl butane) can be used since it has a high thermal conductivity.

Here are some alternative steps for snap freezing tissue. In a styrofoam box with LN2, add a half-full metal canister of isopentane and equilibrate for a few minutes, as shown in the example of the snap freezing of skeletal muscle.[9] A practical way to gauge the coldness of isopentane is to see it boils off when it comes in contact with dry ice; cold isopentane would not. Using long forceps (e.g., 12"), grasp tissue mold and hold them by just immersing the base of the molds in the liquid, but flow over the top of the mold. See Fig. 4.1 for illustration. Temporarily store the snap-frozen blocks by putting them on dry ice. Take precautions to avoid the thaw and refreeze cycle of tissue leading to deforming ice crystal formation in tissues.

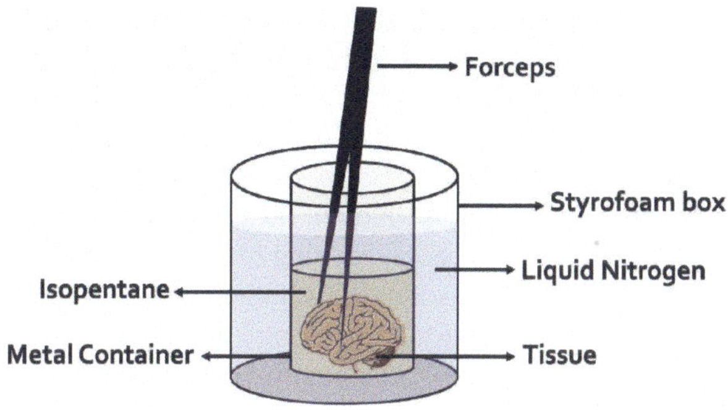

FIG. 4.1 Bulk tissue is placed in a metal container filled with isopentane with a pair of long forceps right after dissection or resection. Isopentane container is cooled in liquid nitrogen. After snap freezing, the tissue can be stored at −80°C or directly placed in a cryostat for sectioning.

Fixing or embedding

In histological imaging, chemical fixatives are often used to preserve and maintain the tissue morphology. Unfortunately, any fixation of tissue can potentially causes loss of metabolites and lipids or, in the case of proteins, leads to its denaturation. The best way to mitigate any loss or change of biomolecules is to prepare frozen sections from unfixed or fresh frozen tissues.

For general histology, tissues are often embedded in a harder medium, such as paraffin wax or optimal cutting temperature (OCT) compound, to allow easier cutting. In paraffin and resin embedding, tissues are dehydrated and replaced with intermediary clearing fluids or directly perfused with an embedding medium that solidifies. During OCT embedding, samples are surrounded by OCT media and snap frozen. The OCT compound contains a water-soluble blend of polyvinyl alcohol and polyethylene glycol (PEG) that freezes between −20 and −80°C for convenient mounting and sectioning. For histological imaging, OCT leaves no residue on slides during the staining procedures and compatible with chromogenic immunohistochemistry experiments. However, the PEGs can overwhelm and dominate the mass spectrometry signal suppressing ionization of other molecular species such as lipids and metabolites. PEGs in OCT suppresses ion formation from molecules in the tissue resulting in a mass spectrum dominated by its ions separated by m/z 44 for single charged ions. Ion suppression is a phenomenon where the ionization efficiency of analyte(s) of interest is significantly reduced due to the presence of molecular species in the sample that compete for ionization or inhibit efficient ionization in other ways.

In some cases, an embedding medium may be required to stabilize the tissue and provides a smooth cutting surface, e.g., sectioning small millimeter-sized samples such as a housefly or morphologically unstable tissue. If some sort of embedding media is needed as a support for cutting, carboxymethyl cellulose (CMC) can be used instead of OCT. Other techniques include sucrose infiltration (30% w/v) followed by embedding in gelatin (10% w/v;

no. 4078; Merck, Darmstadt, Germany).[10] Several embedding media have been reported in the literature, including sodium CMC,[11] gelatin,[12] agarose,[13] or poly,N-(hydroxypropyl)-methacrlyamide,[14] polyvinylpyrrolidone, hydroxypropyl-methylcellulose.[15] Table 4.1 has a summary of several common embedding media and their effect of sectioning properties for imaging MS. A more detailed evaluation is given in the referenced publication.[15]

Here are some general steps for embedding biological samples using CMC. Start with making a 2.5% CMC solution (w/v) in deionized water (e.g., CMC sodium salt C4888) by stirring at medium speed using a magnetic stirrer until the CMC dissolves into a clear, viscous liquid. To embed, add a drop of CMC to the bottom of the cryomold, chill, and secure the sample in the correct orientation. Pour CMC to cover the secured tissue completely and freezes when it will turn from clear to cloudy. The whole CMC block containing the tissue is removed, mounted, and sectioned between about −20°C to −25°C.

For typical imaging MS workflow, fresh-frozen sections are preferred for analysis of small soluble molecules such as lipids. Formalin-fixed paraffin-embedded tissues have been used for imaging glycans attached to asparagine or N-linked glycans, and on-tissue trypsin digested peptides. Detail protocols describing sample preparations steps such as dewaxing, antigen retrieval, application of enzymes of formalin-fixed paraffin-embedded for analysis of tryptic peptides, and N-glycans are referenced herein.[16-17]

Tissue orientation

Tissue must be mounted in a chuck at with the right alignment and orientation. For the embedded sample, tissue must be correctly oriented when it is placed in cryomold for embedding. For fresh-frozen samples, tissue must be correctly oriented when the sample is placed in a sectioning chuck. For a few samples, the orientation of tissue may not be critical, but for the vast majority of specimens and studies, the scientific conclusion depends on the orientation and position of the section.

In human and animal anatomy, the orientation of tissue is defined by a hypothetical anatomical plane that divides the body. Three principal planes are used; sagittal plane (longitudinal) divides the body into left and right, the coronal plane (vertical) divides the body into back and front) portions, and the transverse plane (axial, lateral, and horizontal) divides the body into the head and tail portions. The cardinal or principal plane divides the body or tissue in the middle into two equal segments. An exact location of the section can be defined with respect to the distance to the cardinal plane. A typical example of such anatomical plane and sections is shown in Fig. 4.2. In discussing the anatomy of animals, a simplistic convention has been to name the sections according to the homologous human sections to preserve comparison with a human. For example, a transverse section with respect to the body axis of a rat, i.e., dividing anterior and posterior, is often be referred to as a coronal section and a coronal section with respect to the animal's body, i.e., dividing ventral

TABLE 4.1 Summary of sectioning properties for the embedding media for imaging mass spectrometry.

Embedding medium	Preparation	Viscosity	Possible to section	Adherence to tissues	Matrix interference	Interference with histological stains
OCT medium	Ready to use	Acceptable	Yes	Very good	PEG/PVA introduce polymer peaks into spectra, smearing causing ion suppression	Not observed
10% gelatin	Fast preparation	Acceptable	Yes	Poor	Chemical background for small molecules (e.g., choline, arginine), tissues thaw in hot gelatin	Staining of medium with eosin and anticollagen-1 antibodies resulted in strong background masking tissue outlines
10% HPMA	Fast preparation	Too low for precise sample positioning	Yes	Acceptable	Background for sodium adducts observed	Not observed
12.5% Na-CMC	Swelling of polymer overnight	Very good	Yes	Very good	Shift toward sodium adducts, smearing resulting in ion suppression	Staining of medium with eosin resulting in strong background masking tissue outlines
10% HPMC	Swelling of polymer overnight	Very good	Yes	Acceptable	Not observed	Not observed
15% PVP	Swelling of polymer (hours)	Good	No	Very good	Not evaluated	Not evaluated
7.5% HPMC + 2.5% PVP	Swelling of polymer overnight	Very good	Yes	Very good	Not observed	Not observed

OCT, optimal cutting temperature; Na-CMC, sodium carboxymethyl cellulose; HPMA, poly,N-(hydroxypropyl)-methacrylamide; PVP, polyvinylpyrrolidone; HPMC, hydroxypropyl-methylcellulose.
Reprinted and adapted with permission from Dannhorn A, Kazanc E, Ling S, et al. Universal sample preparation unlocking multimodal molecular tissue imaging. Anal Chem. 2020;92:11080–11088. Copyright 2020 American Chemical Society.[15]

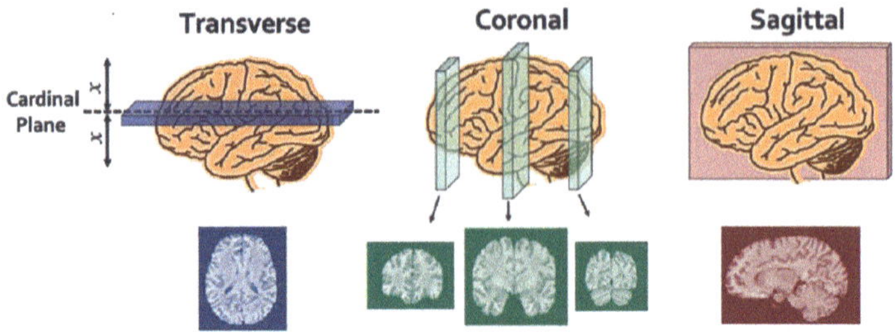

FIG. 4.2 The orientation of the tissue section is based on the direction anatomical plane used to cut the tissue, namely sagittal, coronal, or transverse. The cardinal plane divides the tissue in the middle into two equal segments.

and dorsal, is referred to as transverse. Often anatomical terminology based on anatomical landmarks is used with confusing unique terms and suffixes/prefixes derived from ancient Greek and Latin. In order to be more precise and reduce ambiguity, the international standard on human anatomic terminology *Terminologia Anatomica* was developed by the Federative Committee on Anatomical Terminology and the International Federation of Associations of Anatomists in 1998.

The correct orientation becomes challenging during the embedding and sectioning of especially in small millimeter size samples. To address these problems during people have used using embedding well bar, placing the tissue in embedding well, and use other strategies such as face-down cryoembedding technique.[18]

Preparing a thin fresh-frozen section from large samples (e.g., whole-body) or calcified tissue (e.g., tooth, bone) samples can be more challenging. Decalcification and fixation steps performed for easier sectioning usually result in alteration of the molecular makeup of tissue and not amenable to mass spectrometry workflows. Often heavy-duty sled cryostat with a more robust tungsten carbide blade is needed for sectioning.[19] In general, methods used for sectioning large samples include (a) coating the frozen surface with a polymeric solution before each cutting,[20] (b) covering the frozen surface with a sheet of paper,[21-22] (c) covering the frozen cutting surface with adhesive tape.[23]

The desired portion of the tissue section should not be from the very edge of the mounted tissue. It is prudent to leave some tissue margins on the top because one may need to trim away a significant portion of tissue to get an even section. Similarly, the bottom portion of tissue may be unusable due to malformation or contamination by sticking with embedding, such as OCT.

Cryosectioning

For sectioning in a cryostat, follow the manufacture's recommendation for preparing samples for histopathological analysis. The best way to section tissue for

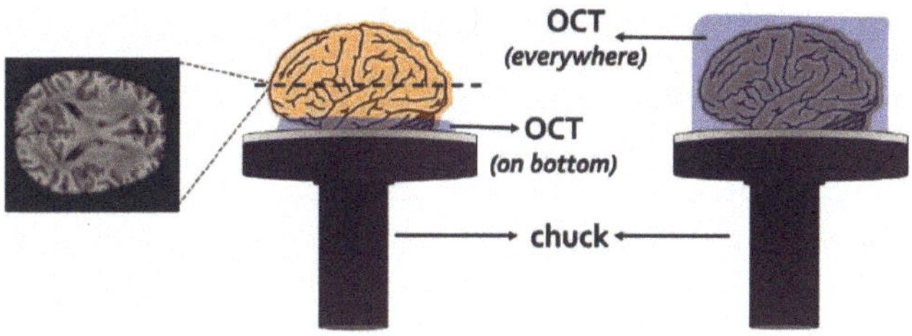

FIG. 4.3 The left chuck has a fresh-frozen brain section is placed on a chuck using OCT compound as a glue, while the right chuck has submerged brain tissue with OCT. In the left chuck, tissue sections can be cut from the portion that never came in contact with the OCT—making it a viable method of tissue mounting for imaging MS. The geometry of placement in chuck has predetermined the type of anatomical plane for section; in this case, it will be axial sections.

imaging MS is to section the fresh-frozen tissues directly on to imaging target plate or slides. OCT compound may be used to glue to adhere the tissue to the chuck of the cryostat instead of submerging the entire tissue with the OCT. See Fig. 4.3 for an illustrative example. Tissue sections can be cut from parts away from OCT, making sure the imaging target plate, blade, and sectioned tissue never came into contact with the OCT. The effect of OCT on matrix-assisted laser desorption/ionization signal of rat liver between *m/z* 4500 and 10,500 is shown in Fig. 4.4. When OCT is used to adhere the tissue to the sample stage without any contact with the sectioned tissue showed more signal than tissue that came in contact with OCT.[7]

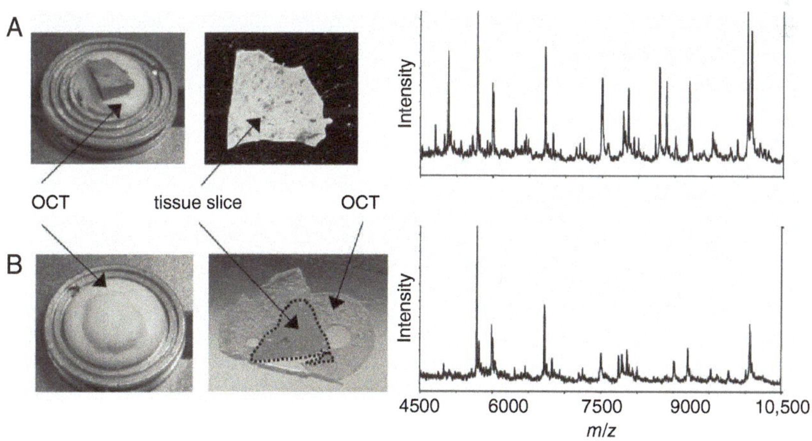

FIG. 4.4 (A) Optimal cutting temperature (OCT) is adhered to the tissue to the sample stage but does not come into contact with the sectioned tissue shows a more robust signal than (B) the tissue that was completely embedded in OCT. Reprinted with permission from Schwartz SA, Reyzer ML, Caprioli RM. Direct tissue analysis using matrix-assisted laser desorption/ionization mass spectrometry: practical aspects of sample preparation. J Mass Spectrom. 2003;38:699–708.[7]

Mounting

The traditional thaw-mounting method is used to adhere to a frozen section of tissue onto the slide for imaging MS data acquisition. Typically, after the section is excised, it is transferred and thaw-mounted on the imaging MS sample plate or glass slide by often dragging with a small paintbrush. The sample plate is maintained at a cold temperature in the cryostat chamber, and the section is thaw-mounted by warming the bottom of the plate using a hand.[32] A few micron-thick frozen tissues is cut in a cryostat and attached to a glass slide by applying heat from the finger at the opposite side of the slide. The thaw-mounting method may be vulnerable to change in analyte distributions due to thermal effects or undergo tissue shrinkage or tearing. Larger tissue (e.g., whole-body) or harder tissue (e.g., bone or plant organs such as seed or rice) are difficult to section and mount. Those tissues can be frozen at 80°C, attached to the double-sided adhesive film, sliced by a cryostat, and mounted using the adhesive tape. For example, sectioning and mounting a delicate heat-stabilized tissue while maintaining tissue integrity is a challenging task. A double-sided adhesive tape can be used to mount and transfer a thin section of the whole-body embedded in CMC or heat-stabilized tissue. A double-sided conductive carbon can be used for time-of-flight matrix-assisted laser desorption/ionization imaging that requires voltage application on the metal plate.[24] A comparison between thaw-mounting and use of conductive tape for sample preparation in secondary ion mass spectrometry imaging of lipids in adult flies found that the tape-transfer method gave a better result.[25] Snap-frozen adult flies embedded in 10% gelatin were cryosectioned and either thaw-mounted or attached to a double-sided adhesive carbon tape. A commercial, cryoJane tape-transfer system that uses adhesive coated slides and adhesive tapes to capture sections.[26] A workflow of two mounting methods, a thaw-mounting, and tape-mounting, on the same gelatin-embedded samples is shown in Fig. 4.5.[25]

Tissue section storage

The cut tissue sections on slides can be processed immediately or stored inside an air-tight container in a −80°C freezer. There are no standardized storage and handling procedures, and it varies a lot. There are a few studies that investigated the storage of aliquots for metabolomics and lipidomic. According to a study, high abundant metabolites and lipids in unfractionated serum were unaffected by repeated freeze–thaw cycles with coefficients of variation close to zero.[27] In another study, prolonged storage of biospecimens led to a change in metabolite and lipid concentrations. A pooled human plasma sample was aliquoted and stored at −80°C and analyzed nearly 200 time points over 5 years. Half of the metabolites remained stable, while oth-

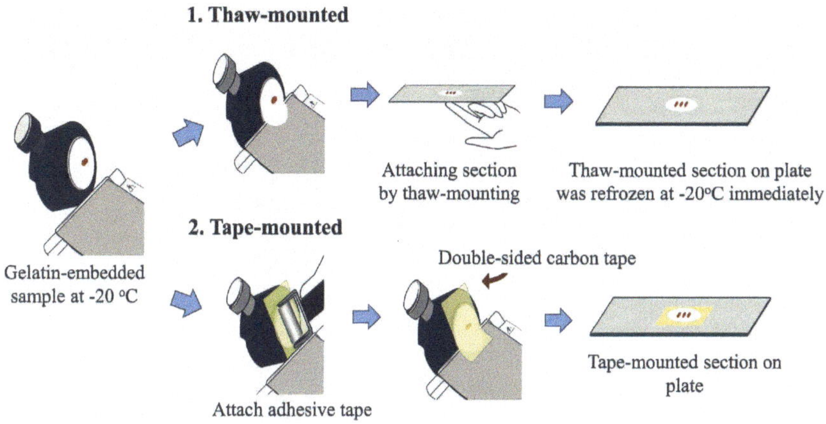

FIG. 4.5 A schematic representation of two mounting methods, a thaw-mounting and tape-mounting show for the same gelatin-embedded samples. Reprinted with permission from Le MUT, Son JG, Shon HK, Park JH, Lee SB, Lee TG, Comparison between thaw-mounting and use of conductive tape for sample preparation in ToF-SIMS imaging of lipids in *Drosophila* microRNA-14 model. Biointerphases. 2018;13:03B414.[25] Creative commons attribution.

ers changed by about a dozen percentage, while lipids decreased concentration levels.[28]

The degradation of molecular species within the tissue is inevitable irrespective of storage conditions. The degree of molecule change will depend on the type of tissue, molecules, and storage conditions. The storage effects on the phospholipid composition of sectioned thaw-mounted mouse tissues (brain, kidney, and liver) showed the global signal intensity decreased as a function of time and temperature while oxidized phospholipid and lysophospholipid species increased within 2 hours and 24 hours, respectively, at ambient conditions.[29] Lipid remains unaffected up to a week after storage in −80°C freezer under nitrogen. Freezing tissue sections in an inert gas, such as nitrogen, at low temperatures of −80°C can minimize changes in molecules.[30] Kidney tissue sections were analyzed for six condition—fresh frozen and openly stored at −80°C; vacuumed-sealed and stored at −80°C; matrix applied and stored at −80°C; under a nitrogen atmosphere and stored at −80°C; and at room temperature in a desiccator. The section stored under nitrogen and −80°C showed the least changes compared to the fresh tissue, and one stored at room temperature showed the most degradation, as illustrated in Fig. 4.6. In another study, optimum stabilization protocols for endogenous metabolite imaging included embedding frozen tissue in 10% w/v gelatin solution, sectioning followed by immediate desiccation in a stream of dry nitrogen, then vacuum packing and storage at −80°C until imaging.[31]

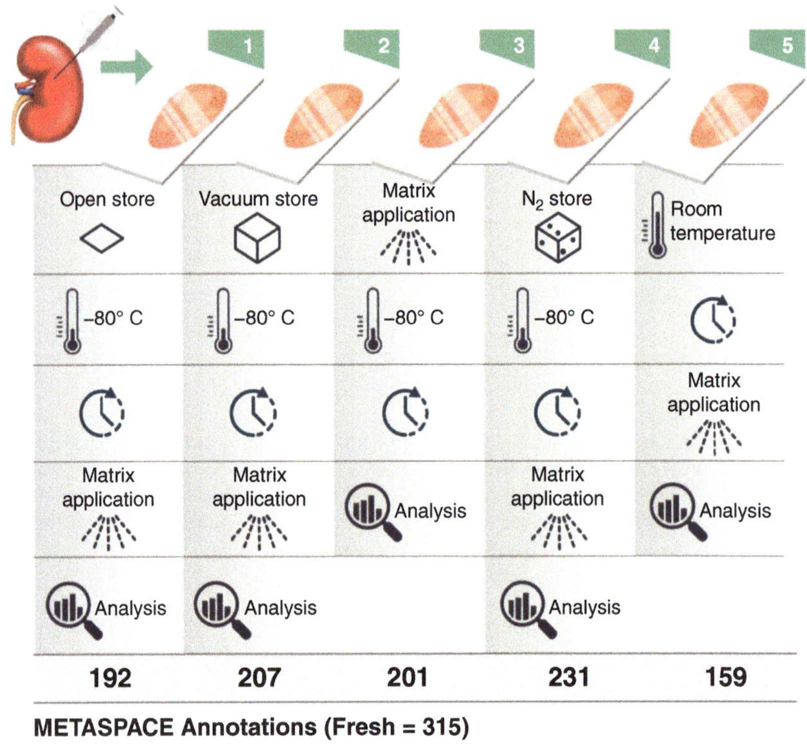

FIG. 4.6 A schematic showing similar slides containing kidney tissue sections were processed and stored in five ways—openly stored at −80°C; vacuumed-sealed and stored at −80°C; matrix applied and stored at −80°C; under a nitrogen atmosphere and stored at −80°C; and at room temperature in a desiccator. The bottom of the image shows the number of annotations by METASPACE for each storage condition has decreased after 144 hours of storage, compared to the fresh tissue section with 315 annotations. Reprinted and adapted with permission from Lukowski J, Pamreddy A, Velickovic D, et al. Storage conditions of human kidney tissue sections affect spatial lipidomics analysis reproducibility. J Am Soc Mass Spectrom. 2020. Copyright 2020 American Chemical Society.[30]

Conclusions

The sectioning of a tissue is the critical first step of imaging MS workflow. Often this step is prepared by scientists or technicians unfamiliar with technical aspects of the mass spectrometer. The sectioned tissue can be used for many types of imaging MS research to map the distribution of any detected endogenous (e.g., lipids, metabolites, peptides, and proteins) and exogenous molecules (drugs, pollutants). Fresh-frozen sections without direct contact with OCT are recommended as the starting point for imaging MS analysis. A good section is the first critical step in performing imaging MS research. Garbage in garbage out is very applicable here—*optimal tissue sections in, useful molecular images out.*

References

1. ISBER Best Practices. Recommendations for Repositories. 4th ed.: International Society for Biological and Environmental Repositories (ISBER); 2018:1–92.
2. NCI Best Practices for Biospecimen Resources. Biorepositories and Biospecimen Research Branch: National Cancer Institute, National Institutes of Health; 2016:4–48.
3. Guide to Building Biospecimen Collections for Study and Future Research Use. The National Heart, Lung, and Blood Institute (NHLBI) Biorepository: National Institutes of Health; 2020:3–24.
4. The Biospecimen Research Database (BRD). National Cancer Institute (NCI), Biorepositories and Biospecimen Research Branch (BBRB). Available at: https://brd.nci.nih.gov/brd/. Accessed on January 1, 2021.
5. McDonnell LA, Römpp A, Balluff B, et al. Discussion point: reporting guidelines for mass spectrometry imaging. *Anal Bioanal Chem.* 2015;407:2035–2045.
6. Stephenson JL. Ice crystal growth during the rapid freezing of tissues. *J Biophys Biochem Cytol.* 1956;2:45.
7. Schwartz SA, Reyzer ML, Caprioli RM. Direct tissue analysis using matrix-assisted laser desorption/ionization mass spectrometry: practical aspects of sample preparation. *J Mass Spectrom.* 2003;38:699–708.
8. Moline S, Glenner G. Ultrarapid tissue freezing in liquid nitrogen. *J Histochem Cytochem.* 1964;12:777–783.
9. Meng H, Janssen PM, Grange RW, et al. Tissue triage and freezing for models of skeletal muscle disease. *J Vis Exp.* 2014:e51586. Accessed on January 1, 2021.
10. Niehoff A-C, Kettling H, Pirkl A, Chiang YN, Dreisewerd K, Yew JY. Analysis of Drosophila lipids by matrix-assisted laser desorption/ionization mass spectrometric imaging. *Anal Chem.* 2014;86:11086–11092.
11. Stoeckli M, Staab D, Schweitzer A. Compound and metabolite distribution measured by MALDI mass spectrometric imaging in whole-body tissue sections. *Int J Mass Spectrom.* 2007;260:195–202.
12. Gill EL, Yost RA, Vedam-Mai V, Garrett TJ. Precast gelatin-based molds for tissue embedding compatible with mass spectrometry imaging. *Anal Chem.* 2017;89:576–580.
13. Nelson KA, Daniels GJ, Fournie JW, Hemmer MJ. Optimization of whole-body zebrafish sectioning methods for mass spectrometry imaging. *J Biomol Tech.* 2013;24:119.
14. Strohalm M, Strohalm J, Kaftan F, et al. Poly[N-(2-hydroxypropyl)methacrylamide]-based tissue-embedding medium compatible with MALDI mass spectrometry imaging experiments. *Anal Chem.* 2011;83:5458–5462.
15. Dannhorn A, Kazanc E, Ling S, et al. Universal sample preparation unlocking multimodal molecular tissue imaging. *Anal Chem.* 2020;92:11080–11088.
16. Drake RR, Powers TW, Norris-Caneda K, Mehta AS, Angel PM. In situ imaging of N-glycans by MALDI imaging mass spectrometry of fresh or formalin-fixed paraffin-embedded tissue. *Curr Protoc Protein Sci.* 2018;94:e68.
17. Angel PM, Norris-Caneda K, Drake RR. In situ imaging of tryptic peptides by MALDI imaging mass spectrometry using fresh-frozen or formalin-fixed, paraffin-embedded tissue. *Curr Protoc Protein Sci.* 2018;94:e65.
18. Peters SR. The art of embedding tissue for frozen section. Part I: a system for precision face down cryoembedding of tissues using freezing temperature-embedding wells. *J Histotechnol.* 2003;26:11–19.
19. Kawamoto T, Shimizu M. A method for preparing 2-to 50-µm-thick fresh-frozen sections of large samples and undecalcified hard tissues. *Histochem Cell Biol.* 2000;113:331–339.

20. Fitz-William WG, Jones GS, Goldberg B. Cryostat techniques: methods for improving conservation and sectioning of tissue. *Stain Technol.* 1960;35:195–204.
21. Gillberg P, Jossan S, Askmark H, Aquilonius S. Large-section cryomicrotomy for in vitro receptor autoradiography. *J Pharmacol Methods*. 1986;15:169–180.
22. Watanabe M. Preparation and stain of whole-body sections. *Cell Mol Bio Include Cyto Enzymol.* 1978;23:311–315.
23. Kawamoto T. Use of a new adhesive film for the preparation of multi-purpose fresh-frozen sections from hard tissues, whole-animals, insects and plants. *Arch Histol Cytol.* 2003;66:123–143.
24. Goodwin RJA, Nilsson A, Borg D, et al. Conductive carbon tape used for support and mounting of both whole animal and fragile heat-treated tissue sections for MALDI MS imaging and quantitation. *J Proteomics*. 2012;75:4912–4920.
25. Le MUT, Son JG, Shon HK, Park JH, Lee SB, Lee TG. Comparison between thaw-mounting and use of conductive tape for sample preparation in ToF-SIMS imaging of lipids in Drosophila microRNA-14 model. *Biointerphases*. 2018;13:03B414.
26. Golubeva YG, Smith RM, Sternberg LR. Optimizing frozen sample preparation for laser microdissection: assessment of CryoJane tape-transfer system®. *PloS one*. 2013;8:e66854.
27. Zivkovic AM, Wiest MM, Nguyen UT, Davis R, Watkins SM, German JB. Effects of sample handling and storage on quantitative lipid analysis in human serum. *Metabolomics*. 2009;5:507–516.
28. Haid M, Muschet C, Wahl S, et al. Long-term stability of human plasma metabolites during storage at -80 °C. *J Proteome Res*. 2018;17:203–211.
29. Patterson NH, Thomas A, Chaurand P. Monitoring time-dependent degradation of phospholipids in sectioned tissues by MALDI imaging mass spectrometry. *J Mass Spectrom.* 2014;49:622–627.
30. Lukowski J, Pamreddy A, Velickovic D, et al. Storage conditions of human kidney tissue sections affect spatial lipidomics analysis reproducibility. *J Am Soc Mass Spectrom.* 2020; 31(12):2538–2546.
31. Swales JG, Dexter A, Hamm G, et al. Quantitation of endogenous metabolites in mouse tumors using mass-spectrometry imaging. *Anal Chem*. 2018;90:6051–6058.
32. Chaurand P, Norris JL, Cornett DS, Mobley JA, Caprioli RM. New Developments in Profiling and Imaging of Proteins from Tissue Sections by MALDI Mass Spectrometry. *J Proteome Res*. 2006;5(11):2889–2900.

Chapter 5

Matrix for matrix-assisted laser desorption/ionization (MALDI)

Chapter Outline

A matrix, usually a small organic molecule, is required in matrix-assisted laser desorption/ionization (MALDI) for ionization of the analyte molecules. Matrix has also been used in secondary ion mass spectrometry (MS) for three-dimensional visualization of the incorporation of peptides in submicron resolution.[1] Matrix molecules, which are in surplus in comparison to analyte molecules, absorb the laser energy shielding the molecular breakdown of analytes. In MALDI, primary ionization occurs shortly after the irradiance of a laser pulse and trailed by secondary reactions in the desorbed plume.[2] Spatial resolution in MALDI can be influenced by matrix due to delocalization during matrix application, uniformity of matrix crystallization, and matrix crystal size.

The choice of matrix influences what type of analyte is ionized. In addition to the selection of matrix, its concentration, composition of matrix solvent, or application technique affect the ion extraction and production in tissue. In a typical experiment, a milligram per millimeter of the organic matrix is dissolved in a mixture of organic and aqueous solvents. A tiny fraction of acid, such as 0.1% trifluoroacetic acid, is added for positive ion mode. The solution is deposited or sprayed on a tissue section, dried, and analyzed by MALDI mass spectrometer.

Matrix application

In an aliquot analysis, a solution with a dissolved matrix is mixed with aliquot before spotting on the target or deposited separately with sample aliquot on the

Introduction to Spatial Mapping of Biomolecules by Imaging Mass Spectrometry. DOI: 10.1016/C2018-0-03962-6

target. In imaging analysis, a matrix is sprayed on the sample surface as a solution or deposited using sublimation. The application of matrix on tissue extracts analyte from the sample leading to the formation of cocrystallized analyte and matrix crystals. The cocrystallized mixture is irradiated by laser pulse leading to ions that are analyzed by a mass spectrometer. The application of matrix can affect analyte extraction and the extent of analyte delocalization.

An optimal matrix application would be homogeneous throughout the sample, have a small crystal size compatible with high-resolution imaging, and above all, efficient analyte extraction without any delocalization. In general, a wetter matrix application results in more analyte extraction, while the drier application may have lower molecular diffusion and small crystal size suitable for high spatial resolution imaging. Even the matrix drying mechanism after may influence the homogenate of the application needed for accurate and reproducible imaging MS. For example, long natural air drying could result in heterogeneous crystal formation, while inducing Marangoni flows within drying droplets can improve homogeneity. Homogeneous distribution can also obtain by decreasing sample surface temperature relative to ambient conditions during droplet drying processes.[3]

Three matrix application techniques, i.e., airbrush, automatic sprayer, and sublimation, were evaluated using the two most widely used matrices, 2,5-dihydroxybenzoic acid (DHB) and α-cyano-4-hydroxycinnamic acid (CHCA) for MALDI imaging MS of metabolites.[4] The study showed an automatic matrix sprayer was successful at analyzing about twice more metabolites than sublimation without humidification and airbrush method. The automated sprayer was also more reproducible than the manual and had less analyte diffusion. Placing the samples in a humidified condition after sublimation enhanced detection similar to the automated sprayer, but also increased analyte diffusion. The image of matrix coating by three different techniques, i.e., airbrush, automatic sprayer, and sublimation, for 2,5-DHB and α-CHCA is shown in Fig. 5.1.

In another study, four methods of matrix deposition were evaluated for analyte migration or delocalization versus analyte integration or extraction, as shown in Fig. 5.2. The goal was to get the least analyte delocalization by minimizing its migration and maximum analyte extraction by incorporation into matrix crystals. The standard dried-droplet preparation by depositing matrix solvent had optimal incorporation of analyte into matrix crystals but also had high delocalization. The electrospray deposition of the matrix had minimum analyte extraction and delocalization. Pneumatic sprayed matrix had high analyte incorporation and relatively low migration. The hybrid approach of decoupling the preparation in two steps, vapor-depositing matrix and recrystallization in humidified condition, had low analyte delocalization with acceptable sensitivity.[5] Matrix recrystallization can also be optimized by varying solvent and incubation time, improving lipid ion signals without any analyte delocalization.[6]

There are many different types of equipment proposed for automated matrix deposition from the nebulizer, thin-layer chromatography sprayer, inkjet or

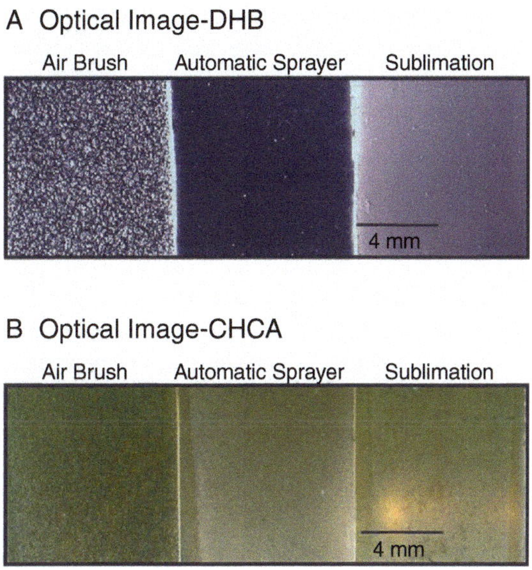

FIG. 5.1 A comparison of matrix application using an airbrush, automatic sprayer, and sublimation is shown using optical images of matrix crystal size for (A) 2,5-DHB and (B) α-CHCA matrix. Automatic sprayer and sublimation have relatively smaller crystal size than airbrushed method. Reprinted with permission from Gemperline E, Rawson S, Li L. Optimization and comparison of multiple MALDI matrix application methods for small molecule mass spectrometric imaging. Anal Chem. 2014;86:10030-10035. Copyright 2014 American Chemical Society.[4] Creative commons.

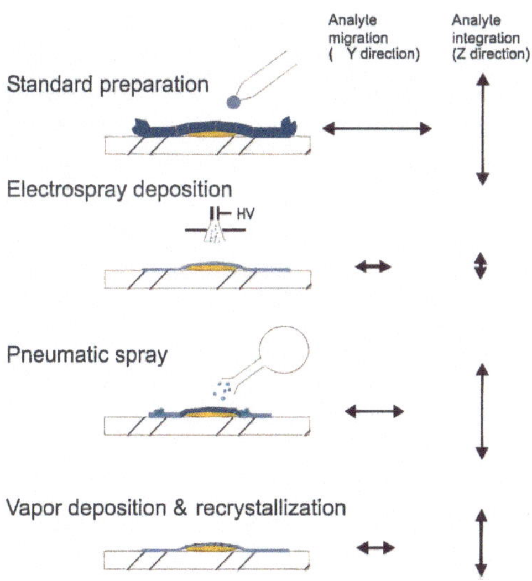

FIG. 5.2 Relative analyte migration (delocalization) versus analyte integration(extraction) evaluated for four matrix deposition methods—standard dried droplet, electrospray deposition, pneumatic sprayed. Reprinted with permission from Bouschen W, Schulz O, Eikel D, Spengler B. Matrix vapor deposition/recrystallization and dedicated spray preparation for high-resolution scanning microprobe matrix-assisted laser desorption/ionization imaging mass spectrometry (SMALDI-MS) of tissue and single cells. Rapid Commun Mass Spectrom. 2010;24:355–364.[5]

chemical printer, acoustic droplet depositor, etc. With such a diversity in tools and protocols, it is critical to include details of the matrix application process in the publication for more reproducible results. Some of the matrix application techniques are discussed below.

Manual matrix application by airbrushing

In earlier days of MALDI, due to lack of commercial solution, many researchers sprayed a matrix solution on tissue using the airbrush, a pneumatic sprayer used in painting. Manual matrix deposition by airbrush is often considered an art form instead of the scientific method with substantial variability. Some matrix compound or solvent used has a deleterious health effect even when sprayed, even in a hood. For example, 1,5-diaminonaphthalene (DAN) is considered an aspiration hazard irritant with carcinogenic and reproductive toxicity.[7]

Automated matrix deposition

Automated matrix depositors can give reproducible results, have a homogeneous coating, are generally safer, prove multisite repeatability, and, once set up, are easier to operate. Due to the advent of robust commercial matrix sprayer with flexible parameters for spraying many different types of matrix composition, many labs have moved toward automated systems for day-to-day matrix application. In most commercial matrix sprayer, a small volume of matrix solution is sprayed over in a uniform pattern with a sprayer mounted on an actuator. The sprayed tissue is placed horizontally at a right angle a few inches away from the sprayer nozzle. The spraying is repeated multiple times to achieve uniform deposition. The use of pneumatically assisted heated matrix sprayer seems to be more prevalent in the literature. These automated sprayers allow users to control matrix application parameters, such as temperature, solvent and gas flow rates, the velocity of spray application or stage movement, number of spraying cycles, spray pattern, and solvent composition for matrix application. Optimizing all these multiple parameters iteratively one at a time is severely time-consuming and may even be unproductive. Alternatively, multiple matrix application parameters can be optimized using response surface methodology.[8] Response surface methodology is a statistical and mathematical technique used for optimizing a process with many independent parameters that influence performance, which is measured by response.[9]

There are some reports of the homebuilt automated matrix deposition system. A desktop inkjet printer compatible with disc printing was modified to deposit matrix on microscope slides by filling empty ink cartridges with MALDI matrix solutions.[10] Researchers have described a custom-built matrix sprayer that can be made in the lab for a tenth of the cost of a commercial solution.[11] In other examples, a commercially available three-dimensional printer was converted to an automated matrix deposition platform,[12] and a mini-humidifier that cost less than five dollars was used for depositing matrix.[13] Electrospray deposition technique has been utilized for coat matrix, as well as, homogeneous analyte over the matrix layer.[14] Electric field-assisted scanning-spraying, used for coating small crystal sizes matrix, was found to improve sensitivity and coverage of small molecules, such as metabolites, in mammalian tissue compared to airbrush and sublimation methods. An automated acoustic spotter has also been used for reproducible matrix deposition for imaging of tissue section.[15] A vapor deposition/recrystallization system for matrix application has also been used for high-resolution MALDI imaging.[5] A portable ultrasonic sprayer has also been used for depositing several matrices, such as 2-[(2E)-3-(4-*tert*-butylphenyl)-2-methylprop-2-enylidene] malononitrilefor triacylglycerol analysis, with a crystal size of 1–15 μm amenable to high spatial resolution imaging.[16]

Homebuilt systems are affordable and customizable, and if set up correctly, can give uniform crystals compatible with high spatial resolution imaging. However, they are time-consuming to set up and program in the beginning. In contrast, commercial systems are ready-to-go, relatively more straightforward, but users need to pay a premium for it. In a market, there are many MALDI matrix deposition systems from different vendors. Besides spraying mechanism, they come with software to define the spraying parameters and actuators to move sprayer for consistent uniform patterns. In both cases, a reproducible matrix application will require proper documentation so that the experiments can be repeated with all the parameters.

Application of matrix by sublimation

Sublimation is a solvent-free dry matrix application technique that can produce homogeneous small matrix crystal compatible with high spatial resolution imaging MS. Some common organic matrices, such as DHB and α-CHCA, undergo solid to vapor-phase transition and sublime without any decomposition in elevated temperature and reduced pressure.[17] Currently, there are commercial solutions for matrix sublimation in the market. However, many publications have reported an affordable apparatus to build a sublimator for the deposition matrix.[18-19] The combination of doping sodium salt, such as sodium carbonate with the sublimation of the matrix 2,5-DHB, enhanced the detection of some of the neutral lipids by 10–140 folds.[20]

A comparison of matrix grain size, internal ion energy, initial velocity, analyte intensity, analyte incorporation depth between dried droplet, and vacuum sublimation methods for a few matrices was done shows dependence of matrix crystal size on those properties.[21] Scanning electron microscopy images of matrix crystal sizes for 2,5-DHB matrix applied by a commercial sprayer and sublimator are shown in Fig. 5.3.[22] The matrix crystal size for sprayer is at least in tens of microns, while the crystal size for a homebuilt sublimation system is in the low micron range. The sublimated matrix is also more uniformly deposited.

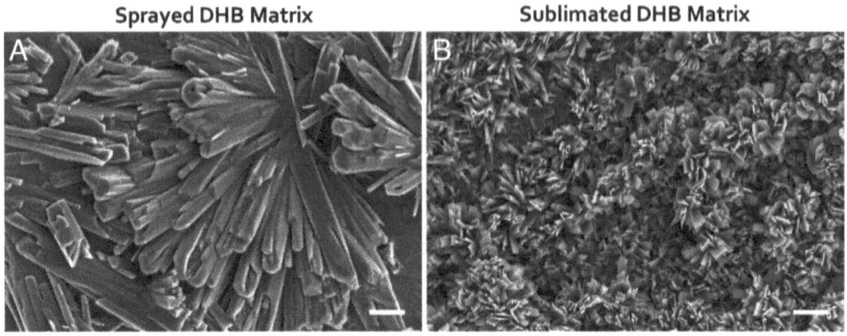

FIG. 5.3 Scanning electron microscope (SEM) images of matrix crystal size obtained from (A) a commercial sprayer compared with (B) a homebuilt matrix sublimation device. SEM images show sublimation produced visibly smaller and more uniform matrix crystals than the sprayer. *Scale bars* represent 10 μm size. Adapted with permission from Ly A, Schöne C, Becker M, et al. High-resolution MALDI mass spectrometric imaging of lipids in the mammalian retina. Histochem Cell Biol. 2015;143:453–462.[22]

Precoated matrix slides

Some applications require speedy turnaround time for imaging. In such a high throughput application, a matrix can be precoated on MALDI targets, and tissue sections can be mounted onto the precoated target. A homogenous layer of a matrix with uniform micron-sized crystals is coated on a slide for precoated slides. Matrix precoated slides have been used for imaging of lipids, peptides, drugs in tissues.[23] For protein analysis, gold-coated slides were layered with sinapinic acid treated with diisopropylethylamine-H_2O vapor.[24] Some precoated matrix slides have shown stability for over 6 months when stored properly. Precoated ready-made target plates coated with thin-layer matrix has also been used for piezoelectric microdispenser deposition of high-density array samples.[25]

Dry-coating matrix

A dry coat of matrix can be deposited on tissue using manual processes. For example, a solvent-free matrix dry matrix-coating method involves filtering a finely ground solid matrix on the tissue through a stainless steel sieve (micron). This dry-coating method was found useful for phospholipid analysis in tissues.[26] Layers of finely ground α-CHCA matrix was applied on thaw-mounted tissue sections that enabled the detection of drug molecules, which could not be detected when the same matrix dissolved in aqueous–organic solvents was sprayed on the tissue.[27] In another study, MALDI was able to perform the quantitative distribution of the positron emission tomography ligands was possible after the solvent-free deposition of finely ground dry α-CHCA matrix on rat brain tissue.[28]

Types of matrices for MALDI

There are several types of matrix molecules used in imaging MS that are empirically found useful or designed for a specific application. Commonly used matrices are organic molecules that can be dissolved in either aqueous or organic solvents, such as DHB, α-CHCA, DAN, and 9-aminoacridine (9-AA). Many commonly used matrices are used for general usage or very broader usage, such as phospholipid imaging in negative ion mode. These common matrices are often used as springboards for initial evaluation or development of an imaging assay because their application methods are well established. Some matrices are found to favor a specific class of molecules, such as 5-chloro-2-mercaptobenzothiazole matrix for imaging gangliosides,[29] endogenous cardiolipins can be detected by using norharmane matrix,[30] DAN[31] or 9-AA matrix for analyzing small molecules in negative ion mode,[32] sinapinic acid matrix for imaging proteins. Some recent developments in MALDI matrices for lipid analysis have been reviewed here.[33] Unfortunately, there is no universal matrix.

Most widely used matrices, such as α-CHCA, were found empirically. Since then, many matrices have been proposed or explicitly synthesized for a defined class of molecules. A matrix was systematically designed by varying the functional groups of the α-cyanocinnamic acid core unit, 4-chloro-α-cyanocinnamic acid resulted in an increase in sensitivity and peptide recovery in proteomic analyses due to uniform response to peptides with varying different basicity.[34] Other examples include combination matrix of 3-hydroxycoumarin, and 6-aza-2-thiothymine was found to be capable of ionizing small molecules such as drugs,[35] or 4-chloro-α-cyanocinnamic acid for analysis of labile peptides,[36] *N*-(1-naphthyl) ethylenediamine dihydrochloride (NEDC) for detection of oligosaccharides in aliquots,[37] 1,5-naphthalenediamine (1,5-DAN) hydrochloride for imaging of small metabolites and lipids in tissue,[38] strong base 1,8-bis(dimethyl-amino) naphthalene (or proton sponge) for analysis of anions,[39] etc. A comprehensive list of several dozen organic MALDI matrices for more than a hundred types of applications is tabulated in the referenced review.[40]

Nanomaterial as a matrix

Besides, small organic molecules, from complex nanostructures to water have been used as matrices, such as ice can be used as a matrix in infrared MALDI.[41] Nanomaterials have been used as a matrix for imaging small molecules, such as drugs and metabolites. Many nanomaterials have been used as matrix or in lieu of matrix for imaging MS, such as immobilized carbon nanotubes for analyzing neutral small carbohydrates,[42] gold nanoparticles for neutral small carbohydrates,[43] colloidal graphite as a matrix in graphite-assisted laser desorption/ionization,[44] multifunctional cadmium selenide quantum dots as the matrix for microwave enzymatic digestion for peptide analysis,[45] polymethyl methacrylate nanoparticle with destroyable core of $CaCO_3$ was used to enrich low-abundance peptides and proteins by adsorbed in linear chains of polymethyl methacrylate linked on the $CaCO_3$ core,[46] nanostructured diamond-like carbon-coated digital versatile disk target,[47] nanostructured solid substrates low molecular weight compounds,[48] chemically modified carbon nanotubes as material enhanced laser desorption ionization material in protein profiling,[49] nanophotonic laser desorption ionization from a highly uniform silicon nanopost array,[50] electrospun polystyrene/oxidized carbon nanotubes film as both sorbent for thin film microextraction and matrix for MALDI,[51] highly ordered mesoporous tungsten trioxide–titanium dioxide (WO_3–TiO_2) as matrix,[52] nanostructure imaging initiator with fluorinated gold nanoparticles for metabolites imaging,[53] etc. Many of these matrix-free laser desorption ionization techniques were developed to analyze low molecular weight compounds that were difficult to analyze by MALDI. Application of these novel matrices can enable an analysis of hard to access drugs and molecules within the tissue. However, one should consider a long-term effect of continuous usage of some of those matrices on the ion optics of the mass spectrometer.

Generally, only a single matrix is applied to a tissue section. Sometimes binary matrix mixtures such as 2,5-DHB with α-CHCA, often termed universal matrix, and 2,5-DHB with sinapinic acid have been reported to broaden the applicability.[54] However, the binary matrix often acts differently from each of the single matrix components. Repeated imaging can be performed by removing matrix, but can lead to distortion due to image acquisition or washing, a multigrid MALDI technique, where different matrices are printed in adjoining grids by inkjet printer for imaging.[55] A binary matrix mixture of organic 2,5-DHB and inorganic Fe_3O_4 nanoparticles was used to image triacylglycerol, phosphatidylethanolamine, phosphatidylinositol, digalactosyldiacylglycerol by alleviating ion suppression by phosphatidylcholine lipids.[56]

Reactive matrix

Imaging MS analysis of many compounds within the tissue is challenging due to poor ionization and ion suppression. Chemical derivatization can lead to derivatized molecular entities that can be easily detected and imaged. On-tissue derivatization reaction should occur rapidly in ambient conditions selectively with the targeted compound. Chemical derivatization protocols similar to ones used in gas chromatography have been used to derivatized molecules for enhancing detection. For example, chemical derivatization with Girard's reagent T has been used to detect drug triamcinolone acetonide in human incubated cartilage.[57] A molecule that can act as a derivatizing agent, while also functioning as an efficient matrix, will be ideal for simplifying workflow and reducing any delocalization of analyte. Reactive matrix is a molecule that can be used for derivatizing a class of challenging to ionize molecule, such as steroids, and also used as a matrix for MALDI.[58] For example, primary amines such as naturally occurring dopamine or exogenous amphetamine were reacted with pyrylium salts, such as 2,4-diphenyl-pyranylium, and imaged.[59] In another example, benzophenone was used as a reactive matrix for simultaneously derivatization reagent of unsaturated phospholipids assisted by photochemical Paternò–Büchi reaction using laser irradiance and ionization generating characteristic fragments that elucidated the location of the lipid double bonds.[60] This type of multifunctional reactive MALDI matrix, benzophenone, 2-benzoylpyridine, was successful at assigning 133 and 58 lipid features in positive and negative ion modes, respectively, in mouse cerebellum tissue and double-bond localization in 12 lipid class standards.[61]

Matrix selection

There is no comprehensive guide or database for a selection of matrices due to many matrices, variations in matrix application protocols, and diversity of tissue types. In a study comparing DAN and 9-AA in plant tissue, DAN provided superior ionization for small metabolites below molecular weight 400 Da, while 9-AA was superior for larger metabolites such as uridine diphosphate standard. DAN also gave a more robust signal for natural phospholipid mixture than 9-AA.[62] In a study that uses graphene oxide, NEDC, and 9-AA was used as MALDI matrices to image lipids in adjacent mouse brain section using negative ion mode. In this example shown in Fig. 5.4, the graphene oxide matrix had higher coverage than NEDC and 9-AA.[63] A study comparing about 13 different nanomaterial and inorganic particles with three traditional matrices (2,5-DHB, 9-AA, DAN) analysis of two dozen small metabolites from various classes is shown in Fig. 5.5 and serve as an excellent guide for matrix selection for small molecules.[64] Most users begin the selection process by searching scientific literature or finding use case examples in a database, such as METASPACE.[65] After selecting a few candidates, a pilot experiment is carried out to determine coverage because matrix application protocol and ion suppression within the specific tissue can affect the detection of the molecule-of-interest.

Matrix stability

For high spatial resolution imaging that requires a long acquisition time, the stability of the matrix in a vacuum is a crucial consideration. Volatile matrices can evaporate in the high vacuum condition of the MALDI ion source, slowly producing a distorted image. Even conventional matrices, such as DHB, are susceptible to evaporation during very long imaging acquisition. Some compounds, 2,5-dihydroxyacetophenone, are the excellent matrix for peptides, proteins, and glycoproteins, but is volatile under high vacuum conditions on MALDI ion source and cannot use in imaging acquisition that lasts more than an hour.

Structurally similar ketone-based matrices, 4-(2,5-dihydroxyphenyl)but-3-en-2-one, have high vacuum stability and can be used for longer acquisition needed for high spatial resolution imaging.[66] The effect of matrix evaporation during imaging MS acquisition was demonstrated by imaging sagittal mouse brain section with a volatile matrix dithranol for about 100 minutes and

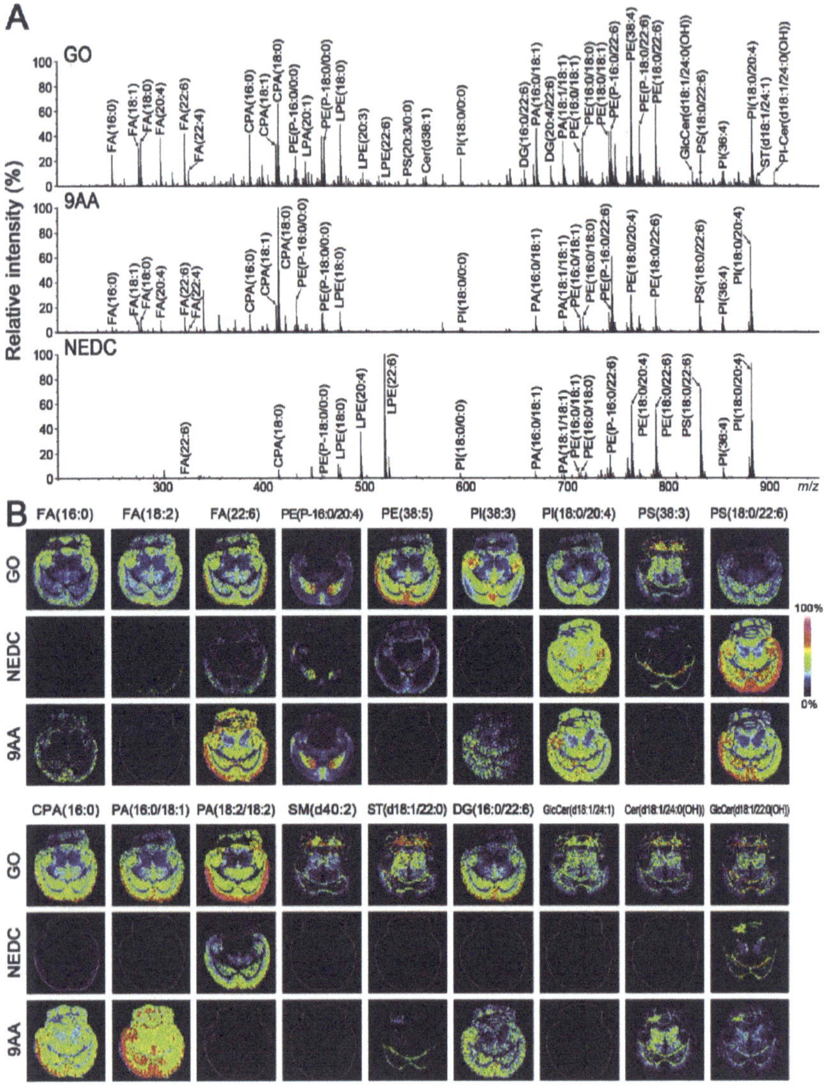

FIG. 5.4 Comparison of three matrices, graphene oxide (GO), N-(1-naphthyl) ethylenediamine dihydrochloride (NEDC), and 9-aminoacridine (9-AA), for imaging lipids in adjacent mouse brain tissue sections, (A) single-pixel mass spectra from the hippocampal area for each matrix, (B) image showing selectivity of different classes of lipids depending on the matrix used. Reprinted with permission from Zhou D, Guo S, Zhang M, Liu Y, Chen T, Li Z. Mass spectrometry imaging of small molecules in biological tissues using graphene oxide as a matrix. Anal Chim Acta. 2017;962:52–59.[63]

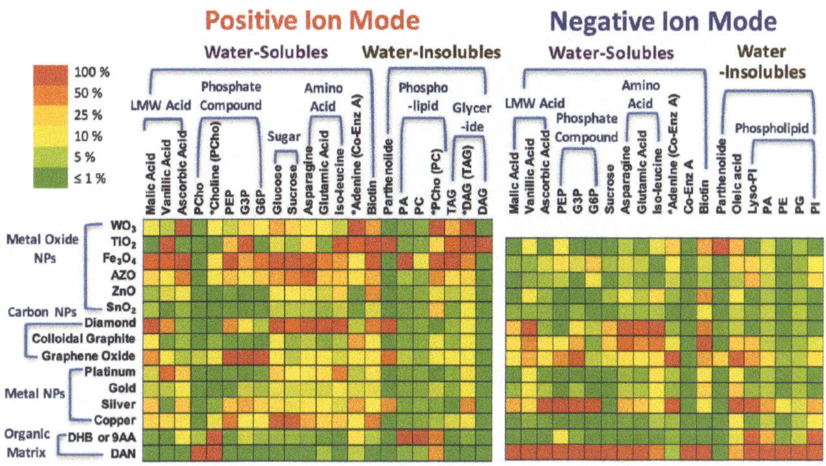

FIG. 5.5 A heatmap evaluation of 13 nanomaterials and three traditional organic matrices for analyzing metabolites and lipids in positive and negative ion modes. Ion intensities are normalized to the highest intensity for the each analyte. Reprinted with permission from Yagnik GB, Hansen RL, Korte AR, Reichert MD, Vela J, Lee YJ. Large scale nanoparticle screening for small molecule analysis in laser desorption ionization mass spectrometry. Anal Chem. 2016;88:8926-8930. Copyright 2016 American Chemical Society.[64] *Notes:* WO3 was excluded from negative ion mode due to significant matrix background, the asterisk (*) indicates a fragment ion with the precursor shown in parentheses. DHB was used for positive ion mode, and 9-AA was used in negative ion mode. The abbreviations include: DAG, diacylglycerol; G3P, glycerol 3-phosphate; G6P, glucose 6-phosphate; PA, phosphatidic acid; PC, phosphatidylcholine; PCho, phosphocholine; PE, phosphatidylethanolamine; PG, phosphatidylglycerol; PI, phosphatidylinositol; PEP, phosphoenolpyruvic acid; TAG, triacylglycerol.

two-and-half times longer. Noticeable loss of signal was observed after 2 hours of image acquisition in high vacuum condition as shown in Fig. 5.6.[67]

Conclusions

For MALDI imaging, matrix selection and application are the two most critical sample preparation parameters. The ionization and thus detection of analyte depends mainly on the selection and application of matrix. The application of matrix influences spatial resolution and delocalization of analyte. An optimal MALDI imaging experiment can be performed only after choosing a correct matrix for the desired analyte and optimizing its application parameters by balancing extraction with delocalization. In the last two decades, several matrices have been reported in the literature claiming the affinity for certain analyte classes or applications. However, due to a lack of independent and preferably multicenter validation studies, it is difficult to compare their performances. The lack of a database for matrix selection depending on analyte also makes the task of selecting a matrix daunting for new researchers. For matrix application parameters, users generally have to rely on the application notes from vendors

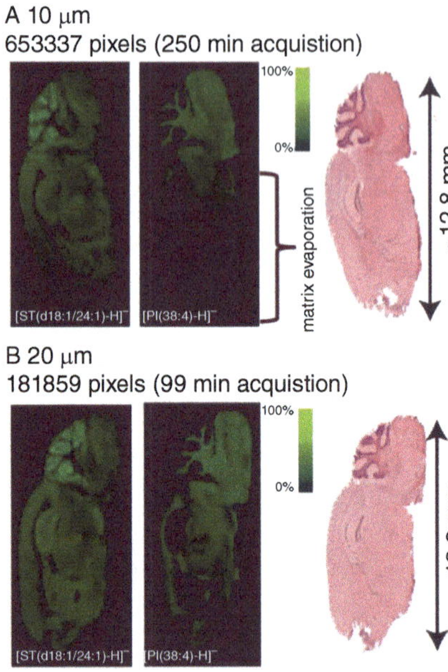

FIG. 5.6 The difference in ion images of two lipids, ST(d18:1/24:1) and PI(38:4), shows the effect of evaporation of the dithranol matrix in vacuum condition. (A) A drop-off in MS signal at the bottom of the image is attributed to matrix evaporation occurring during the acquisition of a 250-minute long image acquisition compared with (B) 99-minute long acquisition, where significant loss of signal due to matrix evaporation is not noticed. The right inset has corresponding H&E images. Reprinted with permission from Ogrinc Potočnik N, Porta T, Becker M, Heeren RMA, Ellis SR. Use of advantageous, volatile matrices enabled by next-generation high-speed matrix-assisted laser desorption/ionization time-of-flight imaging employing a scanning laser beam. Rapid Commun Mass Spectrom. 2015;29:2195–2203.[67]

or use reported parameters in the literature. The control of many matrix application parameters has given flexibility in application development but also created unique parameters depending on laboratory or even users. All the parameters of newly developed methods should be included in the publication for improving reproducibility.

References

1. Körsgen M, Pelster A, Dreisewerd K, Arlinghaus HF. 3D ToF-SIMS Analysis of Peptide Incorporation into MALDI matrix crystals with sub-micrometer resolution. *J Am Soc Mass Spectrom*. 2016;27:277–284.
2. Knochenmuss R. Ion formation mechanisms in UV-MALDI. *Analyst*. 2006;131:966–986.
3. Lai Y-H, Cai Y-H, Lee H, et al. Reducing spatial heterogeneity of MALDI samples with Marangoni flows during sample preparation. *J Am Soc Mass Spectrom*. 2016;27:1314–1321.
4. Gemperline E, Rawson S, Li L. Optimization and comparison of multiple MALDI matrix application methods for small molecule mass spectrometric imaging. *Anal Chem*. 2014;86:10030–10035.

5. Bouschen W, Schulz O, Eikel D, Spengler B. Matrix vapor deposition/recrystallization and dedicated spray preparation for high-resolution scanning microprobe matrix-assisted laser desorption/ionization imaging mass spectrometry (SMALDI-MS) of tissue and single cells. *Rapid Commun Mass Spectrom*. 2010;24:355–364.
6. Dueñas ME, Carlucci L, Lee YJ. Matrix recrystallization for MALDI-MS imaging of maize lipids at high-spatial resolution. *J Am Soc Mass Spectrom*. 2016;27:1575–1578.
7. Millipore Sigma Compound Catalog; 2020.
8. Veličković D, Zhang G, Bezbradica D, et al. Response surface methodology as a new approach for finding optimal MALDI matrix spraying parameters for mass spectrometry imaging. *J Am Soc Mass Spectrom*. 2020;31:508–516.
9. Myers RH, Montgomery DC, Anderson-Cook CM. *Response Surface Methodology: Process and Product Optimization Using Designed Experiments*: John Wiley & Sons; 2016:1–12.
10. Baluya DL, Garrett TJ, Yost RA. Automated MALDI matrix deposition method with inkjet printing for imaging mass spectrometry. *Anal Chem*. 2007;79:6862–6867.
11. Iloro I, Bueno A, Calvo J, Urreta H, Elortza F. Langartech: a custom-made MALDI matrix sprayer for MALDI imaging mass spectrometry. *J Lab Autom*. 2016;21:260–267.
12. Tucker LH, Conde-González A, Cobice D, et al. MALDI matrix application utilizing a modified 3D printer for accessible high resolution mass spectrometry imaging. *Anal Chem*. 2018;90:8742–8749.
13. Huang X, Zhan L, Sun J, et al. Utilizing a mini-humidifier to deposit matrix for MALDI imaging. *Anal Chem*. 2018;90:8309–8313.
14. Malys BJ, Owens KG. Improving the analyte ion signal in matrix-assisted laser desorption/ionization imaging mass spectrometry via electrospray deposition by enhancing incorporation of the analyte in the matrix. *Rapid Commun Mass Spectrom*. 2017;31:804–812.
15. Aerni H-R, Cornett DS, Caprioli RM. Automated acoustic matrix deposition for MALDI sample preparation. *Anal Chem*. 2006;78:827–834.
16. Pei X-L, Liu X-N, Du J-L, Gong C, Xu X. MALDI-MS imaging of lipids in corn using a flexible ultrasonic spraying device as matrix deposition method. *Int J Mass Spectrom*. 2020;455:116373.
17. Murphy RC, Hankin JA, Barkley RM, Zemski Berry KA. MALDI imaging of lipids after matrix sublimation/deposition. *Biochim Biophys Acta*. 2011;1811:970–975.
18. Hankin JA, Barkley RM, Murphy RC. Sublimation as a method of matrix application for mass spectrometric imaging. *J Am Soc Mass Spectrom*. 2007;18:1646–1652.
19. Fernández R, Garate J, Martín-Saiz L, Galetich I, Fernández JA. Matrix sublimation device for MALDI mass spectrometry imaging. *Anal Chem*. 2019;91:803–807.
20. Dufresne M, Patterson NH, Norris JL, Caprioli RM. Combining salt doping and matrix sublimation for high spatial resolution MALDI imaging mass spectrometry of neutral lipids. *Anal Chem*. 2019;91:12928–12934.
21. Jaskolla TW, Karas M, Roth U, Steinert K, Menzel C, Reihs K. Comparison between vacuum sublimed matrices and conventional dried droplet preparation in MALDI-TOF mass spectrometry. *J Am Soc Mass Spectrom*. 2009;20:1104–1114.
22. Ly A, Schöne C, Becker M, et al. High-resolution MALDI mass spectrometric imaging of lipids in the mammalian retina. *Histochem Cell Biol*. 2015;143:453–462.
23. Grove KJ, Frappier SL, Caprioli RM. Matrix pre-coated MALDI MS targets for small molecule imaging in tissues. *J Am Soc Mass Spectrom*. 2011;22:192–195.
24. Yang J, Caprioli RM. Matrix pre-coated targets for high throughput MALDI imaging of proteins. *J Mass Spectrom*. 2014;49:417–422.

25. Miliotis T, Kjellström S, Nilsson J, Laurell T, Edholm L-E, Marko-Varga G. Ready-made matrix-assisted laser desorption/ionization target plates coated with thin matrix layer for automated sample deposition in high-density array format. *Rapid Commun Mass Spectrom.* 2002;16:117–126.
26. Puolitaival SM, Burnum KE, Cornett DS, Caprioli RM. Solvent-free matrix dry-coating for MALDI imaging of phospholipids. *J Am Soc Mass Spectrom.* 2008;19:882–886.
27. Goodwin RJA, Scullion P, MacIntyre L, Watson DG, Pitt AR. Use of a solvent-free dry matrix coating for quantitative matrix-assisted laser desorption ionization imaging of 4-bromophenyl-1,4-diazabicyclo(3.2.2)nonane-4-carboxylate in rat brain and quantitative analysis of the drug from laser microdissected tissue regions. *Anal Chem.* 2010;82:3868–3873.
28. Goodwin RJA, Mackay CL, Nilsson A, Harrison DJ, Farde L, Andren PE, Iverson SL. Qualitative and quantitative MALDI imaging of the positron emission tomography ligands raclopride (a D2 dopamine antagonist) and SCH 23390 (a D1 dopamine antagonist) in rat brain tissue sections using a solvent-free dry matrix application method. *Anal Chem.* 2011;83:9694–9701.
29. Whitehead SN, Chan KHN, Gangaraju S, Slinn J, Li J, Hou ST. Imaging mass spectrometry detection of gangliosides species in the mouse brain following transient focal cerebral ischemia and long-term recovery. *PLoS One.* 2011;6:e20808 e20808.
30. Yang H, Jackson SN, Woods AS, Goodlett DR, Ernst RK, Scott AJ. Streamlined analysis of cardiolipins in prokaryotic and eukaryotic samples using norharmane matrix by MALDI-MSI. *J Am Soc Mass Spectrom.* 2020;31(12):2495–2502.
31. Strnad Š, Pražienková V, Sýkora D, et al. The use of 1,5-diaminonaphthalene for matrix-assisted laser desorption/ionization mass spectrometry imaging of brain in neurodegenerative disorders. *Talanta.* 2019;201:364–372.
32. Vermillion-Salsbury RL, Hercules DM. 9-Aminoacridine as a matrix for negative mode matrix-assisted laser desorption/ionization. *Rapid Commun Mass Spectrom.* 2002;16:1575–1581.
33. Leopold J, Popkova Y, Engel KM, Schiller J. Recent developments of useful MALDI matrices for the mass spectrometric characterization of lipids. *Biomolecules.* 2018;8:173.
34. Jaskolla TW, Lehmann W-D, Karas M. 4-Chloro-α-cyanocinnamic acid is an advanced, rationally designed MALDI matrix. *Proc Natl Acad Sci.* 2008;105:12200–12205.
35. Shanta SR, Kim TY, Hong JH, et al. A new combination MALDI matrix for small molecule analysis: application to imaging mass spectrometry for drugs and metabolites. *Analyst.* 2012;137:5757–5762.
36. Leszyk JD. Evaluation of the new MALDI matrix 4-chloro-alpha-cyanocinnamic acid. *J Biomol Tech.* 2010;21:81–91.
37. Wang J, Qiu S, Chen S, et al. MALDI-TOF MS imaging of metabolites with a N-(1-naphthyl) ethylenediamine dihydrochloride matrix and its application to colorectal cancer liver metastasis. *Anal Chem.* 2015;87:422–430.
38. Liu H, Chen R, Wang J, et al. 1,5-diaminonaphthalene hydrochloride assisted laser desorption/ionization mass spectrometry imaging of small molecules in tissues following focal cerebral ischemia. *Anal Chem.* 2014;86:10114–10121.
39. Shroff R, Svatoš A. Proton sponge: a novel and versatile MALDI matrix for the analysis of metabolites using mass spectrometry. *Anal Chem.* 2009;81:7954–7959.
40. Calvano CD, Monopoli A, Cataldi TRI, Palmisano F. MALDI matrices for low molecular weight compounds: an endless story?. *Anal Bioanal Chem.* 2018;410:4015–4038.
41. Berkenkamp S, Karas M, Hillenkamp F. Ice as a matrix for IR-matrix-assisted laser desorption/ionization: mass spectra from a protein single crystal. *Proc Natl Acad Sci.* 1996;93:7003–7007.

42. Ren S-f, Zhang L, Cheng Z-h, Guo Y-l. Immobilized carbon nanotubes as matrix for MALDI-TOF-MS analysis: applications to neutral small carbohydrates. *J Am Soc Mass Spectrom.* 2005;16:333–339.
43. Su C-L, Tseng W-L. Gold nanoparticles as assisted matrix for determining neutral small carbohydrates through laser desorption/ionization time-of-flight mass spectrometry. *Anal Chem.* 2007;79:1626–1633.
44. Zhang H, Cha S, Yeung ES. Colloidal graphite-assisted laser desorption/ionization MS and MS n of small molecules. 2. Direct profiling and MS imaging of small metabolites from fruits. *Anal Chem.* 2007;79:6575–6584.
45. Shrivas K, Kailasa SK, Wu H-F. Quantum dots laser desorption/ionization MS: multifunctional CdSe quantum dots as the matrix, concentrating probes and acceleration for microwave enzymatic digestion for peptide analysis and high resolution detection of proteins in a linear MALDI-TOF MS. *Proteomics.* 2009;9:2656–2667.
46. Jia W, Chen X, Lu H, Yang P. $CaCO_3$–poly(methyl methacrylate) nanoparticles for fast enrichment of low-abundance peptides followed by $CaCO_3$-core removal for MALDI-TOF MS analysis. *Angew Chem Int Ed.* 2006;45:3345–3349.
47. Najam-ul-Haq M, Rainer M, Huck CW, Hausberger P, Kraushaar H, Bonn GK. Nanostructured diamond-like carbon on digital versatile disc as a matrix-free target for laser desorption/ionization mass spectrometry. *Anal Chem.* 2008;80:7467–7472.
48. Silina YE, Volmer DA. Nanostructured solid substrates for efficient laser desorption/ionization mass spectrometry (LDI-MS) of low molecular weight compounds. *Analyst.* 2013;138:7053–7065.
49. Najam-ul-Haq M, Rainer M, Schwarzenauer T, Huck CW, Bonn GK. Chemically modified carbon nanotubes as material enhanced laser desorption ionisation (MELDI) material in protein profiling. *Anal Chim Acta.* 2006;561:32–39.
50. Stopka SA, Rong C, Korte AR, et al. Molecular imaging of biological samples on nanophotonic laser desorption ionization platforms. *Angew Chem.* 2016;128:4558–4562.
51. He X-M, Zhu G-T, Yin J, Zhao Q, Yuan B-F, Feng Y-Q. Electrospun polystyrene/oxidized carbon nanotubes film as both sorbent for thin film microextraction and matrix for matrix-assisted laser desorption/ionization time-of-flight mass spectrometry. *J Chromatogr A.* 2014;1351:29–36.
52. Yuan M, Shan Z, Tian B, Tu B, Yang P, Zhao D. Preparation of highly ordered mesoporous WO_3–TiO_2 as matrix in matrix-assisted laser desorption/ionization mass spectrometry. *Microporous Mesoporous Mater.* 2005;78:37–41.
53. Palermo A. Charting metabolism heterogeneity by nanostructure imaging mass spectrometry: from biological systems to subcellular functions. *J Am Soc Mass Spectrom.* 2020; 31(12):2392–2400.
54. Laštovičková M, Chmelik J, Bobalova J. The combination of simple MALDI matrices for the improvement of intact glycoproteins and glycans analysis. *Int J Mass Spectrom.* 2009;281:82–88.
55. Urbanek A, Hölzer S, Knop K, Schubert US, von Eggeling F. Multigrid MALDI mass spectrometry imaging (mMALDI MSI). *Anal Bioanal Chem.* 2016;408:3769–3781.
56. Feenstra AD, O'Neill KC, Yagnik GB, Lee YJ. Organic–inorganic binary mixture matrix for comprehensive laser-desorption ionization mass spectrometric analysis and imaging of medium-size molecules including phospholipids, glycerolipids, and oligosaccharides. *RSC Adv.* 2016;6:99260–99268.
57. Barré FPY, Flinders B, Garcia JP, et al. Derivatization strategies for the detection of triamcinolone acetonide in cartilage by using matrix-assisted laser desorption/ionization mass spectrometry imaging. *Anal Chem.* 2016;88:12051–12059.

58. Brombacher S, Owen SJ, Volmer DA. Automated coupling of capillary-HPLC to matrix-assisted laser desorption/ionization mass spectrometry for the analysis of small molecules utilizing a reactive matrix. *Anal Bioanal Chem.* 2003;376:773–779.
59. Shariatgorji M, Nilsson A, Källback P, et al. Pyrylium salts as reactive matrices for MALDI-MS imaging of biologically active primary amines. *J Am Soc Mass Spectrom.* 2015;26:934–939.
60. Wäldchen F, Spengler B, Heiles S, Reactive matrix-assisted laser desorption/ionization mass spectrometry imaging using an intrinsically photoreactive paternò–büchi matrix for double-bond localization in isomeric phospholipids. *J Am Chem Soc* 2019;141:11816–11820.
61. Wäldchen F, Mohr F, Wagner AH, Heiles S. Multifunctional reactive MALDI matrix enabling high-lateral resolution dual polarity MS imaging and lipid C=C position-resolved MS2 imaging. *Anal Chem.* 2020;92(20):14130–14138.
62. Korte AR, Lee YJ. MALDI-MS analysis and imaging of small molecule metabolites with 1,5-diaminonaphthalene (DAN). *J Mass Spectrom.* 2014;49:737–741.
63. Zhou D, Guo S, Zhang M, Liu Y, Chen T, Li Z. Mass spectrometry imaging of small molecules in biological tissues using graphene oxide as a matrix. *Anal Chim Acta.* 2017;962:52–59.
64. Yagnik GB, Hansen RL, Korte AR, Reichert MD, Vela J, Lee YJ. Large scale nanoparticle screening for small molecule analysis in laser desorption ionization mass spectrometry. *Anal Chem.* 2016;88:8926–8930.
65. Palmer A, Phapale P, Chernyavsky I, et al. FDR-controlled metabolite annotation for high-resolution imaging mass spectrometry. *Nat Methods.* 2017;14:57–60.
66. Yang J, Norris JL, Caprioli R. Novel vacuum stable ketone-based matrices for high spatial resolution MALDI imaging mass spectrometry. *J Mass Spectrom.* 2018;53:1005–1012.
67. Ogrinc Potočnik N, Porta T, Becker M, Heeren RMA, Ellis SR. Use of advantageous, volatile matrices enabled by next-generation high-speed matrix-assisted laser desorption/ionization time-of-flight imaging employing a scanning laser beam. *Rapid Commun Mass Spectrom.* 2015;29:2195–2203.

Chapter 6

Molecule identification approaches in imaging mass spectrometry

Chapter Outline

Mapping localization of molecules on the tissue by imaging mass spectrometry (MS) has a wide array of utility in biomedical research and drug discovery. One of the critical aspects of imaging MS analysis is the need to identify the mapped ions confidently. A mass spectrometer reports intensity data for ions as a mass-to-charge (*m/z*) ratios. The *m/z* value needs to be annotated as one or more molecule candidates. One of the most significant benefits of imaging MS is the ability to simultaneously visualize the spatial distribution of several hundred to thousand ions directly from the tissue without any prior knowledge or target-specific reagents such as antibodies. However, without a confident molecule identification, the underlying biochemical process within the tissue cannot be interpreted and studied. The molecule identification step is often straightforward for the imaging of nonbiological samples, because many times, the chemical is known and is placed on a relatively less complicated substrate than tissue.

Molecule identification remains a bottleneck of many omics, such as metabolomics, lipidomics, and proteomics. The annotation of ions into confident molecular entities represents a key challenge in imaging MS as well. The confident identification of molecules is further confounded in imaging MS because of the direct ionization nature of the technique. Lack of chromatographic separation and sample cleanup results in the detection of numerous ions. Isobaric ions with close *m/z* values and isomeric ion with the same *m/z* value due to the same chemical formula interfere with unambiguous detection.

Introduction to Spatial Mapping of Biomolecules by Imaging Mass Spectrometry. DOI: 10.1016/C2018-0-03962-6

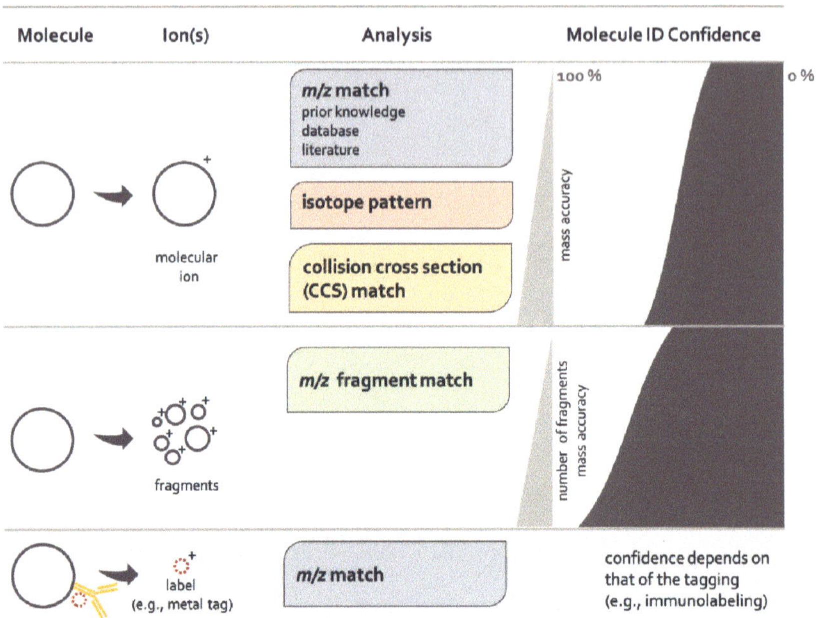

FIG. 6.1 Three workflows for the identification of molecules in tissue during imaging are illustrated. The first approach is *m/z* analysis of the intact ion and annotation using an accurate *m/z* match against previously known data, supplemented by isotope pattern and collision cross section (CCS) match. The confidence of molecular identification increases with a rise in mass accuracy, as well as the incorporation of other orthogonal measurements such as CCS. The second approach is the MS/MS fragmentation of the selected ion. Here, the confidence increases with the number of fragment matches and incorporation of the accurate mass match of fragments. The third approach is antibody tagging of molecules by a metal ion. The confidence of the match depends on the confidence of the tagging, such as immunolabeling with rare earth metal tags.

The molecular annotation of detected ions can be stratified into hierarchical levels depending on the identification's specificity as shown in Fig. 6.1. The specificity of the identification increases with the increase in analytical information. As the level increases, the number of molecules attributed decreases—with the ultimate goal of assigning the signal to a single molecule species. Similar approaches for increasing identification confidence levels for high-resolution mass spectrometric analysis have been proposed elsewhere, starting from accurate mass ion, unequivocal molecular formula, tentative candidate, probable structure, and finally, a confirmed structure.[1-3] Table 6.1 shows the identification level for a glycerophosphocholine molecule based on the analytical information. There are potentially hundreds of molecules that can fit that identification in the first level, while the last level of identification is specifically ascribed to a single type of molecule. Analogous strategies of hierarchical levels of molecular identification for lipids have been described elsewhere.[4-6]

TABLE 6.1 Molecular identification level of lipid.

Identification level	Information	Example
Class	Head group	PC
Species	Chain length	PC (34:1)
Acyl-chain species	Acyl chain group	PC (18:1_16:0)
Subspecies	Sn fatty acyl position	PC (16:0/18:1)
Positional isomer	Double bond position	PC (16:0/18:1(7))
Stereo isomer	Double bond position	PC (16:0/18:1(7Z))

Accurate mass matching

The most straightforward workflow to annotate an ion as a small molecule in imaging MS is based on *m/z* ratio match with known molecules or using a database. High mass accuracy, within a few milli-Daltons, of some mass spectrometers now used for imaging—such as quadrupole time of flight, orbitrap, Fourier transform ion cyclotron resonance—combined with the previous knowledge of the sample's biochemical makeup is capable of annotating ions within a single or a few isomeric candidates with a high degree of certainty. Curated databases, such as HMDB,[7] METLIN,[8] LipidMAPS,[9] can serve as a reference to match the detected *m/z* within a defined mass tolerance in milli-Daltons. Due to the vastness of chemical structures, the selection of a biologically relevant database is essential while matching the detected *m/z* value with the known or predicted metabolites and lipids. For example, we cannot successfully use the human glycan database to search for natural products found in ginseng.

METASPACE is an annotation tool for high accuracy *m/z* imaging data.[10] METAPSACE improves the annotation accuracy by calculating a false discovery rate (FDR) based on a match from a curated database such as the Human Metabolome Database.[11] The FDR is estimated using a decoy set generated using randomly select an implausible adduct from the Commission on Isotopic Abundances and Atomic Weights list of the elements such as B^+, Db^+, or Ag^+.[12]

In addition to *m/z* accuracy, the isotope pattern match also helps to narrow down the molecule candidates. Most elements naturally occur as a mixture of isotopes with different masses. Thus, a molecule will have ions with multiple masses. For example, chlorine has a natural isotope distribution mass of 35 and 37, with an approximate abundance of 76% and 24%. For this molecule, a mass spectrometer will detect two ions with a relative abundance of 100% for *m/z* 35 and 32% (24/76) for *m/z* 37. This isotope pattern can be calculated for each chemical formula and help to strengthen the conformation of the identification for the proposed chemical formula each ion by determining the number and type of elements.[13] In other to study compounds that are not in the database, the software platform, such as CycloBranch 2, can generate the de novo molecular formulas of unknown ions using the fine isotope ratio.[14]

For the early part of imaging MS by matrix-assisted laser desorption/ionization (MALDI) time-of-flight system, detected proteins were annotated using a nominal mass and some preexisting biochemical knowledge. The in situ sequencing of proteins or peptides by tandem MS is discussed in the section below on MS fragmentation. A database match against accurate mass collision cross section (CCS) is often considered the first step in molecule identification. An unambiguous identification often requires a secondary confirmation using tandem MS or MS/MS.

Orthogonal identifier—collision cross section

Imaging MS on an ion mobility separation equipped mass spectrometer can provide two significant advantages. First, ion mobility provides gas-phase separation postionization. Ion mobility coupled to imaging MS ion sources, such as MALDI,[15] desorption electrospray ionization,[16] and laser ablation electrospray ionization,[17] has shown improvement in the signal-to-noise ratio of analytes. Ion mobility separation during MS imaging improved the spectral clarity of detected analytes by separating a molecule-of-interest from interfering isobaric species found in a complex biological matrix of tissue.[18]

In addition to clarifying the image of a single molecular species, the drift times of ions obtained during ion mobility analysis can be used to calculate CCSs. Accurate mass value supplemented with CCS as an orthogonal identifier an ion. CCS offers a structural measurement of size and shape for the molecules, independent of the sample matrix.[19] A database with CCS value is a prerequisite in order to exploit the full utility of CCS values. A CCS database can be populated by experimentally measuring the standards, when possible, and if not, in silico CCS values can be predicted using a computational method such as a machine-learning algorithm.[20]

Fragmentation by ion dissociations

Imaging MS can provide more confident identification of molecules by using the targeted tandem MS (MS/MS) approach. Multiple reaction monitoring (MRM) involves selecting a precursor ion to monitor known product–ion transitions to identify targeted molecules. MRM imaging MS can be used to identify and quantify a few molecules accurately. For instance, zolpidem consumption was monitored by single hair analysis using imaging MS on a MALDI triple quadrupole mass spectrometer showed a good semiquantitative correlation from concentrations obtained from validated routine liquid chromatography (LC)–MS/MS methods.[21] A confident molecular identification can be obtained by matching product ion spectra. In another example, desorption electrospray ionization coupled with a triple quadrupole to perform MRM imaging lipids, a drug candidate, metabolites in mammalian tissue. The MRM transitions were used to strengthen the identification of the drug molecule on tissue.[22]

Protein and peptide identification also utilizes MS fragmentation of protein into amino acid residues. Fundamentally there are two approaches for peptide and protein identification by mass spectrometer that can be used for in situ identification proteins using MS/MS—top-down approach and bottom-up approach. In the top-down approach, an intact protein species undergoes fragmentation using tandem MS (MS/MS). There are a few examples of a top-down identification in imaging MS, such as a small peptide identification on mouse pancreatic tissue.[23] MALDI imaging, which is used for the majority of protein imaging, typically generates singly charged ions lowering the efficiency of the gas-phase fragmentation needed for sequencing. The top-down approach works for smaller proteins less than a couple of thousand Daltons depending on instrumentation due to the difficulty of dissociating singly charged molecule species of higher molecular weight.

The bottom-up approach involves enzymatically digesting the proteins and analyzing the peptides by MS or MS/MS. Imaging MS follows on-tissue digestion of proteins by a proteolytic enzyme such as trypsin. Multiple proteolytic peptides match for a protein is needed for its confident identification. On-tissue digestion of protein results in a complex mixture of peptides, and the lower sensitivity due to ion-suppression limits the robust MS/MS fragmentation and the identification of many moderate-to-low abundant proteins.

In both approaches, the fragmented peptide peaks are searched in peptide sequence databases, such as MASCOT or SEQUEST, to identify proteins. Detailed discussions on protein identification in imaging MS can be found in the referenced reviews.[24-25]

Tandem MS imaging can be performed separately or in concert with a mass spec profile scan. For example, a multiplex approach, imaging raster step, or pixel is composed of multiple smaller substeps or subpixels. One of the subpixel or step is used for acquiring high mass resolution while others can be used to acquire MS/MS data on specific molecules. This workflow allows users to obtain accurate mass data and MS/MS data from a single imaging experiment but at the cost of spatial resolution. A workflow and example of the utility of the discussed simultaneous MS and MS/MS imaging is presented in Fig. 6.2.[26] A similar multiplex imaging MS workflow can be utilized to incorporate accurate mass profile with MS/MS and/or MS^3 or to perform positive and negative ion imaging in a single MS experiment using polarity switching—depending on the mass spectrometer.[27]

MRM of precursor ions to monitor a few known product–ion transitions can be a useful approach for targeted imaging of a handful of molecules. Imaging MS coupled with the data-independent acquisition (DIA) collects MS/MS data without a predefined list of ions In DIA, two sister scans, one at lower collision energy-generating precursor molecular ion spectrum and another at high collision energy with fragmentation information, are acquired for each pixel or successive pixels. Two scans are aligned by matching the precursor ion in the low energy function with the remnant precursor ion in the high energy function, spatial correlation of fragments, and the precursor or any other quadrupole of ion mobility alignment based on the technique.

For example, in SONAR, a resolving quadrupole slides over a defined mass range with a selected mass window during two alternating scans with low and high collision energy. Multistep correlation, scanning quad correlation, and pixel-by-pixel spatial correlation (ρ) are used to align the precursor ions with

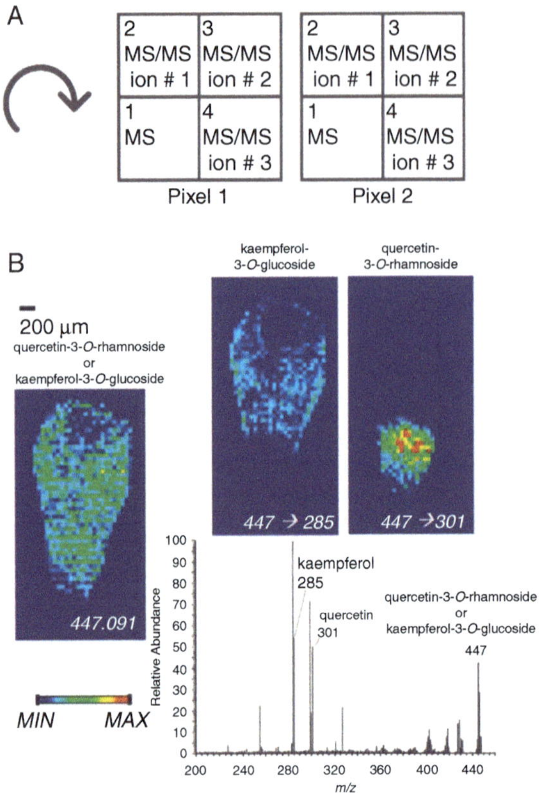

FIG. 6.2 (A) Top shows a conceptual illustration of MS imaging workflow where a sample is analyzed in a clockwise pattern into multiple (four here) subpixels generating MS, and several MS/MS images (B) the left MS image of *m/z* 447.091 ion on *Arabidopsis* petals can represent the distribution of either quercetin-3-*O*-rhamnoside or kaempferol-3-*O*-glucoside, the MS/MS imaging of fragments kaempferol ion (*m/z* 285), and the quercetin ion (*m/z* 301) differentiated their localization at either top or bottom petal. Adapted with permission from Perdian DC, Lee YJ. Anal Chem. 2010;*82*:9393–9400. Copyright 2010 American Chemical Society.[26]

the fragment ions. The spatial correlation (ρ) is based on the Pearson product–moment correlation, where a perfect spatial match gives a correlation factor R of 1, and the opposite gives an R value of −1. Another DIA method, HDMS^E is based on ion mobility separation and analyze isobaric species with very close mass, but dissimilar structure. In HDMS^E, two functions are aligned by matching its drift time and pixel-by-pixel spatial correlation (ρ). A schematic description of HDMS^E for imaging is given in Fig. 6.3. The utility of data-independent analysis workflow in imaging MS is still in infancy. Due to the complex molecular makeup of tissue, the DIA method often suffers from ion suppression and alignment issues for lowly abundant molecular species.

The molecule fragmentation in imaging MS is not limited to conventional gas-phase collision-induced dissociation or collisionally activated dissociation.

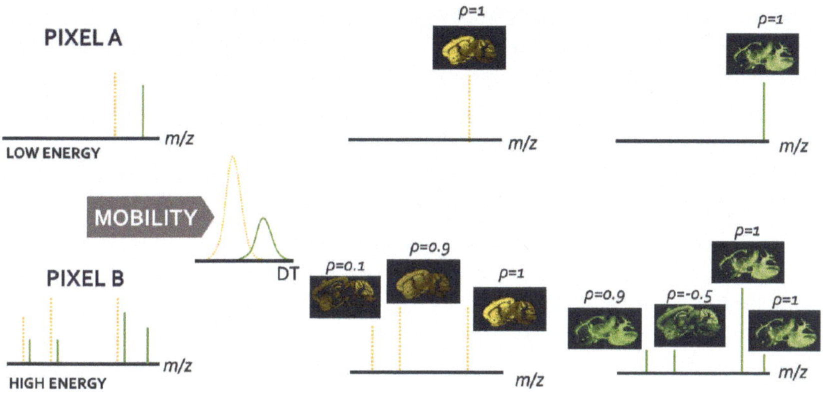

FIG. 6.3 HDMS^E is a data-independent acquisition method, where two sister functions, one at low energy giving molecular profile and another at high energy gives fragmentation information. Both are acquired for each pixel. Two functions are aligned by matching the precursor ion in the low energy function with the remnant precursor ion in the high energy function, drift time window of ion mobility separation, and spatial correlation (ρ) of fragments with their precursors.

Alternative fragmentation mechanism, such as electron capture dissociation, electron transfer dissociation, surface-induced dissociation, ion–ion reactions, ozone-induced dissociation, photodissociation, has been developed for other applications.[28] For example, ultraviolet photodissociation at 193 nm utilizes photons to activate and produce information-rich fragments to identify double bond positional isomers of lipids.[29] The gas-phase ion–molecule reactions of lipid ions and ozone vapor have been used to produce a separate image of isomeric lipids with different double bond and *sn* positions.[29] In ozone-induced dissociation, unsaturated lipid ions react with ozone vapor, producing fragments indicative of the position of saturation within the precursor ion. There are other innovative approaches to do protein digestion in research, such as one based on plasmon surface resonance absorption and heating for on-surface photothermal decomposition and digestion of protein. In this method, a continuous wave laser excitation and gold nanoparticles induce predictable decomposition reactions and cleave amino acid chains at the N-terminus of cysteine and the C-terminus of aspartic acid.[30]

Secondary MS analysis

One strategy of molecule annotation is by using established LC-coupled tandem MS workflow of extracted sample counterparts. The LC–MS/MS molecular annotation will provide complementary molecular identification to the spatial information provided by imaging MS. Some workflows utilize a multimodal approach, where imaging MS is supplemented with LC–MS/MS analysis. LC–MS/MS can be performed on a chunk of data or spatially directed to a region of interest (ROI) in tissue using a microextraction on the same slide or an adjacent serial section. A straightforward way of microextraction of analyte from a region in tissue using a pipet tip. A few microliters of extraction solvent drops are placed onto the tissue for a few seconds, and the extract is recovered for LC–MS/MS. Solvent selection can be made to maximize the extraction of molecules-of-interest from the tissue. The detection limits for such 1-μL extraction were found sufficient to detect 50–100 proteins.[31] The extracts were analyzed by LC–MS/MS electron transfer dissociation to identify proteins from three regions of tissue confidently. Another approach of extraction includes a hydrogel extraction of proteins on the tissue. In this method, an SDS-PAGE gel is washed, dehydrated, and rehydrated in trypsin solution. A small trypsin-reconstituted band is placed on an ROI of tissue. The bands are rehydrated and incubated. A supernatant collected from washing in a series of gradient dehydration solvents was analyzed by LC/MS for peptide sequence. Another approach involves tissue microarrays or TMA excised from laser capture microdissection. TMA consists of array of core biopsy each few millimeters in diameter. The tissue microarrays were analyzed by LC–MS/MS after antigen retrieval and trypsin digestion leading to identification of 3844 distinct peptide sequence from 840 proteins from ovarian cancer TMA. The identified peptide sequence can be used to identify proteins in future MALDI imaging studies. A conceptual workflow of LC–MS/MS analysis TMA followed by MALDI imaging is shown in Fig. 6.4.[32]

Similarly, liquid microextraction using liquid extraction surface analysis (LESA) can be used to extract a few microliters of solvent placed on the tissue section in an automated manner. The extracts from LESA can be used for infusion electrospray MS/MS or LC–MS/MS to confidently identify a broad range of molecules within ROIs. LESA can be used in concert with other imaging modalities to obtain congruent multimodal imaging. For example, MALDI can provide a high spatial resolution image of a molecule while LESA MS/MS can

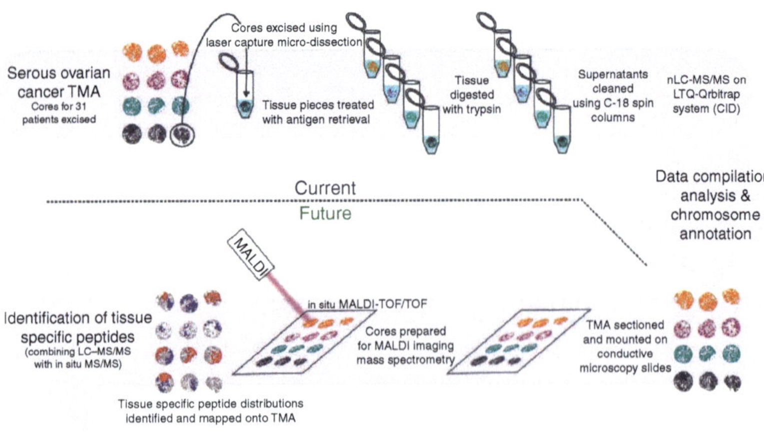

FIG. 6.4 A workflow showing the creation of accurate peptide identification data using LC–MS/MS from tissue microarrays of serous ovarian cancer (top) and future application of the generated peptide data for identification in MALDI imaging (bottom). Reprinted with permission from Meding S, Martin K, Gustafsson OJR, Eddes JS, Hack S, Oehler MK, Hoffmann P. Tryptic peptide reference datasets for MALDI imaging mass spectrometry on formalin-fixed ovarian cancer tissues. J Proteome Res. 2013;12:308–315. Copyright 2013 American Chemical Society.

provide highly informative structural information with ROIs.[33] Moss, cyanobacteria, and fungus were reanalyzed by LESA-MS/MS, leading to the confident annotation of ions as metabolites involved in a tripartite symbiosis system. The workflow used is shown in Fig. 6.5. Generally, any liquid microextraction

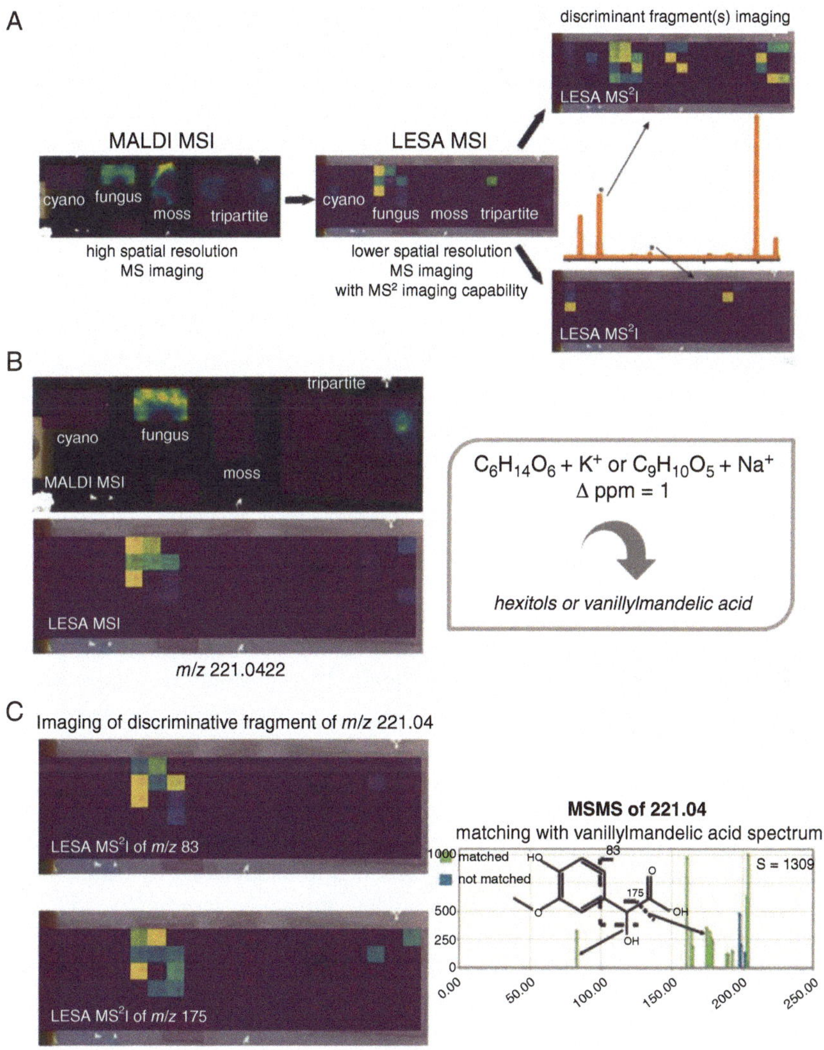

FIG. 6.5 (A) A workflow illustrating multimodal imaging of sample by MALDI followed by LESA analysis, (B) distribution of *m/z* 221 in MALDI-MS and LESA-MS imaging in fungi can be annotated to either hexitols or vanillylmandelic acid within 1 ppm mass accuracy, (C) fragmentation of the *m/z* 221 ion onto discriminative fragments of vanillylmandelic acid from fungi support its identification as vanillylmandelic acid. Adapted with permission from Veličković D, Chu RK, Carrell AA, Thomas M, Paša-Tolić L, Weston DJ, Anderton CR. Anal Chem. 2018;90:702–707. Copyright 2018 American Chemical Society.[33]

techniques have a relatively larger ROI and will not be able to resolve any features smaller than a fraction of a millimeter.

Instead of microextraction, a serial section of tissue placed on the aluminum foil can be punched out at the ROI and analyze by conventional LC–MS/MS.[34] Laser capture microdissection can be utilized to isolate a smaller ROI in tissue and analyzed separately by MALDI[35] or extracted for conventional LC–MS/MS analysis. In laser capture microdissection, a tissue section is placed on a microscope and excised off using a laser beam. A cell subgroup or single cells or any relevant ROI in a tissue section can be selected.[36] Imaging MS and laser capture microdissection are usually performed on serial sections, but recently the same tissue section was imaged by atmospheric pressure MALDI and used for ROI sampling by laser capture microdissection, followed by the analysis of several protein/peptides identified by LC–MS/MS.[37]

Immunolabeling

Identification of unambiguous antigens *m/z* value after immunoaffinity capture of target antigens has been used to overcome ion-suppression effects in MALDI since mid-1990s.[38] Recently, multiplexed ion beam imaging and imaging mass cytometry or CyTOF analyze metal mass tags conjugated with antibodies bound to protein in tissue or to identify cells or its components such as the nucleus. Lanthanides and other rare earth metals are used as elemental mass tags for immunolabeling since they do not occur naturally in typical biological samples. Metal isotopes with unique *m/z* values without spectral overlapping chemical signals are selected for the labeling. Thus, the detected *m/z* values can be confidently assigned to the metal tags, which in turn can be used for the identification of larger biomolecules, such as proteins. The detection of metal tag ions by MS results in discrete signals and not prone to overlapping compared with fluorescence-based detection.

The confidence in the identification of molecules during imaging depends on the confidence of metal tags and antibodies used. The preparation of metal tag must address any impurities in the enriched isotopes to avoid any potential signal overlap. The metal tags are attached to an antibody by bifunctional chelating agents consisting of two functional groups—a coordinative chelator that enables chelation of metal cations and a maleimide-functionalized group that coupled to the antibody.[39] The molecule identification also depends on the specificity of the antibody that binds to the chosen target and its cross reactivity. For example, the presence of metal tags may interfere with the specificity of the conjugated antibody to recognize the antigen.[40] A handful of validated metal-conjugated antibodies for phenotyping or functional profiling is available commercially. Although a limited number of antibodies have undergone rigorous validation, thousands of antibodies are available for research use for immunohistochemistry.[41] Potentially, those that antibodies can be potentially used with mass tags for imaging MS. Nevertheless, a validation will be needed to get target-specific metal-conjugated antibodies for confident identification.

Conclusions

One of the critical barriers to the adoption of imaging MS by the broader biomedical research community is the lack of confidence in the identity of the imaged molecules. Molecular annotation has improved due to the advancements in the hardware of mass spectrometers, such as increased mass resolution, high mass accuracy, and incorporation of the CCS. The development of new software tools, such as METASPACE, that streamline the annotation with FDR has not only increased the identification but also streamlined the curated database-based annotation. The incorporation of new ion activation techniques, such as ultraviolet photodissociation to obtain more descriptive fragmentation pattern, has greatly improved the unambiguous identification for molecules. Also, the evolution of data-independent methods, such as $HDMS^E$, in imaging MS space has the potential to enhance the confidence in molecular identification in untargeted discovery experiments. The continual improvement in hardware, software, algorithms, and workflows will need to further increase identification confidence. It is critically important to get the correct molecular identification to have an accurate interpretation of biomedical data.

References

1. Schymanski EL, Jeon J, Gulde R, et al. Identifying small molecules via high resolution mass spectrometry: communicating confidence. *Env Sci Technol*. 2014;48:2097–2098.
2. Schrimpe-Rutledge AC, Codreanu SG, Sherrod SD, McLean JA. Untargeted metabolomics strategies—challenges and emerging directions. *J Am Soc Mass Spectrom*. 2016;27:1897–1905.
3. May JC, McLean JA. Advanced multidimensional separations in mass spectrometry: navigating the big data deluge. *Annu Rev Anal Chem*. 2016;9:387–409.
4. Foster JM, Moreno P, Fabregat A, et al. LipidHome: a database of theoretical lipids optimized for high throughput mass spectrometry lipidomics. *PLOS One*. 2013;8:e61951.
5. Tu J, Zhou Z, Li T, Zhu Z-J. The emerging role of ion mobility-mass spectrometry in lipidomics to facilitate lipid separation and identification. *Trends Anal Chem*. 2019;116:332–339.
6. Ni Z, Goracci L, Cruciani G, Fedorova M. Computational solutions in redox lipidomics – current strategies and future perspectives. *Free Radical Biol Med*. 2019;144:110–123.
7. Wishart DS, Feunang YD, Marcu A, et al. HMDB 4.0: the human metabolome database for 2018. *Nucl Acids Res*. 2018;46:D608–D617.
8. Guijas C, Montenegro-Burke JR, Domingo-Almenara X, et al. METLIN: a technology platform for identifying knowns and unknowns. *Anal Chem*. 2018;90:3156–3164.
9. O'Donnell V, Dennis EA, Wakelam MJ, Subramaniam S, MAP: LIPID. Serving the next generation of lipid researchers with tools, resources, data and training. *Sci Signal*. 2019:12.
10. METASPACE. Available at: https://metaspace2020.eu. Accessed on January 23, 2021.
11. Palmer A, Phapale P, Chernyavsky I, et al. FDR-controlled metabolite annotation for high-resolution imaging mass spectrometry. *Nat Methods*. 2017;14:57–60.
12. Wieser ME, Coplen TB. Atomic weights of the elements 2009 (IUPAC Technical Report). *Pure Appl Chem*. 2010;83:359–396.
13. Böcker S, Letzel MC, Lipták Z, Pervukhin A. Decomposing metabolomic isotope patterns. In: Bücher P, Moret BME, eds. Decomposing metabolomic isotope patterns. *Algorithms in Bioinformatics, Berlin, Heidelberg, 2006*. 2006:12–23.

14. Novák J, Škríba A, Havlíček V. CycloBranch 2: Molecular formula annotations applied to imzML data sets in bimodal fusion and LC-MS data files. *Anal Chem*. 2020;92:6844–6849.
15. Ridenour WB, Kliman M, McLean JA, Caprioli RM. Structural characterization of phospholipids and peptides directly from tissue sections by MALDI traveling-wave ion mobility-mass spectrometry. *Anal Chem*. 2010;82:1881–1889.
16. Škrášková K, Claude E, Jones EA, Towers M, Ellis SR, Heeren RM. Enhanced capabilities for imaging gangliosides in murine brain with matrix-assisted laser desorption/ionization and desorption electrospray ionization mass spectrometry coupled to ion mobility separation. *Methods*. 2016;104:69–78.
17. Shrestha B, Vertes A. High-throughput cell and tissue analysis with enhanced molecular coverage by laser ablation electrospray ionization mass spectrometry using ion mobility separation. *Anal Chem*. 2014;86:4308–4315.
18. Midey A, Olivos H, Shrestha B. Spatial mapping of cellular metabolites using DESI ion mobility mass spectrometry*Single Cell Metabolism*: Springer; 2020:181–190.
19. Levy AJ, Oranzi NR, Ahmadireskety A, Kemperman RHJ, Wei MS, Yost RA. Recent progress in metabolomics using ion mobility-mass spectrometry. *Trends Anal Chem*. 2019;116:274–281.
20. Zhou Z, Shen X, Tu J, Zhu Z-J. Large-scale prediction of collision cross-section values for metabolites in ion mobility-mass spectrometry. *Anal Chem*. 2016;88:11084–11091.
21. Poetzsch M, Steuer AE, Roemmelt AT, Baumgartner MR, Kraemer T. Single hair analysis of small molecules using MALDI-triple quadrupole MS imaging and LC-MS/MS: investigations on opportunities and pitfalls. *Anal Chem*. 2014;86:11758–11765.
22. Lamont L, Eijkel GB, Jones EA, et al. Targeted drug and metabolite imaging: desorption electrospray ionization combined with triple quadrupole mass spectrometry. *Anal Chem*. 2018;90:13229–13235.
23. Minerva L, Clerens S, Baggerman G, Arckens L. Direct profiling and identification of peptide expression differences in the pancreas of control and ob/ob mice by imaging mass spectrometry. *Proteomics*. 2008;8:3763–3774.
24. Ryan DJ, Spraggins JM, Caprioli RM. Protein identification strategies in MALDI imaging mass spectrometry: a brief review. *Curr Opin Chem Biol*. 2019;48:64–72.
25. Mascini NE, Heeren RM. Protein identification in mass-spectrometry imaging. *Trends Anal Chem*. 2012;40:28–37.
26. Perdian DC, Lee YJ. Imaging MS methodology for more chemical information in less data acquisition time utilizing a hybrid linear ion trap–orbitrap mass spectrometer. *Anal Chem*. 2010;82:9393–9400.
27. Korte AR, Lee YJ. Multiplex mass spectrometric imaging with polarity switching for concurrent acquisition of positive and negative ion images. *J Am Soc Mass Spectrom*. 2013;24:949–955.
28. Brodbelt JS. Photodissociation mass spectrometry: new tools for characterization of biological molecules. *Chem Soc Rev*. 2014;43:2757–2783.
29. Klein DR, Feider CL, Garza KY, Lin JQ, Eberlin LS, Brodbelt JS. Desorption electrospray ionization coupled with ultraviolet photodissociation for characterization of phospholipid isomers in tissue sections. *Anal Chem*. 2018;90:10100–10104.
30. Zhou R, Basile F. Plasmonic thermal decomposition/digestion of proteins: a rapid on-surface protein digestion technique for mass spectrometry imaging. *Anal Chem*. 2017;89:8704–8712.
31. Schey KL, Anderson DM, Rose KL. Spatially-directed protein identification from tissue sections by top-down LC-MS/MS with electron transfer dissociation. *Anal Chem*. 2013;85:6767–6774.

32. Meding S, Martin K, Gustafsson OJR, et al. Tryptic peptide reference data sets for maldi imaging mass spectrometry on formalin-fixed ovarian cancer tissues. *J Proteome Res.* 2013;12:308–315.
33. Veličković D, Chu RK, Carrell AA, et al. Multimodal MSI in conjunction with broad coverage spatially resolved MS2 increases confidence in both molecular identification and localization. *Anal Chem.* 2018;90:702–707.
34. A Masucci J, D Mahan A, D Kwasnoski J, W Caldwell G. A novel method for determination of drug distribution in rat brain tissue sections by LC/MS/MS: functional tissue microanalysis. *Curr Topics Med Chem.* 2012;12:1243–1249.
35. Chaurand P, Fouchécourt S, DaGue BB, et al. Profiling and imaging proteins in the mouse epididymis by imaging mass spectrometry. *Proteomics.* 2003;3:2221–2239.
36. Datta S, Malhotra L, Dickerson R, Chaffee S, Sen CK, Roy S. Laser capture microdissection: big data from small samples. *Histol Histopathol.* 2015;30:1255–1269.
37. Dilillo M, Pellegrini D, Ait-Belkacem R, de Graaf EL, Caleo M, McDonnell LA. Mass spectrometry imaging, laser capture microdissection, and LC-MS/MS of the same tissue section. *J Proteome Res.* 2017;16:2993–3001.
38. Nelson RW, Krone JR, Bieber AL, Williams P. Mass spectrometric immunoassay. *Anal Chem.* 1995;67:1153–1158.
39. Han G, Spitzer MH, Bendall SC, Fantl WJ, Nolan GP. Metal-isotope-tagged monoclonal antibodies for high-dimensional mass cytometry. *Nat Protoc.* 2018;13:2121–2148.
40. Majonis D, Ornatsky O, Kinach R, Winnik MA. Curious results with palladium- and platinum-carrying polymers in mass cytometry bioassays and an unexpected application as a dead cell stain. *Biomacromolecules.* 2011;12:3997–4010.
41. Howat WJ, Lewis A, Jones P, et al. Antibody validation of immunohistochemistry for biomarker discovery: recommendations of a consortium of academic and pharmaceutical based histopathology researchers. *Methods.* 2014;70:34–38.

Chapter 7

Strategies for quantitative imaging mass spectrometry

Chapter Outline

Imaging mass spectrometry (MS) is a label-free analytical technique that is used to visualize detected molecules on the sample surface. In a typical imaging MS experiment, a unique mass spectrum is generated at each pixel by moving a focused ion beam on a two-dimensional surface. An MS image for an ion is created by plotting signal intensities of that ion at every pixel using a false-color heatmap representing its relative distribution on the *x–y* plane. Qualitative MS imaging already gives some sort of quantitative information. Improving quantitative workflow in imaging MS is needed expand the utility of technology in pharmaceutical and biomedical research. Due to the inherent nature of the direct analysis of tissue instead of extracted and purified samples. All imaging MS modalities will need to address analyte and tissue-specific ion suppression, ion extraction. Consistent data preprocessing and appropriate normalization routines are also crucial to get more accurate values. In this chapter, we discuss various aspects of quantitative imaging MS, such as normalization, calibration strategy with internal standards.

Most of the challenges associated with quantitative analysis in MS also exist in imaging MS and more so, such as sample preparation workflow, ion suppression, detection bias, spectral interferences, calibration nonlinearity. In principle, the goal of any quantitative analysis by MS is to obtain a concentration value from the measured ion signal by comparing it with the ion signal produced by an internal standard. In conventional MS workflows, such as electrospray liquid

Introduction to Spatial Mapping of Biomolecules by Imaging Mass Spectrometry. DOI: 10.1016/C2018-0-03962-6

chromatography (LC) MS, quantitation specificity is increased by separating the molecule-of-interest by chromatography separation and quadrupole mass isolation, such as multiple reaction monitoring (MRM) experiments. Data normalization is done using an internal standard to correct any matrix effects such as inconsistent extraction, ionization efficiencies, and even instrumentation-related variability such as injection-to-injection stability. A significant challenge for quantitative MS imaging is the direct in situ nature of the ionization used for analysis that does not account for heterogeneous ion suppression and inhomogeneous extraction effects of a different part of a tissue. Accounting and compensating for matrix effects in imaging MS are essential for accurately quantifying spatial localization of molecules in biological samples.[1]

In addition to the instrumental and experimental variance that may influences quantification, MS imaging also has to account for biological variation within the tissue and pixel-to-pixel variation. One can think of each pixel as a separate mass spectrum measurement. Many excellent reviews on the quantitative imaging MS can be found in the literature, some of them listed in the referenced publications.[2–7,54]

In a simple ideal sample, such as pure molecule standard, the ion intensities denoted by the heatmap scale of the image should correlate to a semiquantitative trend. However, in a real-world application, many factors complicate the quantitation. For instance, in the most widespread application of MS imaging, tissue analysis would need to account for ion suppression for the inherent molecular heterogeneity of the tissue section.

Quantitative imaging MS analysis is hindered by—(a) undulation in the analytical beam (e.g., shot-to-shot variability in laser fluence, electrospray instability due to fluidics); (b) nonideal analytical workflow such as manual sample preprocessing steps (e.g., sectioning, storage), matrix crystal inhomogeneity; or (c) inherent complexity brought in by heterogeneous nature of tissue sections (e.g., tissue-specific differential ionization efficiencies or ion suppression, variance in extraction efficiency). The issues related to analytical tools can be resolved by designing or using instrumentation with more precision and adopting consistent protocols. Additionally, to avoid signal from other overlapping chemical species, either imaging of a product ion using a tandem MS (MS/MS) or high-resolution mass spectral measurements able to resolve analytes is needed. The unavoidable matrix effect due to heterogeneous tissue needs to be addressed by appropriate normalization schemes and calibration routines constituting the tissue matrix.

In a mass spectrometer, an ion signal of a molecule may decrease when analyzed as part of a complex mixture, such as biological tissue with numerous molecules, compared to when analyzed by itself. This is known as ion suppression, where an MS signal of a specific compound is suppressed by competition for charge with other compounds in tissue due to instrumental factors such as fragmentation, complex ion formation, low mass resolution. When a molecular ion is suppressed, the measured intensities response does not reflect the true concentration. Due to the heterogeneous nature of the tissue, ion suppression of molecule-of-interest may change depending on its location. Quantification of molecules using imaging MS can be affected by matrix and substrate effects. For instance, three illicit drugs (cocaine, methamphetamine, and heroin) residues in fingerprint were quantitatively imaged on silicon and paper by secondary ion mass spectrometry and desorption electrospray ionization (DESI) MS. All drugs were successfully imaged on silicon, but except cocaine, other drugs were entirely suppressed by the matrix and substrate effect of the paper.[8] In addition to assessing for the matrix effect, it is imperative to understand the molecule being analyzed. Comprehending its physicochemical properties, in vitro and in vivo metabolism, stability, protein binding, or any other biochemical interactions are critical during imaging many molecules in tissue, such as drugs. Due to critical biochemical functional considerations, it is often necessary to quantitatively determination of the distribution of drugs and metabolites in tissues.

The broad definition of quantitative imaging is an objective numerical measurement of tissue visualization. Here, quantitative imaging with respect to imaging MS is defined as quantitative information obtained on a molecule based on the ion intensity signal of the imaging MS experiment. Imaging MS measurements can be either perform (a) relative quantitation or (b) absolute quantitation. Multivariate analysis, statistical, and segmentation aspects of quantitative imaging MS are not discussed here. For more information on these topics, please refer to the referenced articles here.[9,10]

The data processing routine is very critical in quantitative MS imaging. Careful attention should be paid during baseline subtraction, denoising, spectral realignment, peak peaking, bin definition. One of the most important aspects of data processing for quantitative analysis is normalization. There are many imaging software that supports various normalization scheme and quantitation tools.[11–17]

Normalization

Normalization is applied correct variance due to instrument, methods, or sample microenvironment, and transforms measured intensities to intensities that accurately depict the scale of molecular concentration. Normalization of MS imaging data encompasses correcting intensities of the ion at each pixel by applying a specific correction factor. In the literature, many normalization methods are proposed for imaging MS, and many are readily accessible in software. The normalization strategies swing from a global normalization scheme based on a single component such as total ion count (TIC) to a normalization scheme based on a single isotopically labeled internal standards. The MS imaging normalization methods differ in their assumptions, signal correction, ease of application, etc.

Normalization may need to change due to ion source or mass analyzer. For example, in matrix-assisted laser desorption/ionization (MALDI), normalization of the signal intensity needs to compensate for additional effects such as uneven MALDI matrix deposition, laser fluence fluctuation.[18] Similarly, in DESI-MS imaging normalization strategies can be used to account for fluidic stability, surface morphology, sprayer instability, etc.[19]

The choice of normalization may depend on the type of mass analyzer. For example, normalizing MALDI imaging from fourier transform ion cyclotron resonance mass spectrometer (FTICRMS or FTMS), where it has a uniform electronic baseline may be different from one done on time of flight, which has a large chemical noise due to a lower mass resolution.[20] Some commonly used normalization methods used in imaging MS have been discussed here.[21,22]

Imaging MS simply cannot copy the normalization procedures adopted in other imaging fields because MS is a fundamentally different technology. Additionally, other modality often has only a few channels of information, whereas MS has thousands. Normalization may also produce misleading results by skewing or biasing the distributions of ions. Normalization is well suited for systematic artifacts.

Global normalization

Global normalization performs a spectrum-wide scaling of ion intensity. The three most common global normalization strategies include; the TIC, the root mean square, and the median intensity.[11,20,23] Global normalization is the most commonly used method due to its simplicity and ease-of-use. TIC normalization could adequately address the minor differences in matrix coating and laser power for MALDI time-of-flight analysis. However, the TIC normalization is prone to high-intensity peaks, excessive background peaks, significantly changing peaks. For instance, it has been shown that TIC normalization can benefit from omitting intense peaks.[20]

A single metric global normalization also cannot account for ionization bias and tissue extraction efficiency. Global normalization is a good starting point for qualitative or semiquantitative analysis but cannot adequately address molecule-specific ionization biases for all molecules present in a complex mixture. Global normalization has been often widely used to get semiquantitative information in discovery biomedical research, such as spatial metabolomics. Many of the detected compounds may not have affordable and readily available commercial stable isotope standards or, more so, it is impracticable to use matched internal standards for each of the thousands of molecules detected. At the discovery stage, the semiquantitative information may be enough to plan and move forward with more targeted imaging experiments. Factor like ion suppression that depends on the microenvironment of tissue cannot be corrected by application of the spectrum-wide global normalization approaches.

Internal standard normalization

Internal standard normalization is molecule-specific normalization using a reference standard for accounting for ion suppression, and maybe ion extraction. The internal standard-based normalization strategy provides the most accurate quantitative results. However, the sample preparation for this method is most demanding. It requires the use of suitable reference standards that will depend on the type of molecule being studied. For example, the pharmaceutical compound can be most quantitatively analyzed by using isotopically labeled counterpart.

The use of the internal standard for quantitative imaging MS has been demonstrated multiple times, most commonly with small pharmaceutical drugs such as cocaine in brain tissue using a deuterated internal standard using MALDI,[24,25] clozapine quantitative DESI-MS imaging in brain sections[26] or LAESI.[27,28] An internal standard can be added on the tissue prior to the experiment, or during the sample preparation step (e.g., with matrix), or online during acquisition, depending on the imaging modality. For example, nano-DESI solvent was doped with internal standards that were used to obtain online quantification of several phospholipids in brain tissue.[29]

Background ions, such as ubiquitous matrix peak in MALDI, can be used as a surrogate in lack of appropriate internal standards such as stable isotope compounds or structural analogs—especially small molecule drug imaging. For example, a pharmaceutical compound in rat brain normalized against the α-Cyano-4-hydroxycinnamic acid (α-CHCA) matrix found a good correlation with absolute quantification by LC–MS/MS of laser-capture microdissected tissue.[30] Others have used matrix peak normalization to perform quantitative MS imaging of ketoconazole in the skin and raclopride.[31,32] Two peptide isotopologues, differing with only neutrons, have been used as internal standards for quantification of low-abundance proteins using the coupling of immunoenrichment of proteotypic peptides with method called, immuno-MRM using immuno-MALDI.[33]

Tissue-specific normalization coefficient

Compound-specific ionization efficiencies can be addressed by using the internal standard, but tissue-specific ion suppression needs a tissue-specific ionization efficiency normalization factor that considers changing microenvironments of tissues. The ion intensity of the same concentration of the sample could vary significantly between the different areas of the tissue due to local ion suppression in that area. The significant advantage of such normalization is easier to implement because it does not need to require the addition of a reference standard. Tissue extinction coefficient normalization has been proposed to consider tissue-dependent ionization efficiency due to local ion suppression.[21,34,35]

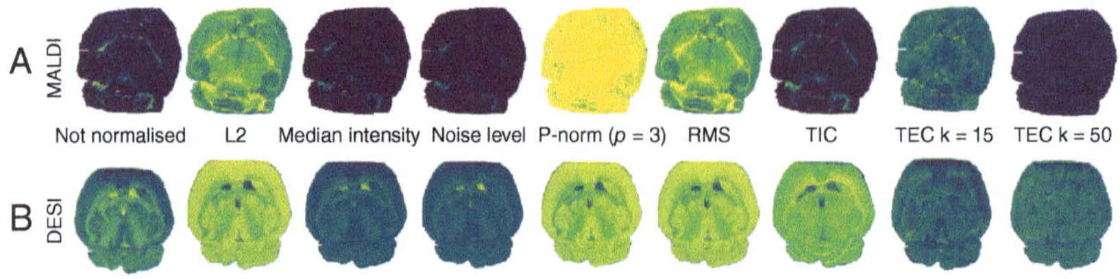

FIG. 7.1 MALDI and DESI images of uniformly coated olanzapine standard on mouse brain after application of various normalization methods—not normalized, L2 normalization, median intensity normalization, noise level normalization, P-norm normalization ($p = 3$), root-mean-square (RMS) normalization, total ion count (TIC) normalization, and tissue extinction coefficient (TEC) normalization. Adapted with permission from Taylor AJ, Dexter A, *Bunch J. Anal Chem.* 2019;90:5637–5645. Copyright 2019 American Chemical Society.[21]

Fig. 7.1 shows various normalization methods, from left to right—not normalized, L2 normalization, median intensity normalization, noise level normalization, P-norm normalization ($p = 3$), root-mean-square normalization, TIC normalization, and tissue extinction coefficient normalization ($k = 15$ and $k = 50$) applied to MALDI and DESI image of uniformly coated olanzapine standard on the mouse brain.[21] An effective normalization would show a homogeneous intensity of olanzapine across the brain section by correcting any ion suppression.

Quantitation level in imaging MS

There are many levels of quantitation, depending on the precision of the accuracy required. The first one is the relative quantitation, which can elude to gross trends or fold change from normalized intensities without any internal standards. The second level is more accurate and specific quantitation obtained

Table 7.1 A general level of quantitation in imaging mass spectrometry.

RELATIVE QUANTITATION • without internal standard • fold change, trends
SPECIFIC QUANTITATION • general internal standards • molar concentration range, percentage
ABSOLUTE QUANTITATION • ideally isotopically labeled or other suitable internal standard • molar concentration

using any internal standard that can give molarity level or percentage. The final level is the absolute quantitation that provides accurate molar concentration with isotopically level or other suitable internal standards. The general level of quantitation is summarized in Table 7.1. Relative and absolute quantitation is described more in detail below.

Relative quantitation

Relative quantitation is defined here as a corresponding variation in an ion in a different part of the image, such as fold change or trend. The absolute quantification is defined as an accurate measurement of the amount or concentration of the compound in a well-defined region of interest (ROI) of the tissue. The goal of relative quantitation is to ensure the measured intensities of the molecules are consistent with their real localizations within the tissues, and any increasing or decreasing trends are captured by a similar movement in the ion signal. The relative intensity will not be able to get an exact concentration of the molecule within tissue or region within the tissue, but just inform if it is more or less versus other areas of tissue or other tissues. Relative quantitation would need normalization of data with or without internal standards but would not require any calibration curve.

Absolute quantification

Absolute quantification means knowing the exact concentration of a specific molecule at each pixel of tissue or defined ROI. Many calibration workflows have been defined to obtain absolute quantitation of molecules, particularly for xenobiotic drug molecules during MALDI imaging. The absolute quantitation during imaging MS requires tissue-specific calibration to address the ion suppression due to the endogenous molecules present in the tissue. Additionally, it would also need tissue-specific normalization steps to addresses different chemical microenvironments of tissue.

Absolute quantification of imaged molecules by imaging MS is not straightforward because the measured ion intensity cannot be directly converted to an absolute quantity. A reference standard is needed in order to obtain absolute quantification as with any other MS analysis because several molecules-specific properties can affect the ion intensity, including ionization efficiency, ion suppression, extraction efficiency, molecular stability, etc.

The specificity of the quantification is increased by measuring fragmentation after isolating the parent ion and fragmenting by MS/MS. The method of measuring the transition of parent to fragment ion is called selected reaction monitoring or MRM, depending on the number of transitions or molecules.[36] MALDI analysis has been utilized in selected reaction monitoring and MRM mode for the quantification of pharmaceutical compounds in triple quadrupole (QqQ) mass spectrometers.[37] MALDI MRM imaging MS has been used as a sensitive method for quantifying the distribution of an antituberculosis drug in rabbit lung biopsies.[38] The peptide quantitation is measured using a ratio of MALDI MRM transitions of the endogenous unlabeled proteolytic peptides to the isotopically labeled standard peptides. MALDI MRM imaging MS showed protein concentrations over two orders of magnitude after applying reference standard to reduces the ion suppression or any matrix localization effects.[39]

Absolute quantification calibration strategies

In most of the absolute quantitative workflow, in order to get the absolute concentration of drug from the tissue section, an appropriate calibration curve is created. The calibration curve is created by analyzing a series of standard solutions at different concentration levels. Ideally, the calibration curve has to take into account the tissue biochemical matrix to accurately obtain an intensity response similar to dosed molecules. The use of analyte-matched internal standards, such as stable isotope label surrogates, are necessary for all the calibration strategies. These calibration strategies differ in their correction to tissue matrix effects, analyte extraction efficiency, and ease of sample preparation. The three broad calibration strategies for absolute quantitation are shown in Fig. 7.2 and a comparative summary of three absolute quantitation strategies is presented in Table 7.2. The off-tissue, on-tissue, and mimetic calibration strategies are discussed in detail below.

Off-tissue calibration

The off-tissue quantitation strategy involves constructing the calibration curve by spotting various concentrations of the standard molecule on a target plate

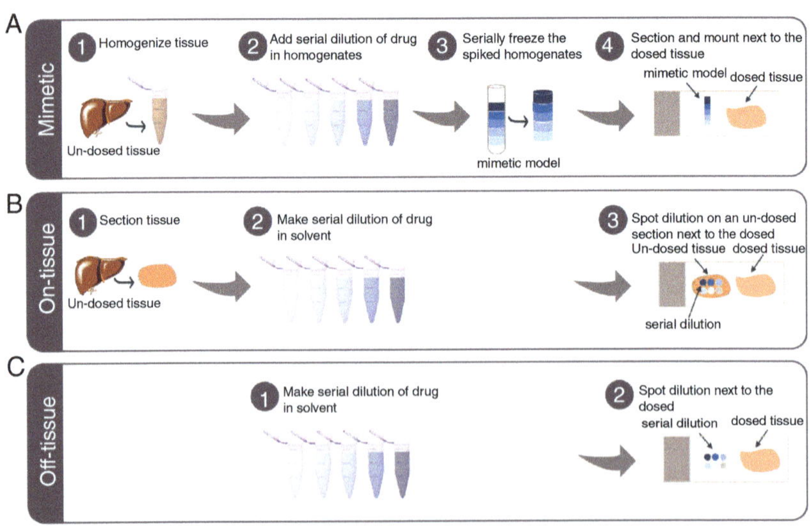

FIG. 7.2 Three broad calibration strategies for quantitation of molecule in imaging mass spectrometry are shown. Details are elaborated in Table 7.2.

Table 7.2 Comparison of three absolute quantitation strategies used in imaging MS.

	Mimetic	On-tissue	Off-tissue
			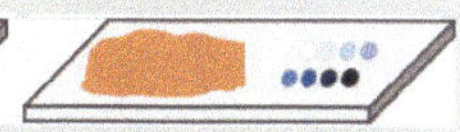
Ion suppression	Homogenate represents "average" tissue suppression	Homogenate represents "local" tissue suppression	Tissue suppression is not taken into account
Extraction	Extraction efficiency most similar to the tissue	Extraction efficiency may differ due to drug spotted on top of the tissue	The extraction efficiency is not taken into account
Reproducibility	Replicates most consistent	Reproducibility depends on spotting	Reproducibility depends on spotting
Preparation	Preparation is more involved for a few samples, but less effort for a larger number of samples	Easy prep steps	Easy prep steps
Suitability	A large cohort of samples with the same level of drug	Quick answer suitable for homogeneous tissue	For quick LOD and proof-of-principle experiment

outside of the tissue section. Typically, all the spots are imaged together as a single image, with or without tissue. The analysis is done by outlining ROI of each spot to get absolute or normalized ion intensity values, which are used to create the calibration curve. This strategy has the most straightforward sample preparation step but does not compensate for the ion suppression effect of tissue or extraction efficiency. This is a suitable first step quick quantitation strategy, as well as an alternate where the nondosed (blank) tissue sample is not available to spike standard and stable isotope label molecule is inaccessible.

On-tissue calibration

The on-tissue quantitation strategy involves spotting the standard analyte on a second tissue section instead of a target plate. The second tissue section is usually a serial section with similar characteristics. Usually, the entire spotted tissue is imaged together with the tissue being examined. Spotting the analyte onto a blank tissue takes into account ionization effects due to tissue matrix but may not account for the extraction efficiency. On-tissue calibration is a suitable strategy for a quick quantitation strategy where the nondosed (blank) tissue sample is readily available to spike standard, and tissue is homogenous. There are several strategies for adding the standard, either on top of tissue or underneath the

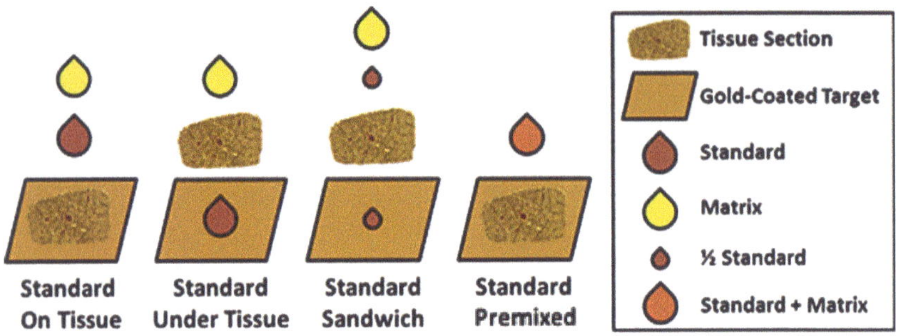

FIG. 7.3 Four approaches of depositing reference standard on the tissue were evaluated; standard applied first, underneath the tissue, tissue sandwich between standards or standard, and matrix mixture applied. The first approach of applying the standard followed by the matrix onto tissue sections resulted in the most consistent result as the HPLC analysis. Reprinted with permission from Chumbley CW, Reyzer ML, Allen JL, et al. Absolute quantitative MALDI imaging mass spectrometry: a case of rifampicin in liver tissues. Anal Chem. 2016;88:2392–2398. Copyright 2016 American Chemical Society.[40]

tissue. In a study, several approaches of depositing reference standards on the tissue were systematically evaluated, as shown in Fig. 7.3. In the first approach, the standard was applied, followed by the matrix on the tissue section. On the second approach, the standard was deposited on the bottom of the tissue before thaw mounting, and the matrix was applied. On the third approach, the standard was applied on top and bottom of tissue, followed by matrix coating on tip. Finally, a mixture of standard and matrix was applied. No statistically significant differences were detected in the concentration of standard analyzed by high-performance liquid chromatography MS and MALDI-MS when standards were deposited on the tissues followed by matrix application.[40] Similarly, three strategies for internal standard application in DESI imaging MS was explored. Applying a thin film of internal standard and depositing micro spots on top of tissues was found statistically similar, while spiking DESI solvent spray with internal standard needs more investigation.[55] Fig. 7.4 shows an example of quantitative imaging of cocaine drugs on the tissue with several orders of magnitude using tandem MS and stable isotope labeling.[24]

Mimetic tissue calibration

In the mimetic model strategy, a calibration curve is created by a series of tissue homogenate mixtures spiked with various concentrations of the molecule-of-interest instead of spotting the standard on or off tissue. Briefly, the initial protocol proposed on the mimetic model was constructed by adding the appropriate concentration of drug standard to the exact amount of control tissue homogenate, freezing the spiked homogenate in a single mold adjacent to each other, and finally cryosectioning the frozen homogenate as tissue section next to tissue.[41] Recently, the mimetic tissue calibration workflow was revised. Briefly,

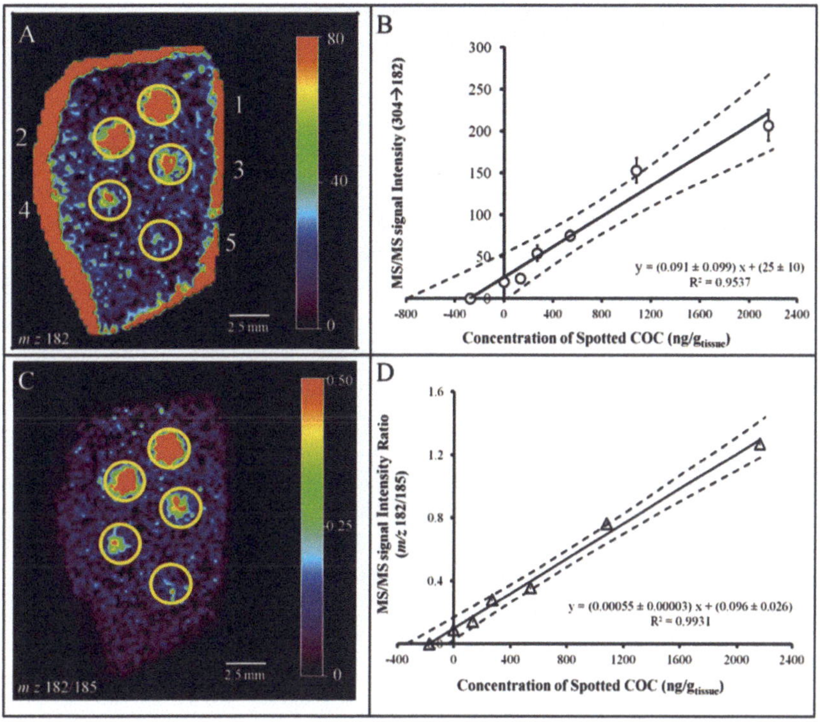

FIG. 7.4 Quantitative analysis of cocaine by image by using ratio of normal by deuterated fragment is shown. Normalization to the deuterated standard improved the linearity and the error of the calibration plot leading to improved quantitative estimation of cocaine. Adapted with permission from Pirman DA, Reich RF, Kiss A, Heeren RMA, Yost RA. Anal Chem. 2013;85:1081–1089. Copyright 2012 American Chemical Society.[24]

a single mold was laid with the serially frozen analyte and tissue homogenate at increasing or decreasing concentration and sectioned adjacent to tissue. A complete protocol can be found in the referenced publication.[42]

Both on-tissue calibration strategy and mimetic model in-tissue calibration strategy account for ion suppression effects, but the mimetic model can account for analyte extraction better. The memetic model is well adapted to single organ imaging, mainly if the organ is homogeneous as the liver. However, heterogeneous tissue, multiple organs, or whole-body imaging MS analysis may be complicated by uneven ion response based on its microenvironment. Additionally, the whole-body study would require a large amount of nontreated tissue material. Mimetic model addressed extraction efficiency, but the extraction of spiked standard in the frozen homogenate section may not accurately reflect the analyte in the tissue section, such as protein-bound drug molecule. The effect of cell disruption due to the homogenization process and refreezing also needs to be addressed.

Offline and other tissue quantitation strategy

Imaging MS is often supplemented by other MS methods for a local quantitative analysis of molecule-of-interest in tissue. These include measuring the molecule quantity in the entire organ or a small region by its tedious dissection. Here quantitation of molecule-of-interest is achieved by conventional analysis by LC–MS/MS. One of the other methods includes laser-capture microdissection, where ablated tissue is captured and analyzed by electrospray ionization.[43] Laser-capture microdissection coupled to LC–MS/MS was able to demonstrate the accumulation of tuberculosis drug ethambutol's accumulation in the lesion.[44] A straightforward approach, called functional tissue microanalysis, involves punching out a region of the tissue section, extract and analyze by conventional LC–MS/MS.[45] Like laser-capture microdissection, many subregions can be sampled from a single section. Liquid extraction surface analysis (LESA) performs efficient sampling by liquid extraction, followed by highly sensitive nanoelectrospray analysis. Its ability to couple with LC–MS/MS also provides it with high selectivity. LESA has been mainly employed to perform profiling at the discrete location of the tissue, but can also be used for imaging MS. Due to its high sensitivity and selectivity, LESA can be useful to map distribution of molecules that lacked sufficient sensitivity or require high selectivity due to isobaric interefences.[46]

Like any other analytic workflow, quantitative measurements using imaging MS require validation by other assays. The validation may include defining limits of detection or limits of quantification. Quantitative imaging MS for pharmaceuticals drugs will need a set of validation method, such as ones used for validation of high-performance liquid chromatography methods as prescribed by the US Food and Drug Administration, United States Pharmacopeia, and the International Council for Harmonisation of Technical Requirements for Pharmaceuticals for Human Use.[47]

Only a handful of strategies are discussed here. Often there is a tradeoff between accurate quantitative information and spatial resolution as broadly illustrated in Fig. 7.5. Research is ongoing to find innovative ways to obtain precise and accurate quantitative information during imaging MS. For example, virtual calibration has been demonstrated for performing a label-free calibration of tissue-specific matrix effects and quantitation using air-flow-assisted DESI. The technique employs the analyte response of endogenous metabolite ions from each mass spectrum as an internal standard. Machine-learning segmentation is used to analyze ionization variation within the tissue and for region-level or pixel-level normalization resulting in a label-free calibration and quantitation.[48] Another novel approach termed imprint mass spectrometry imaging uses tissues imprints on slides coated with a dopamine-modified TiO_2 monolith preloaded with analyte internal standards. The incorporation of endogenous molecules onto the coated slide results in similar extraction of endogenous analyte and internal standards, as well as equal incorporation of the MALDI matrix. Endogenous molecules, such as phospholipids, ceramides, cholesterol, were imprinted from the brain without noticeable delocalization. Their imprinting can lead to a reduction in tissue matrix effects and result in more accurate quantitation.[49] For calibration, the imprinting imaging MS used kinetic calibration method, in which

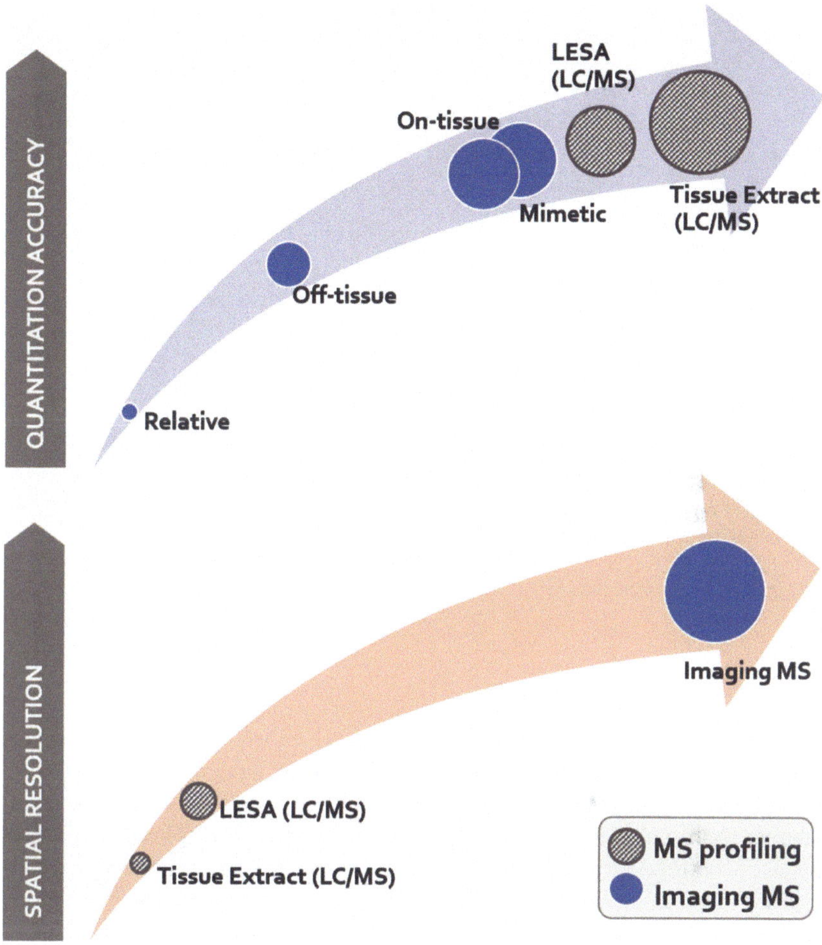

FIG. 7.5 Current tradeoffs in spatial resolution and quantitative information in imaging mass spectrometry.

preloaded internal standards are used to calibrate the extraction of the analytes and was demonstrated with solid-phase microextraction (Fig. 7.6).[50]

Relative quantitation of proteins can be done by a method called stable isotope label-based mass spectrometric imaging. The technique uses stable isotope-labeled chromogens applied using the immunohistochemistry tissue staining protocols.[51] In stable isotope label-based mass spectrometric imaging, primary antibody specific for the biomarker is attached to a secondary antibody, conjugated with alkaline phosphatase, that is specific to the primary antibody. The liberated azo dye precipitate during immunohistochemistry staining absorbs the laser energy and fragments at the amide bond to produce signature reporter ions similar to the targeted multiplex mass spectrometry imaging technique that employs photocleavable discrete mass tags that are released from their respective antibodies.[52] The peak intensities of the reporter ions from imaging MS were used to relatively quantitate proteins in the tissue.

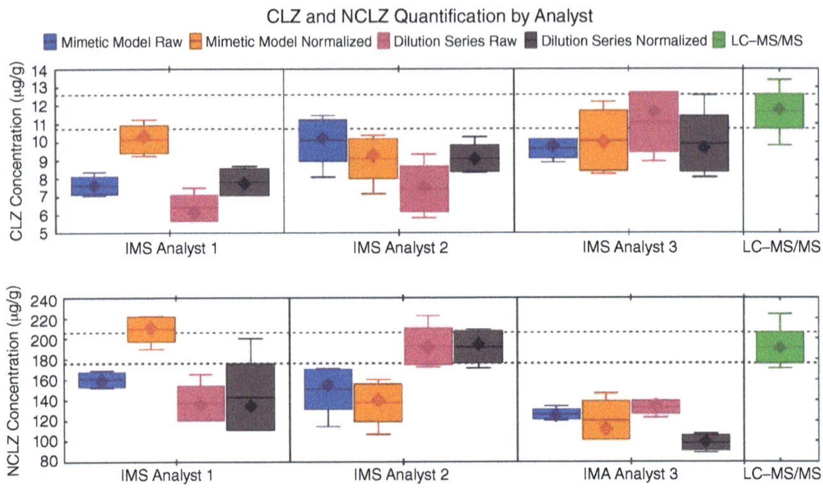

FIG. 7.6 Comparison of quantitation done by mimetic model, dilution series, and LC–MS/MS of clozapine (CLZ) and norclozapine (NCLZ) by three different analysts. Adapted with permission from Barry JA, Ait-Belkacem R, Hardesty WM, et al. Anal Chem. 2019;91:6266–6274. Copyright 2019 American Chemical Society.[53]

Conclusions

It is given that an excellent quantitative measurement would need precise instrumentation from mass spectrometers to pipettes that have minimum variation. Quantitative imaging MS would also need a consistent workflow. Only a few quantitations approach for imaging MS have been discussed above. Currently, the most popular workflow for absolute quantification seems to a variant of on-tissue calibration with internal standards. Multicenter and multianalyst validation studies as shown in Fig. 7.6 showed that the quantification accuracy of the on-tissue dilution series and mimetic model are comparable. Thus, the choice of the method depends on the nature of the sample, such as tissue heterogeneity, and resources available. Both calibration method normalized to the internal standard gave about 80% accuracy with a relative standard deviation of roughly 35% compared to corresponding LC–MS/MS homogenate samples.[53] These workflow developments in quantitative imaging have enabled quantification of detected analyte often down to a handful of pixels. Despite many advancements, precise absolute quantification from tissue sections by imaging MS still require additional research, such as dilution series to that mimics the behavior of that analyte in tissue, solving ion suppression/extraction in tissue microenvironment due to heterogeneity, robust validation efforts, and more.

References

1. Lanekoff I, Stevens SL, Stenzel-Poore MP, Laskin J. Matrix effects in biological mass spectrometry imaging: identification and compensation. *Analyst*. 2014;139:3528–3532.

2. Rzagalinski I, Volmer DA. Quantification of low molecular weight compounds by MALDI imaging mass spectrometry–a tutorial review. *Biochim Biophys Acta Proteins Proteomics*. 2017;1865:726–739.
3. Ellis SR, Bruinen AL, Heeren RM. A critical evaluation of the current state-of-the-art in quantitative imaging mass spectrometry. *Anal Bioanal Chem*. 2014;406:1275–1289.
4. Sun N, Walch A. Qualitative and quantitative mass spectrometry imaging of drugs and metabolites in tissue at therapeutic levels. *Histochem Cell Biol*. 2013;140:93–104.
5. Porta T, Lesur A, Varesio E, Hopfgartner G. Quantification in MALDI-MS imaging: what can we learn from MALDI-selected reaction monitoring and what can we expect for imaging?. *Anal Bioanal Chem*. 2015;407:2177–2187.
6. Prideaux B, Lenaerts A, Dartois V. Imaging and spatially resolved quantification of drug distribution in tissues by mass spectrometry. *Curr Opin Chem Biol*. 2018;44:93–100.
7. Tobias F, Hummon AB. Considerations for MALDI-based quantitative mass spectrometry imaging studies. *J Proteome Res*. 2020;19(9):3620–3630. Publication Date: August 7. doi:10.1021/acs.jproteome.0c00443.
8. Muramoto S, Forbes TP, van Asten AC, Gillen G. Test sample for the spatially resolved quantification of illicit drugs on fingerprints using imaging mass spectrometry. *Anal Chem*. 2015;87:5444–5450.
9. Jones EA, Shyti R, van Zeijl RJM, et al. Imaging mass spectrometry to visualize biomolecule distributions in mouse brain tissue following hemispheric cortical spreading depression. *J Proteomics*. 2012;75:5027–5035.
10. Alexandrov T. MALDI imaging mass spectrometry: statistical data analysis and current computational challenges. *BMC Bioinformatics*. 2012;13:S11.
11. Källback P, Shariatgorji M, Nilsson A, Andrén PE. Novel mass spectrometry imaging software assisting labeled normalization and quantitation of drugs and neuropeptides directly in tissue sections. *J Proteomics*. 2012;75:4941–4951.
12. Källback P, Nilsson A, Shariatgorji M, Andrén PE. MsIQuant–quantitation software for mass spectrometry imaging enabling fast access, visualization, and analysis of large data sets. *Anal Chem*. 2016;88:4346–4353.
13. Bokhart MT, Nazari M, Garrard KP, Muddiman DC. MSiReader v1. 0: evolving open-source mass spectrometry imaging software for targeted and untargeted analyses. *J Am Soc Mass Spectrom*. 2017;29:8–16.
14. Tortorella S, Tiberi P, Bowman AP, et al. LipostarMSI: comprehensive, vendor-neutral software for visualization, data analysis, and automated molecular identification in mass spectrometry imaging. *J Am Soc Mass Spectrom*. 2020;31:155–163.
15. Grey AC, Demarais NJ, West BJ, Donaldson PJ. A quantitative map of glutathione in the aging human lens. *Int J Mass Spectrom*. 2019;437:58–68.
16. Baluya D, Shrestha B, Cressman EN. Semi-Quantitative Mapping of Oncological Therapies with Mass Spectrometry Imaging. Proceedings: AACR Annual Meeting 2019; March 29-April 3, 2019; Atlanta, GA. Published July 2019. doi: 10.1158/1538-7445.AM2019-1411.
17. Suzuki T, Sakata S, Makino Y, et al. iQuant2: software for rapid and quantitative imaging using laser ablation-ICP mass spectrometry. *Mass Spectrom*. 2018;7:A0065 A0065.
18. Steven RT, Race AM, Bunch J. Probing the relationship between detected ion intensity, laser fluence, and beam profile in thin film and tissue in MALDI MSI. *J Am Soc Mass Spectrom*. 2016;27:1419–1428.
19. Dong Y, Guella G, Franceschi P. Impact of tissue surface properties on the desorption electrospray ionization imaging of organic acids in grapevine stem. *Rapid Commun Mass Spectrom*. 2016;30:711–718.

20. Deininger S-O, Cornett DS, Paape R, et al. Normalization in MALDI-TOF imaging datasets of proteins: practical considerations. *Anal Bioanal Chem.* 2011;401:167–181.
21. Taylor AJ, Dexter A, Bunch J. Exploring ion suppression in mass spectrometry imaging of a heterogeneous tissue. *Anal Chem.* 2018;90:5637–5645.
22. Fonville JM, Carter C, Cloarec O, et al. Robust data processing and normalization strategy for MALDI mass spectrometric imaging. *Anal Chem.* 2012;84:1310–1319.
23. Norris JL, Cornett DS, Mobley JA, et al. Processing MALDI mass spectra to improve mass spectral direct tissue analysis. *Int J Mass Spectrom.* 2007;260:212–221.
24. Pirman DA, Reich RF, Kiss A, Heeren RMA, Yost RA. Quantitative MALDI tandem mass spectrometric imaging of cocaine from brain tissue with a deuterated internal standard. *Anal Chem.* 2013;85:1081–1089.
25. Reich RF, Cudzilo K, Levisky JA, Yost RA. Quantitative MALDI-MS n analysis of cocaine in the autopsied brain of a human cocaine user employing a wide isolation window and internal standards. *J Am Soc Mass Spectrom.* 2011;21:564–571.
26. Vismeh R, Waldon DJ, Teffera Y, Zhao Z. Localization and quantification of drugs in animal tissues by use of desorption electrospray ionization mass spectrometry imaging. *Anal Chem.* 2012;84:5439–5445.
27. Deimler RE, Razunguzwa TT, Reschke BR, Walsh CM, Powell MJ, Jackson GP. Direct analysis of drugs in forensic applications using laser ablation electrospray ionization-tandem mass spectrometry (LAESI-MS/MS). *Anal Methods.* 2014;6:4810–4817.
28. Bokhart MT, Rosen E, Thompson C, Sykes C, Kashuba AD, Muddiman DC. Quantitative mass spectrometry imaging of emtricitabine in cervical tissue model using infrared matrix-assisted laser desorption electrospray ionization. *Anal Bioanal Chem.* 2015;407:2073–2084.
29. Lanekoff I, Thomas M, Laskin J. Shotgun approach for quantitative imaging of phospholipids using nanospray desorption electrospray ionization mass spectrometry. *Anal Chem.* 2014;86:1872–1880.
30. Goodwin RJA, Scullion P, MacIntyre L, Watson DG, Pitt AR. Use of a solvent-free dry matrix coating for quantitative matrix-assisted laser desorption ionization imaging of 4-bromophenyl-1,4-diazabicyclo(3.2.2)nonane-4-carboxylate in rat brain and quantitative analysis of the drug from laser microdissected tissue regions. *Anal Chem.* 2010;82:3868–3873.
31. Bunch J, Clench MR, Richards DS. Determination of pharmaceutical compounds in skin by imaging matrix-assisted laser desorption/ionisation mass spectrometry. *Rapid Commun Mass Spectrom.* 2004;18:3051–3060.
32. Takai N, Tanaka Y, Inazawa K, Saji H. Quantitative analysis of pharmaceutical drug distribution in multiple organs by imaging mass spectrometry. *Rapid Commun Mass Spectrom.* 2012;26:1549–1556.
33. Ibrahim S, Froehlich BC, Aguilar-Mahecha A, et al. Using two peptide isotopologues as internal standards for the streamlined quantification of low-abundance proteins by immuno-MRM and immuno-MALDI. *Anal Chem.* 2020;92:12407–12414.
34. Stoeckli M, Staab D, Schweitzer A. Compound and metabolite distribution measured by MALDI mass spectrometric imaging in whole-body tissue sections. *Int J Mass Spectrom.* 2007;260:195–202.
35. Hamm G, Bonnel D, Legouffe R, et al. Quantitative mass spectrometry imaging of propranolol and olanzapine using tissue extinction calculation as normalization factor. *J Proteomics.* 2012;75:4952–4961.
36. James A, Jorgensen C. Basic design of MRM assays for peptide quantification*LC-MS/MS in Proteomics*: Springer; 2010:167–185.
37. Hatsis P, Brombacher S, Corr J, Kovarik P, Volmer DA. Quantitative analysis of small pharmaceutical drugs using a high repetition rate laser matrix-assisted laser/desorption ionization source. *Rapid Commun Mass Spectrom.* 2003;17:2303–2309.

38. Prideaux B, Dartois V, Staab D, et al. High-sensitivity MALDI-MRM-MS imaging of moxifloxacin distribution in tuberculosis-infected rabbit lungs and granulomatous lesions. *Anal Chem*. 2011;83:2112–2118.
39. Clemis EJ, Smith DS, Camenzind AG, Danell RM, Parker CE, Borchers CH. Quantitation of spatially-localized proteins in tissue samples using MALDI-MRM imaging. *Anal Chem*. 2012;84:3514–3522.
40. Chumbley CW, Reyzer ML, Allen JL, et al. Absolute quantitative MALDI imaging mass spectrometry: a case of rifampicin in liver tissues. *Anal Chem*. 2016;88:2392–2398.
41. Groseclose MR, Castellino S. A mimetic tissue model for the quantification of drug distributions by MALDI imaging mass spectrometry. *Anal Chem*. 2013;85:10099–10106.
42. Barry JA, Groseclose MR, Fraser DD, Castellino S. Revised preparation of a mimetic tissue model for quantitative imaging mass spectrometry. *Protoc Exch*. 2018;1:104.
43. Cahill JF, Kertesz V, Van Berkel GJ. Characterization and application of a hybrid optical microscopy/laser ablation liquid vortex capture/electrospray ionization system for mass spectrometry imaging with sub-micrometer spatial resolution. *Anal Chem*. 2015;87:11113–11121.
44. Zimmerman M, Lestner J, Prideaux B, et al. Ethambutol partitioning in tuberculous pulmonary lesions explains its clinical efficacy. *Antimicrob Agents Chemother*. 2017;61:e00924 17.
45. A Masucci J, D Mahan A, D Kwasnoski J, W Caldwell G. A novel method for determination of drug distribution in rat brain tissue sections by LC/MS/MS: functional tissue microanalysis. *Curr Topics Med Chem*. 2012;12:1243–1249.
46. Swales JG, Tucker JW, Spreadborough MJ, et al. Mapping drug distribution in brain tissue using liquid extraction surface analysis mass spectrometry imaging. *Anal Chem*. 2015;87:10146–10152.
47. Shabir GA. Validation of high-performance liquid chromatography methods for pharmaceutical analysis: understanding the differences and similarities between validation requirements of the US Food and Drug Administration, the US Pharmacopeia and the International Conference on Harmonization. *J Chromatogr A*. 2003;987:57–66.
48. Song X, He J, Pang X, et al. Virtual calibration quantitative mass spectrometry imaging for accurately mapping analytes across heterogenous biotissue. *Anal Chem*. 2019;91:2838–2846.
49. Wu Q, Rubakhin SS, Sweedler JV. Quantitative imprint mass spectrometry imaging of endogenous ceramides in rat brain tissue with kinetic calibration. *Anal Chem*. 2020;92:6613–6621.
50. Zhou SN, Zhao W, Pawliszyn J. Kinetic calibration using dominant pre-equilibrium desorption for on-site and in vivo sampling by solid-phase microextraction. *Anal Chem*. 2008;80:481–490.
51. Wang H, DeGnore JP, Kelly BD, True J, Garsha K, Bieniarz C. A technique for relative quantitation of cancer biomarkers in formalin-fixed, paraffin-embedded (FFPE) tissue using stable-isotope-label based mass spectrometry imaging (SILMSI). *J Mass Spectrom*. 2015;50:1088–1095.
52. Thiery G, Shchepinov MS, Southern EM, et al. Multiplex target protein imaging in tissue sections by mass spectrometry–TAMSIM. *Rapid Commun Mass Spectrom*. 2007;21:823–829.
53. Barry JA, Ait-Belkacem R, Hardesty WM, et al. Multicenter validation study of quantitative imaging mass spectrometry. *Anal Chem*. 2019;91:6266–6274.
54. Unsihuay D, Sanchez DM, Laskin J, Quantitative Mass Spectrometry Imaging of Biological Systems. *Ann Rev Phy Chem*. 2021;72:1.
55. Perez CJ, Ifa DR. Internal Standard Application Strategies in Mass Spectrometry Imaging by Desorption Electrospray Ionization Mass Spectrometry. *Rapid Commun. Mass Spectrom*. 2021 Jan;20:e9053.

Chapter 8

Spatial resolution of imaging mass spectrometry

Chapter Outline

"Intuitively, spatial resolution is a measure of the smallest discernible detail in an image." Gonzalez and Woods elegantly defined spatial resolution in Digital Image Processing.[1]

Spatial resolution is important because it influences if we can discern between two objects, as well as, how sharply we see the imaged objects. In biomedical science contexts, spatial resolution refers to the ability to differentiate small cellular or tissue-level structures. Increasing spatial resolution of imaging mass spectrometry (MS) gives us the capability to map molecular distributions at a finer detail from tissue to single cells to the subcellular compartment. In this chapter, spatial resolution or pixel size in imaging MS is discussed.

The spatial resolution of the study should always match the scientific question posed. Fig. 8.1 shows the theoretical minimum spatial resolution required to study various biological features. For example, the whole human body can be studied with a minimum resolution of a centimeter, while a resolution of a millimeter would be needed to differentiate between two adjacent organs, such as liver versus pancreas. Similarly, the resolution of a micron would be needed to distinguish between various types of cells within the pancreas, such as insulin-producing beta cells from glucagon-producing alpha cells in the islet of Langerhans. In the same manner, a resolution of nanometers will be required to map the subcellular distribution of glucose versus insulin in beta cells. At this point, without antibody tagging, imaging MS is excellent at doing tissue-level imaging with thriving research to reach a single-cell level resolving power.

The spatial resolution of the image is conveyed in many ways, often with the size of pixels. In generic digital images, spatial resolution is defined by the size and number of pixels utilized in the construction of the image. An image

Introduction to Spatial Mapping of Biomolecules by Imaging Mass Spectrometry. DOI: 10.1016/C2018-0-03962-6

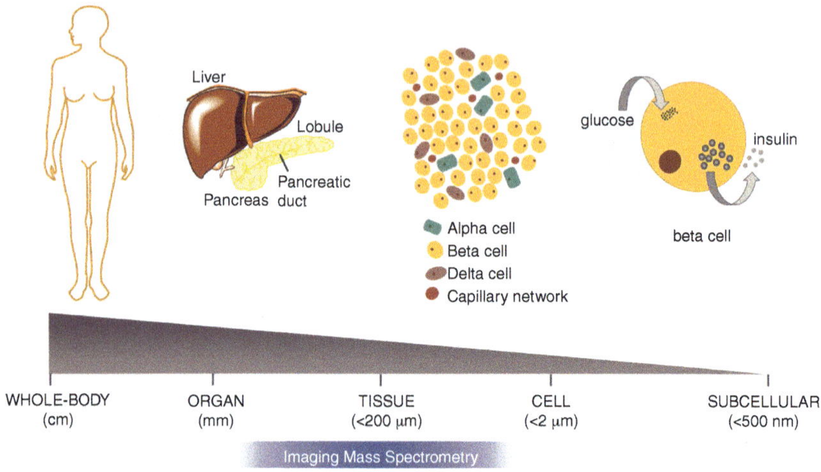

FIG. 8.1 The minimum spatial resolution needed depends on the biological features being studied. Theoretically, in order to image the whole human body, a centimeter-level resolution may be adequate, while organ-level imaging may need at least millimeter-level resolution, the tissue-level MS imaging may need at least a couple of hundred pixel size to be meaningful, single-cells imaging will require at least a couple of micron resolution, while subcellular will need at least a few hundred-nanometer resolutions.

is said to have a higher spatial resolution if it has a higher number of pixels of smaller pixel size than its lower spatial resolution counterpart of the same dimension (area). A higher spatial resolution image has more information and able to resolve smaller features. Spatial resolution should be stated with respect to spatial units, i.e., pixels per unit distance.

Spatial resolution in biomedical imaging

Spatial resolution is defined in many ways in different biomedical imaging fields. In ultrasound imaging, spatial resolution is defined as distinguishing characteristics between two points at a defined depth in tissue. Ultrasound imaging has axial resolution and lateral resolution. The axial resolution, which is also called the longitudinal resolution, is the minimum distance that can be distinguished between two reflectors located parallel to the ultrasound beam. Lateral resolution is the minimum distance needed to differentiate between two reflectors located perpendicular to the ultrasound beam.[2] In positron emission tomography reconstructed image, the spatial resolution is the minimum difference between two points that are distinguished as separate sources. Spatial resolution in positron emission tomography images depends on three factors; the size of the detector elements, the energy of the emitted positrons, and filtering applied during or post image reconstruction.[3]

Imaging MS has complementary spatial resolution and spatial range it can study compared to other biomedical images besides also providing completely orthogonal-specific information on molecules. A comparison of imaging MS

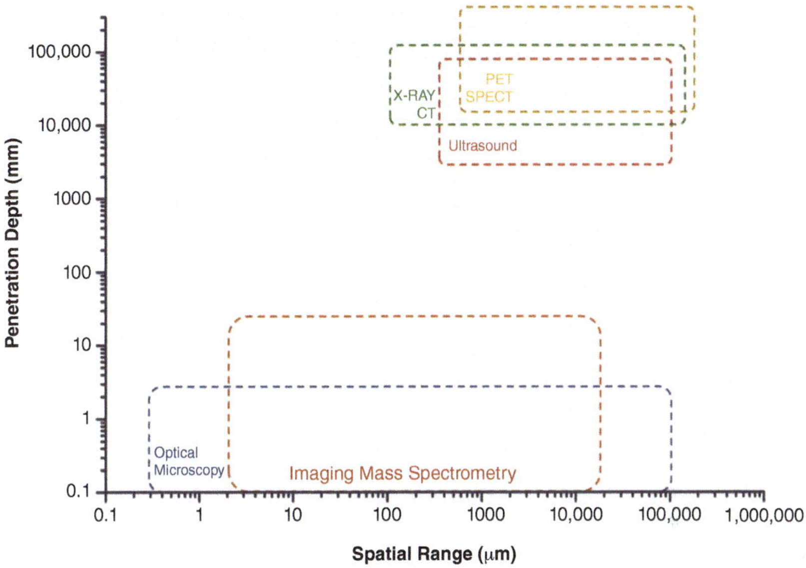

FIG. 8.2 A comparison of common biomedical imaging modalities with imaging MS shows complementary strength in spatial resolution, range, and penetration depth with other in vivo imaging.

with other common biomedical imaging modalities is presented in Fig. 8.2. Imaging MS has better spatial resolution than many other modalities except optical microscopy. Like microscopy, it has lower penetration depth requiring excising of tissue, making it unsuitable for in vivo imaging.

Numbers of pixels for spatial resolution

According to Nyquist–Shannon sampling theorem, any analytical tools, including an imaging device, must utilize a sampling interval at a minimum of half the size of the smallest resolvable feature desired in the image. Thus, there must be at least two pixels per highest spatial feature to accurately preserve the spatial resolution in the image. In other words, if one needs to resolve a single cell of 10 μm diameter, the minimum pixel size must be its half or 5 μm or less. If less than half the pixel size of the feature is utilized to acquire the sample, then all the spatial details comprising the specimen will not be preserved in the final image. In practice, to ensure adequate sampling for imaging, an interval of 2.5–3 times for the smallest resolvable feature is recommended. The additional pixels may not contribute to the spatial resolution but can often help improve the accuracy of the feature.

Spatial resolution in laser desorption imaging MS

In laser desorption imaging MS, such as matrix-assisted laser desorption/ionization (MALDI), the laser beam spot can be defined as the diameter or width

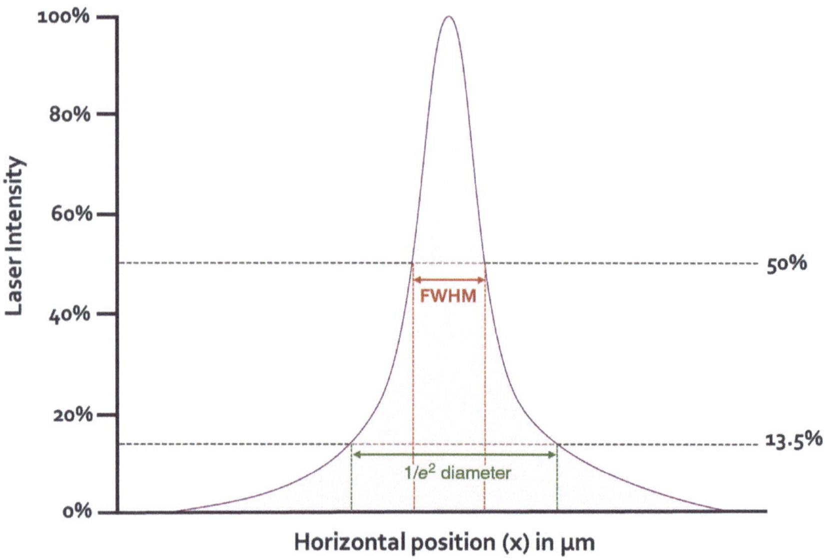

FIG. 8.3 The definition of Gaussian laser beam diameter is shown for full width at half-maximum (FWHM) and at $1/e^2$ or 13.5%. Laser beam diameter is an intrinsic property of a laser beam and should not be confused with ablation spot size.

of the laser beam on the sample. In this text, to avoid any confusion with ablation spot size, the laser beam width is referred to as beam diameter. The laser beam diameter can be defined in many ways, such as full width at half-maximum, where laser power is half or $D4\sigma$ diameter, which is four times the standard deviation of the distribution of intensity.[4] Typically for a Gaussian laser beams, the beam diameter is defined at the $1/e^2$ or 0.135 beam width, which is the distance between the edges where the intensity is $1/e^2$ or 13.4% of the maximum value. The beam diameter measurements are illustrated in Fig. 8.3. The laser beam diameter is independent of number laser shots, laser fluence, or substrate. The desorption or ablation spot is the area of the sample that is removed and/or ionized by laser. The ablation spot is dependent on the laser fluence, number of shots, and material ablated. Many researchers use the term "spot size" to reference ablation spot size.

The pixel size of MALDI and laser ablation electrospray ionization (LAESI) is determined by the distance that the sample is moved between two pixels or the distance between two pixels. At "optimal sampling," the step size is the same as the laser ablation spot size or ablation diameter without any overlapped analyzed areas. The "undersampling" or suboptimal sampling is used if the step size is considerably larger than the ablation diameter, leaving parts of tissue unanalyzed. If the sampling distance is less than the ablation diameter, it is called "oversampling." Oversampling can be used to generate MS images with a pixel size that is smaller than the ablation diameter. In order for oversampling to

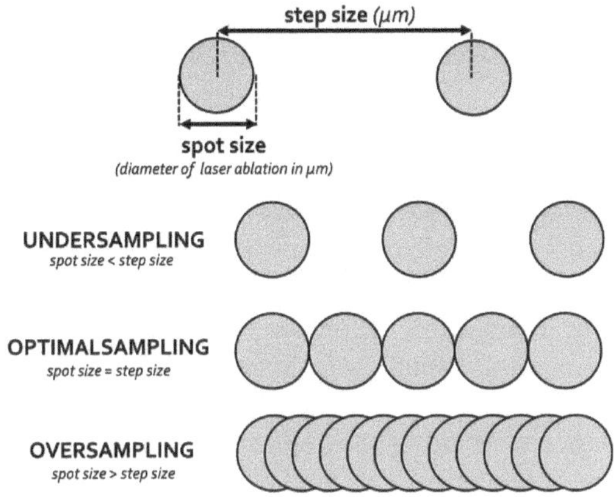

FIG. 8.4 Various sampling mode for discrete laser-based imaging MS ionization, such as MALDI, is shown. Spot size refers to the ablation spot size corresponding to materials removed and ionized from the sample surface.

work, the laser beam must ablate all material at each pixel before moving to the next, as demonstrated for peptides prepared from cells.[5] A schematic diagram of various sampling mode is given in Fig. 8.4.

The spatial resolution in laser-based imaging systems MALDI and LAESI, imaging MS experiments depends on the distance between each pixel, i.e., step size or sampling interval, and focused laser beam size. The minimum sampling interval is the function of the mechanical fidelity of the *x–y* translational stage and any mechanical vibration at the ion source. The spatial resolution of the translational stage is not a limiting factor because the sample plate can be laterally moved by piezoelectric actuators and displacement sensors at a low nanometer resolution, which is orders of magnitude lower than the spatial resolution of any imaging MS system.

The minimum laser focus is often defined by the diffraction limit. Ernst Karl Abbe approximated the diffraction limit of a microscope given below can be used to measure the resolvable features;

$$r = \frac{\lambda}{2n \times \sin\theta} = \frac{\lambda}{2NA}$$

where r is the resolvable feature size, λ is the wavelength of the laser beam, n is the index of refraction of the medium, and θ is the half-angle subtended by the optical objective lens.

For a perfect optical system, an ultraviolet wavelength of 355 nm found in the majority of solid-state ND:YAG laser use in commercial MALDI systems, would have the Abbe limit is roughly $d = \lambda/2 = 176$ nm. The $n \sin\theta$ represents the numerical aperture and can reach about 1.4–1.6 in modern optics, Abbe limit is $d = \lambda/2.8$.

For the MS imaging platform based on laser ablation, the diffraction-limited ablation spot size can also be defined by the following equation[6–7]:

$$S = M^2 \times \frac{4}{\pi} \times \lambda \times \frac{f}{B} \times \frac{1}{\cos\theta}$$

where S is the diffraction-limited spot size, M is the beam quality factor, λ is the wavelength, f is the focal length of the lens, B is the input beam diameter to the final focusing lens, and $\cos\theta$ is the incident angle of the laser beam.

There are three main ways to reduce the laser spot size based on this equation—first by using the smallest focal length focusing lens with minimum working distance, second by improving the beam quality factor, e.g., a perfect Gaussian beam has M closer to 1, and finally maximizing the input beam diameter.

Most of the work on reducing laser focus diameter has been done by using aspherical lenses with small focal distance. However, the reduction in working distance of lens means the lens would be very close to the sample surface blocking the ion transport from the sample surface to the mass analyzer. This challenge has been addressed in many ways. One can use the lens in the transmission-mode geometry, where the sample is ablated from the back of the slide. MALDI and LAESI have used these approaches to reduced laser spot size several times compared to conventional geometry.[8–11] However, transmission geometry may not be appropriate for all samples and may need thinner sections. Another way of using a small focal distance lens is to create a path for ions through the focusing lens to move to the mass analyzer. This can be done by creating a central bore in the middle of the focusing lens. Such coaxial laser optics have been used for 1.4 μm spatial resolution imaging of small molecules in cells.[12] Finally, a smaller laser spot can be generated by delivering the laser beam via an etched optical fiber tip closer to the sample surface, like the approach used in scanning nearfield optical microscopy.[13–14]

Beam quality can be improved by using auxiliary optical setup such as beam expanders or by simply increasing distance. For example, for infrared ablation, by increasing the distance between the laser and the lens from 1 to 18 m, the ablation spot size decreases 10-fold from 440 to 44 μm.[15] Improving the beam expander and a few other components was able to improve the spatial resolution of MALDI down to 5 μm.[6] The beam quality factor depends on the laser and we can anticipate this improvement as the laser with better beam quality is more readily and affordably available in the market.

The majority of MALDI imaging studies are performed at between 10 and 100 μm pixel size. There are a few notable exceptions of low single-digit pixel size, such as obtained at 1 μm or below.[10,16] However, most often they are done using home-built systems for selected samples. The reduction of pixel size comes with another major challenge of decreasing sensitivity. Ion abundances are considerably reduced at small pixel sizes. For example, the reduction in half of laser focus diameter equals to four times reduction in ions assuming imaging as a

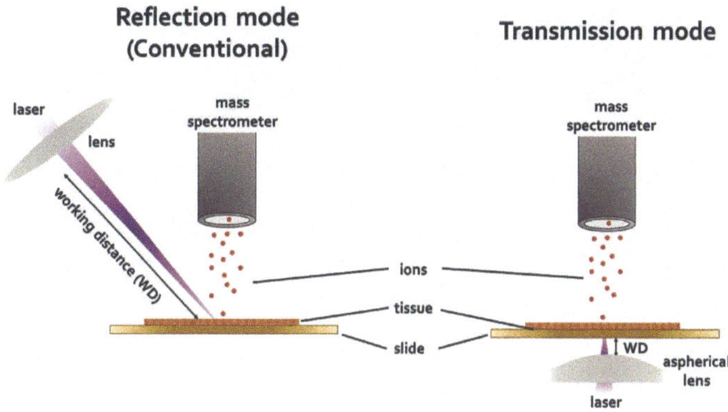

FIG. 8.5 A schematic diagram of the MALDI ion source in reflection versus transmission mode is shown.

two-dimensional surface analysis. The loss is more because a pixel in the imaging MS experiment is in practice three-dimensional pixel or voxel. For example, MALDI in transmission-mode geometry can have a laser focus diameter of less than 1 μm capable of ablating single cells or subcellular components. In order to improve sensitivity, laser-induced postionization, termed as "MALDI-2" was used to boost the sensitivity of ions leading to brain tissue imaging at the pixel size of 600 nm.[16] A schematic representation of MALDI in conventional, i.e., reflection mode and transmission mode is shown in Fig. 8.5.

Measuring spatial resolution in imaging MS

Spatial resolution in imaging MS methods can be measured by resolving features in an object with a defined spatial feature, such as transmission electron microscopy (TEM) meshed grid with defined bar spacing. The object may be used as a mask to cover a homogeneous sample surface or coated with a sample surface. A TEM grid with 92 μm bar spacing, as shown in Fig. 8.6A (left), was used as a mask, a homogeneous toluidine blue dye. The sample imaged by atmospheric pressure infrared (IR) MALDI at 40 μm step size providing two measurements or pixel size for spatial resolution, is shown in Fig. 8.6A (right).[17] Similarly, the achievable lateral spatial resolution of laser desorption/ionization MS using etched silver substrates was estimated using TEM copper grids with a well-defined web width of 25 um and a hexagonal structure as shown in Fig. 8.6B (left). Fig. 8.6B (right) shows the MS image of the grids were coated with hexacosane (C26:0) placed onto a silver substrate indicating spatial resolution of at least in the range of the grid's web width of 25 um.[18] A similar approach with either grid or spaced features can be used to measure the spatial resolution of any imaging MS system. In lieu of the actual measurement of the spatial resolution, many researchers use pixel size as a yardstick of the spatial resolution.

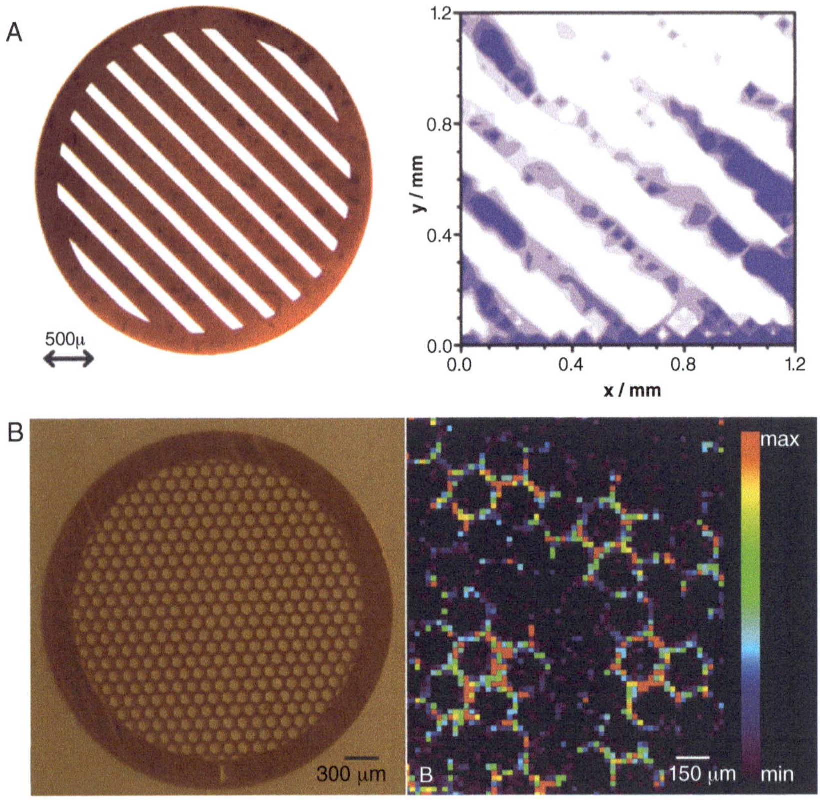

FIG. 8.6 The spatial resolution of imaging MS methods can be measured by resolving features of transmission electron microscopy (TEM) grid with defined spacing. Examples of measure spatial resolution in two imaging MS systems from the referenced article is given here.[17–18] Adapted with permission from Li Y, Shrestha B, Vertes A. *Anal Chem*. 2007;79:523–532. Copyright 2007 American Chemical Society and Schnapp A, Niehoff A-C, Koch A, Dreisewerd K. *Methods*. 2016;104:194–203.

Analyte-dependent pixel size or spatial resolution

In theory, a focused ultraviolet laser beam will have a smaller focus diameter, hence pixel size, than a focused liquid jet in desorption electrospray ionization or a focused midinfrared laser beam in LAESI. In turn, ions, such as used in secondary ion MS (SIMS), are capable of being focused on an even smaller diameter. Thus, in terms of pixel size, SIMS can image tissue at a smaller pixel size. However, SIMS may be only able to detect a select group of chemical species, such as metal ions, some intact molecule, and many fragments ions. On the other hand, desorption electrospray ionization, which is based on different ionization mechanisms, may be able to some metabolites at slightly higher pixel size.

Defining pixel size or spatial resolution of an imaging MS system for any ions informs about the smallest pixel size for that analyte or, more particularly, the smallest area analyzed for any analyte regardless of ion suppression or ionization efficiency. This is similar to defining the limit of detection (LOD) of a system based on the most sensitive analyte, not the analyte of interest. The LOD is always defined in the context of a specific analyte and matrix, not the analytical device. Similarly, the limit of resolution for a specific molecule needs to be defined in the context of the tissue sample matrix. At the smaller laser focus diameter, imaging MS will not be able to detect all the analyte detected at the larger spot size. This is due to the reduction in sensitivity because of the lower amount of material available. At the smaller spot size, MALDI imaging MS still may be able to detect a handful of very abundant and easily ionizable lipid ions. Defining a blanket pixel size of an imaging MS system leads to misleading information about the ability to detect those analytes with a higher LOD for that tissue. Thus, when reporting pixel size, one needs to specify the pixel size for a defined analyte or ion if pixel size information on the specific molecules is unavailable. In order to have a meaningful biomedical study, a random, highly abundant molecule does not convey the capability of imaging biologically relevant molecules.

Conclusions

Spatial resolution is a critical experimental parameter in biomedical imaging because it allows us to differentiate between various tissues, cell populations, and single cells. In all imaging MS experiments, one needs to balance the analytical sensitivity, computational resource, and time versus spatial resolution needed for the study. The better spatial resolution always needs higher sensitivity and longer experimental time, and more significant computational resources. The spatial resolution of the imaging system is affected by several factors depending on ion source (e.g., MALDI vs. SIMS), sample preparation (e.g., matrix coating, washing), hardware specification (e.g., laser beam quality), mass spectrometer sensitivity, and other parameters. Analytical techniques, such as laser capture microdissection followed by electrospray ionization, can perform molecular analysis at cellular and even subcellular resolution. However, currently, it lacks throughput to perform imaging MS experiments with thousands of pixels.[19] The fundamental bottleneck of imaging at cellular resolution is due to the lower number of molecules available for detection in smaller pixels. Thus, there is often a compromise between the sensitivity and spatial resolution driven by the application need.

References

1. Gonzales RC, Woods RE. *Digital Image Processing*. New Jersey: Prentice hall; 2002:1–31.
2. Ng A, Swanevelder J. Resolution in ultrasound imaging. *Contin Educ Anaesth Crit Care Pain*. 2011;11:186–192.
3. Knaapen P, Lubberink M. Positron emission tomography. In: Nieman K, Gaemperli O, Lancellotti P, Plein S, eds. *Advanced Cardiac Imaging*: Woodhead Publishing; 2015:71–95.

4. Siegman AEEDDM. How to (maybe) measure laser beam quality, *DPSS (Diode Pumped Solid State) Lasers: Applications and Issues, Washington, D.C., 1998/01/01*. Washington, D.C.: Optical Society of America; 1998:MQ1. doi:10.1364/DLAI.1998.MQ1.
5. Jurchen JC, Rubakhin SS, Sweedler JV. MALDI-MS imaging of features smaller than the size of the laser beam. *J Am Soc Mass Spectrom*. 2005;16:1654–1659.
6. Feenstra AD, Dueñas ME, Lee YJ. Five micron high resolution MALDI mass spectrometry imaging with simple, interchangeable, multi-resolution optical system. *J Am Soc Mass Spectrom*. 2017;28:434–442.
7. Zavalin A, Yang J, Caprioli R. Laser beam filtration for high spatial resolution MALDI imaging mass spectrometry. *J Am Soc Mass Spectrom*. 2013;24:1153–1156.
8. Zavalin A, Yang J, Hayden K, Vestal M, Caprioli RM. Tissue protein imaging at 1 μm laser spot diameter for high spatial resolution and high imaging speed using transmission geometry MALDI TOF MS. *Anal Bioanal Chem*. 2015;407:2337–2342.
9. Ovchinnikova OS, Bhandari D, Lorenz M, Van Berkel GJ. Transmission geometry laser ablation into a non-contact liquid vortex capture probe for mass spectrometry imaging. *Rapid Commun Mass Spectrom*. 2014;28:1665–1673.
10. Zavalin A, Todd EM, Rawhouser PD, Yang J, Norris JL, Caprioli RM. Direct imaging of single cells and tissue at sub-cellular spatial resolution using transmission geometry MALDI MS. *J Mass Spectrom*. 2012;47:1473–1481.
11. Jacobson RS, Thurston RL, Shrestha B, Vertes A. In situ analysis of small populations of adherent mammalian cells using laser ablation electrospray ionization mass spectrometry in transmission geometry. *Anal Chem*. 2015;87:12130–12136.
12. Kompauer M, Heiles S, Spengler B. Atmospheric pressure MALDI mass spectrometry imaging of tissues and cells at 1.4-μm lateral resolution. *Nat Methods*. 2017;14:90.
13. Stöckle R, Setz P, Deckert V, Lippert T, Wokaun A, Zenobi R. Nanoscale atmospheric pressure laser ablation-mass spectrometry. *Anal Chem*. 2001;73:1399–1402.
14. Shrestha B, Vertes A. In situ metabolic profiling of single cells by laser ablation electrospray ionization mass spectrometry. *Anal Chem*. 2009;81:8265–8271.
15. Hieta J-P, Vaikkinen A, Auno S, et al. A simple method for improving the spatial resolution in infrared laser ablation mass spectrometry imaging. *J Am Soc Mass Spectrom*. 2017;28: 1060–1065.
16. Niehaus M, Soltwisch J, Belov M, Dreisewerd K. Transmission-mode MALDI-2 mass spectrometry imaging of cells and tissues at subcellular resolution. *Nat Methods*. 2019;16: 925–931.
17. Li Y, Shrestha B, Vertes A. Atmospheric pressure molecular imaging by infrared MALDI mass spectrometry. *Anal Chem*. 2007;79:523–532.
18. Schnapp A, Niehoff A-C, Koch A, Dreisewerd K. Laser desorption/ionization mass spectrometry of lipids using etched silver substrates. *Methods*. 2016;104:194–203.
19. Cahill JF, Kertesz V, Van Berkel GJ. Characterization and application of a hybrid optical microscopy/laser ablation liquid vortex capture/electrospray ionization system for mass spectrometry imaging with sub-micrometer spatial resolution. *Anal Chem*. 2015;87: 11113–11121.

Chapter 9

Visualization in imaging mass spectrometry

Chapter Outline

Biomedical sciences have relied on effective visualization to infer the relationship of anatomical features with biological functions. Scientific data visualization is the process of converting numbers to a visual representative image. In imaging mass spectrometry (MS), visualization is a process of plotting intensities of ions onto two- or three-dimensional space to make a pictorial representation of the distribution of that ion on the surface, such as tissue section. An effective molecular visualization allows scientists to gain insights into the accurate spatial distribution of the ions or molecules.

Several imaging MS toolkits give their users various options to plot raw or processed intensities, such as normalized intensities, of ions. Aside from data processing, the user interpretation of visualization can be fundamentally altered by the selection of colormaps, scale, and interpolation. In this chapter, we will discuss those parameters that can help us to make an effective visualization without processing the data.

Colormaps

One of the most effective ways of visualizing scientific data is using a sequence of color hue or luminance representing a value, also known as colormaps. The definition of colormaps is one of the most fundamental features of any scientific image visualization. A colormap should intuitively convey the information on the difference in the distribution of the molecule(s) being plotted in a molecular image. A wrong colormap will misconstrue the data being communicated, extenuate nonexisting difference, introduce artifacts. For instance, many scientific visualization tools, by default, use a colormap that is famous for its ineffectiveness, i.e., the rainbow colormap, which cycles through saturated colors.[1] The rainbow colormap was used as the default in the majority of scientific visualization toolkits and used about half of the time.[2]

Introduction to Spatial Mapping of Biomolecules by Imaging Mass Spectrometry. DOI: 10.1016/C2018-0-03962-6

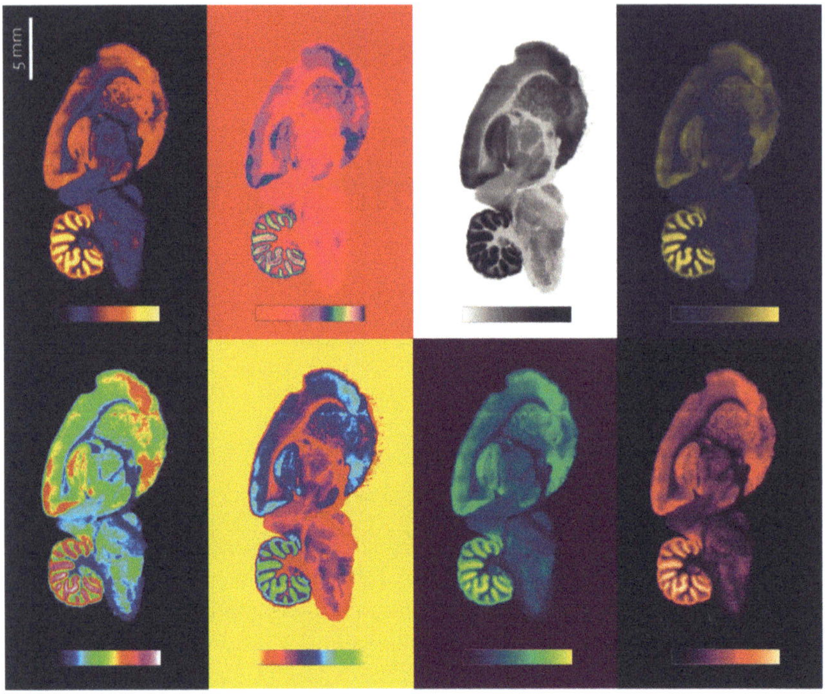

FIG. 9.1 The same mass spectrometry imaging data plotted with different colormaps; from top right-to-left are HotMetal, Abstract, Grayscale, and Cividis. The colormaps used in the bottom right-to-left are Jet-black, HueWheel, Viridis, and Magma.

The definition of colormap influences the representation of data to image and hence its interpretation. Fig. 9.1 shows the collage of the same imaging MS data plotted using various colormaps. The most important aspect of the image visualization process is the choice of pseudocolor(s) or their shades and their order corresponding to the numerical data. There are four broad classes of colormaps—sequential, diverging, cyclic, and miscellaneous. A brief description of classes of colormaps is given in Table 9.1, and representative examples are shown in Fig. 9.2.

An ideal colormap for imaging MS data should have the following characteristics. It should be (a) colorful, spanning a wide palette making differences in values easy to discernable to an observer; (b) perceptually uniform, i.e., close values have similar-appearing colors and values farther away have more different appearing colors; (c) robust to individuals with a common form of color vision deficiency; (d) grayscale printing friendly since many people will print the images in black and white, and finally (e)aesthetically pleasing—no one likes awkward color combination. Currently, Viridis was created for the Matplotlib library (Python) is considered a gold standard for scientific heatmap visualization consisting of all these attributes. Similarly, Cividis derived from

Table 9.1 Definition of classes of colormap.

Sequential: Sequential colormaps change in lightness and often saturation of color incrementally. Sequential colormaps often using a single hue and suited for representing info that has order.
Diverging: Diverging colormaps change in lightness and possibly saturation of two different colors that meet in the middle at an unsaturated color. Diverging colormaps are used when data have a critical middle value, e.g., topography or data that deviate around zero.
Cyclic: Cyclic colormaps change in the lightness of two different colors that meet in the middle and beginning and end. Cyclic colormaps are used for values that wrap around at the endpoints, e.g., time of day, wind direction.
Qualitative: Qualitative colormaps are miscellaneous colors and used to represent info that does not have ordering or relationships.

the Viridis is considered a choice of colors taking account for people with color vision deficiency, which affects 8% of men and 0.5% of women and reduces the ability to distinguish between certain colors, causing misinterpretation of data.[3] Grayscale is not considered an ideal colormap because human eyes can distinguish 10 million colors but only about 30 different shades of gray. Grayscale colormaps can provide linear luminance for people with color vision deficiency but have a smaller dynamic range.

Many different colormaps have been used to plot MS images. There are no set standards or guidelines. The ideal colormap for any given dataset will depend on several factors. A perceptually uniform colormap, in which equal numerical

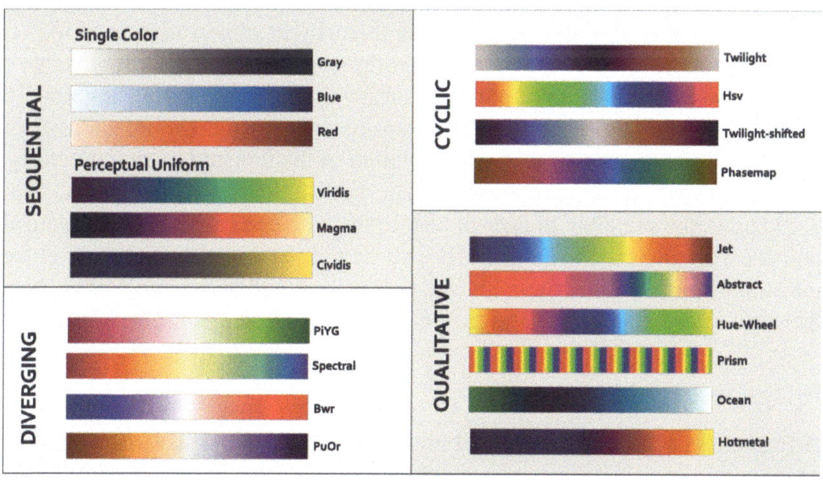

FIG. 9.2 Colormap classes are shown here, and the detailed descriptions are provided in Table 9.1.

value in data is perceived as equal steps in color, is the ideal choice for many MS images. Research has shown that humans can perceive changes in the lightness better than changes in hue. Thus, an ideal colormap has a gradual change in lightness through the colormap will be better interpreted by the viewer.

Colormap scale

Multiple images in a dataset can be plotted using the same colormaps scale or scale tailored to that individual ion. The same colormap scale for all the ions does not take into account the intensity of ions and misrepresenting the spatial ion distribution of the image. For example, Fig. 9.3 shows two ions with low and high intensities plotted with scale suited for high-intensity ion versus low-intensity ion. The low-intensity ion is barely visible at the colormap scale for higher intensity ion, while high-intensity ion is oversaturated in the colormap scale suited for the lower intensity ion. Thus, it is important to choose a scale that is correctly suited for the ion being plotted. Generally, the colormap is scaled from the lowest ion intensity value, typically near zero, to the highest ion intensity value for each ion. Even this default scaling can suffer from the presence of a hot spots pixel due to an artifact, where the intensity of ion in one pixel significantly exceeds other pixels skewing the colormap scale. Hot spots can be a data processing issue, hardware error, or introduced during the sample preparation step. Similarly, avoiding cold spots is important in low-intensity images. Cold spots can be defined as background noise, such as chemical noise that distorts the spatial distribution of the low-intensity ions.

Image plotting may require a definition of thresholds to extract or avoid hot spots or cold spots in images. Some other biomedical imaging techniques widely use automatic techniques to determine a threshold that will separate the imaged object from the background.[4] In imaging MS, hot spots are often removed by plotting ions from first to 99th percentile intensity or some variation of it. Fig. 9.4 shows images of an ion plotted with original data without hot spot, as well as, two artificially introduced hot spot consisting of a single pixel with 10× (10 times) or 100× (100 times) more ion intensities than the maximum intensity of that ion. Fig. 9.4A has all of the ions plotted to the hundred percent intensity. The image with the artificially introduced hot spots is barely registered. Fig. 9.4B shows the same ion images plotted to the 99th percentile. For all practical purposes, all the images of Fig. 9.4B with or without the artificial hot pixel are indistinguishable to the original image in Fig. 9.4A and, more importantly, give the same information to the researchers irrespective of the hot spot.

Image interpolation

A fundamental issue in biomedical imaging, including imaging MS, is linking a large discrete acquisition steps that do not match the continuous nature of the biological samples. Images obtained from imaging MS are often interpolated to improve their appearance by giving a continuous image from discrete pixel

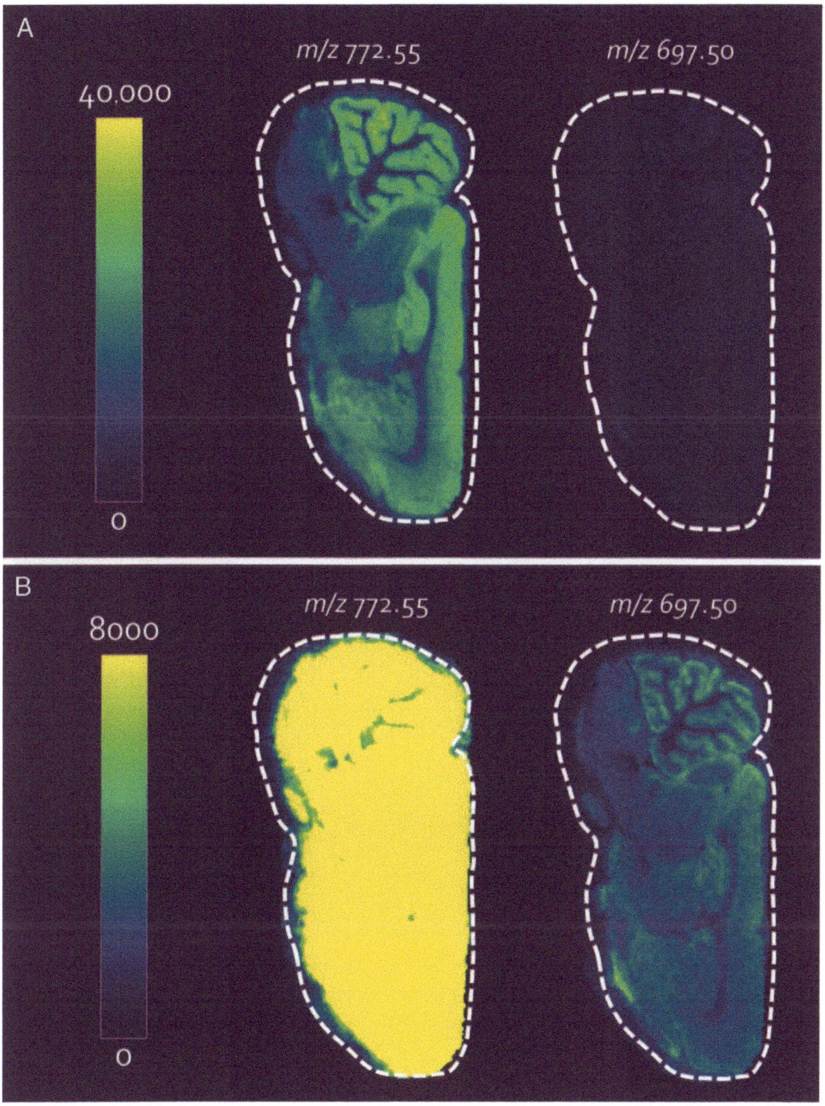

FIG. 9.3 Effect of colormap scale is demonstrated here, (A) has colormap scale suited for higher intensity ion at *m/z* 772, while (B) as colormap scale suited for lower intensity ion at *m/z* 697, which is about five times less intense.

data. Fig. 9.5A shows RAW (unprocessed) images without interpolation, and Fig. 9.5B shows the same image after a bilinear interpolation.

Image interpolation is a commonly used image processing step in many biomedical imaging applications. For instance, in magnetic resonance imaging, a sequence sliced image of an organ is interpolated to make three-dimensional construction. Image interpolation is often needed because the unimaged space

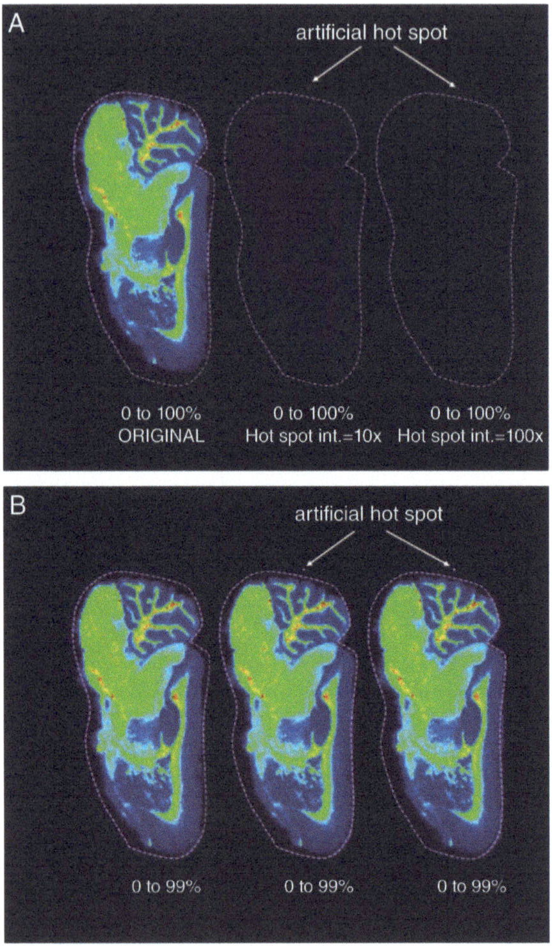

FIG. 9.4 An effect of plotting scale of colormap is demonstrated using artificially created hot spot pixel, (A) shows images plotted to hundred percent or their maximum intensities leads images with artificially induced pixel failing to account for hot spots, (B) which can be easily rectified by plotting images just below hundred percent.

between neighboring slices is generally larger than the acquisition pixel size due to either inherent pixel-size limitation of the imaging modality or other limitations imposed by time or processing. All image interpolation step resamples and transforms discrete matrix values into a continuous image.[5] There are several image interpolation methods, starting from simple algorithm-based methods, such as linear interpolation or nearest neighbor to neural network-based algorithms.[6]

Pixel interpolation can be changed in many MS imaging MS software. Fig. 9.6A–C shows the example of three types of interpolations (linear,

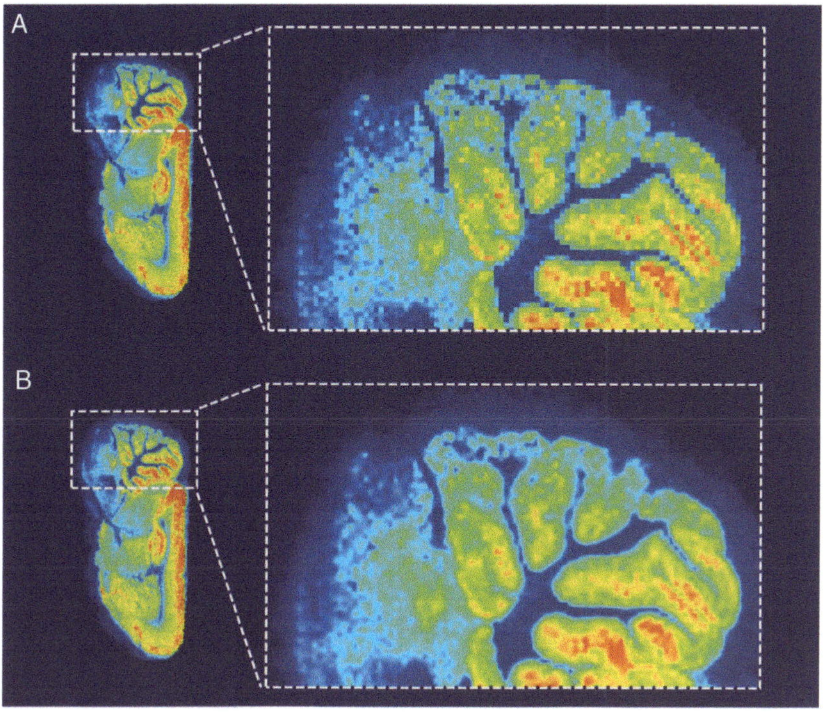

FIG. 9.5 The same ion image plotted without (A) and with (B) bilinear interpolation is shown with zoomed inset on the right.

spline, and cubic) applied to the first-order in MSIReader software.[7] Fig. 9.6D shows the same image with linear interpolation applied to the fifth order. Due to the current lack of guidelines or consensus on interpolation, it is recommended that researchers disclose the type of interpolation used in the publication. Image interpolation can provide smooth, pleasant images, but overinterpolation can misconstrue the spatial data.

Conclusions

Visualization tasks in imaging MS are similar to many other biomedical images, but also differ from them in many aspects. The primary goal of these scientific imaging techniques is to extract spatially relevant biomedical information from the sample. We can all communicate that biomedical information better by carefully considering some fundamental aspects of visualization, such as colormap choice, colormap scale, and interpolation. Currently, there is no standardization for the visualization scheme for imaging MS since a lot of work is still confined to basic research. As we move toward more clinical translational space, we can expect some standardization or at least some consensus for a particular application.

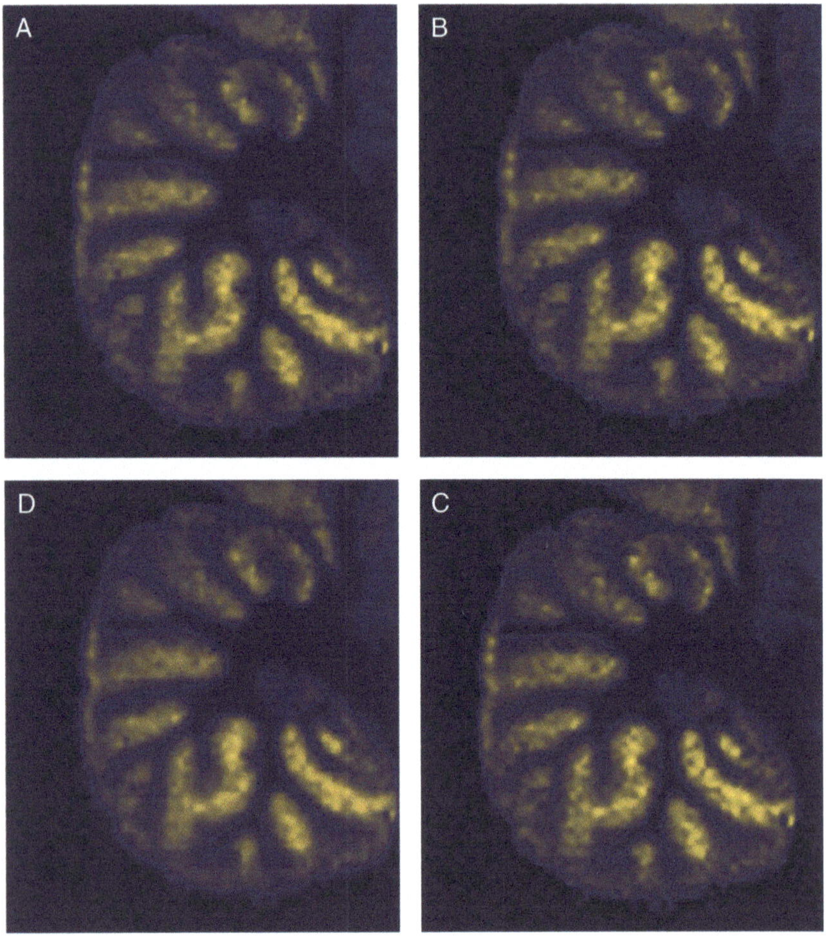

FIG. 9.6 (A) Linear, (B) spline, and (C) cubic interpolation applied to the first order in MSIReader software, (D) shows the same image with linear interpolation applied to the fifth order.

References

1. Moreland K. Why we use bad color maps and what you can do about it. *Electron Imaging*. 2016;2016:1–6.
2. Borland D, Ii RMT. Rainbow color map (still) considered harmful. *IEEE Comput Gr Appl*. 2007;27:14–17.
3. Nuñez JR, Anderton CR, Renslow RS. Optimizing colormaps with consideration for color vision deficiency to enable accurate interpretation of scientific data. *PLOS One*. 2018;13:e0199239.
4. Reyes-Aldasoro CC. *Biomedical Image Analysis Recipes in MATLAB: For Life Scientists and Engineers*: John Wiley & Sons; 2015:1–400.

5. Lehmann TM, Gonner C, Spitzer K. Survey: interpolation methods in medical image processing. *IEEE Trans Med Imaging*. 1999;18:1049–1075.
6. Umehara K, Ota J, Ishida T. Application of super-resolution convolutional neural network for enhancing image resolution in chest CT. *J Digit Imaging*. 2018;31:441–450.
7. Bokhart MT, Nazari M, Garrard KP, Muddiman DC. MSiReader v1. 0: evolving open-source mass spectrometry imaging software for targeted and untargeted analyses. *J Am Soc Mass Spectrom*. 2017;29:8–16.

Chapter 10

Data analysis and computation for imaging mass spectrometry

Chapter Outline

Imaging mass spectrometry (MS) provides a spatial molecular characterization of samples by employing data acquired from a mass spectrometer. Imaging MS obtains the spatial distribution of molecules by detecting ions pixel-by-pixel, a pixel at a time, on a sample using a mass spectrometer. The intensity plot of a specific ion is aligned with the spatial coordinate. Data analysis is utilized to render this spatial information of the ion as a false-color heatmap image. A two-dimensional imaging MS data can be considered as a three-dimensional data matrix containing hundreds of thousands of x,y positions with mass spectra containing ion intensities of thousands of ions. Imaging MS data analysis workflow has most of the requirements of mass spectrometer data analysis and some more due to added extra dimension of space and visualization requirements.

The fundamental imaging MS data analysis includes spectral processing and visualization. Other functionality may include molecule identification, quantitation, statistical analysis, segmentation, multimodal capabilities, the region-of-interest (ROI) analysis, etc. Some publications have done an excellent job on reviewing the various aspect of imaging MS data analysis, as indicated in the references here.[1–4]

Introduction to Spatial Mapping of Biomolecules by Imaging Mass Spectrometry. DOI: 10.1016/C2018-0-03962-6

Spectral processing

Spectral processing is an essential data processing procedure before creating images using mass spectral data. Many spectral processing steps are often similar to MS data processing for other ion sources and applications. Fig. 10.1 illustrates spectral processing steps using simulated data for matrix-assisted laser desorption/ionization (MALDI) imaging. Depending on the experimental design, some more steps are modified, removed, or added.[4]

Spectral alignment and mass correction

In imaging MS, peaks may shift across different spectra or pixels due to instrumental drift or sample inhomogeneity. For example, in linear time-of-flight MALDI, *m/z* values can change because of different starting points of the ions due to random topographical heights within the imaged tissue or a systematic tilt of sample plate.[5] Spectral alignment or each spectrum or pixel can be performed using typical endogenous peaks or added reference peaks across the acquisition. Without

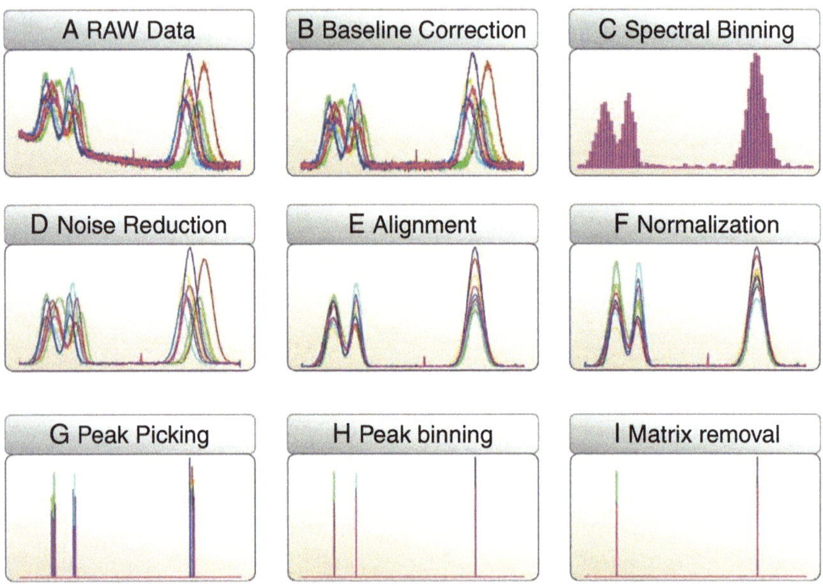

FIG. 10.1 Data processing steps in imaging MS illustrated by simulated data. (A) RAW MS spectra before processing, (B) MS spectra after the baseline correction, here using the top-hat subtraction, (C) spectral binning for data reduction, here *m/z* value is represented as a bar, (D) spectral noise reduction or smoothing, here using Savitzky–Golay method, (E) mass alignment or correction for possible mass drifts, (F) intensity normalization, here using TIC normalization, (G) peak picking or centroiding data to a list of *m/z* peaks, deisotoping steps can be added here, (H) peak binning to eliminate slight mass drift between pixels, (I) matrix or background peak removal. Reprinted with permission from Ràfols P, Vilalta D, Brezmes J, et al. Signal preprocessing, multivariate analysis and software tools for MA(LDI)-TOF mass spectrometry imaging for biological applications. Mass Spectrom Rev. 2018;37:281–306.[4]

spectral alignment, many other processing steps, such as binning and statistical analysis, would be error prone. Spectral alignment is a data processing step that optimally shifts the peaks for better alignment of the same compound peaks across multiple spectra or pixels in case of imaging. Spectral alignment is often done by using an alignment with reference peak. A mass correction can also be introduced at the same time. Spectral alignment is widely used for other MS applications, such as aligning peptides for proteomics.[6–8] For example, spectral alignment was used to discover positional isomers from spectra of mixtures of isobaric protein forms with numerous modifications and efficient database search for identification of mutated and modified proteins.[7–8] Spectral alignment is extensively used in nuclear magnetic resonance spectral analysis to correct chemical shift data.[9]

Baseline correction and noise reduction

Baseline correction is a spectral processing method important for removing spectral artifacts due to electronic distortions, imperfect digital sampling, or insufficient digital filtering. Baseline correction may be a prerequisite for correct spectral alignment and integration peak area for quantitation. Tiny errors in the baseline may lead to a more massive error during quantitation of low abundance ions. Visually, baseline correction makes a spectrum where the region without signal is flat with zero intensity, and one with a signal starts at zero intensity. There are many baseline correction methods, such as a nonlinear filter known as the top hat, leading to the enhancement of minor signals.[10] Other baseline correction mechanisms include monotone minimum, linear interpolation, Loess, continuous wavelet transform, moving average of minima, etc.[11] Smoothing is a noise reduction step that aims to capture the approximate data pattern. Smoothing filters include moving average filter, Savitzky–Golay filter, Gaussian filter, Kaiser window, continuous wavelet transform, discrete wavelet transform.[11]

Peak detection and binning

Peak detection is an indispensable data processing step for obtaining a list of accurate *m/z* values for subsequent molecule identification and quantification. The peak detection process falls after smoothing and baseline correction to mitigate any noise as shown in Fig. 10.2. There are several peak detection algorithms. A study looking into various peak detection algorithms using simulated and experimental MALDI data showed that the continuous wavelet-based algorithm provides the optimal performance.[11] Other peak detections include signal-to-noise ratio, intensity threshold, slopes of peaks, local maximum shape ratio, ridgelines, peak width, etc.[11]

Background subtraction

Background subtraction is a common data processing step for MS data, where analyte acquisition is subtracted from a control acquisition with background

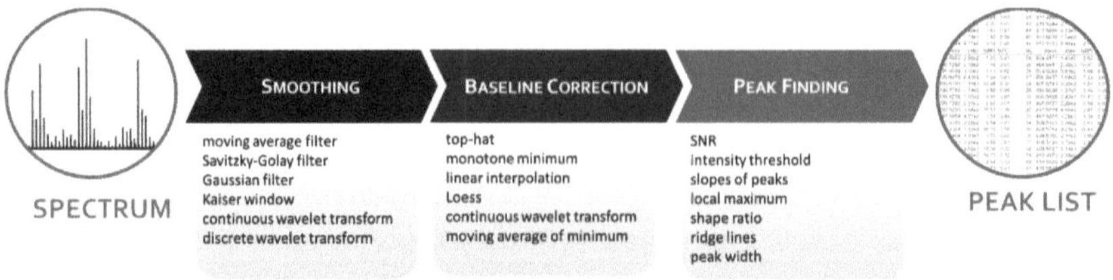

FIG. 10.2 Peak detection framework from the mass spectrum to the peak list. The mass spectrum is transformed into a list of peaks after smoothing, baseline correction, and peak finding. Based and adapted with permission on data from Yang C, He Z, Yu W. Comparison of public peak detection algorithms for MALDI mass spectrometry data analysis. BMC Bioinformatics 2009;10:4.[11]

ions.[12] Noise from random electric noise generated by the mass spectrometer or adjacent equipment or ubiquitous chemical species such as plasticizer may distort mass spectra interfering with the detection of analytes. Stochastic signal processing techniques can be applied for analyzing electrical noise.[13]

During imaging MS, background subtraction of data can improve the detection of molecules in tissue after eliminating the interfering background peaks. The nature of background noise depends on the sample, application preparation steps, or instruments. For example, small molecule imaging could have interference from matrix ion peaks in MALDI, background electrospray solvent peaks in desorption electrospray ionization (DESI), or polyethylene glycol peaks if tissue was prepared with optimal cutting temperature. Robust background subtraction can improve signal-to-noise ratio and reduced file sizes without losing any data fidelity. The simplest way to perform background subtraction is to do a manual background subtraction using spreadsheets such as Microsoft Excel.[14] Adaptive background subtraction can utilize signal processing algorithms to reduce background chemical noise that has short periodicity with a more extended range of intensity variation.[15–16]

Normalization

Normalization is an essential data processing step to render an accurate distribution of the ions over the tissue. There are several normalization strategies used for normalization. For example, MALDI imaging MS dataset commonly uses normalization techniques based on total ion count.[17] However, inappropriate normalization algorithms can create a misleading spatial distribution of ion, leading to the erroneous conclusion. Normalization can correct signal suppression due to the ion suppression effect on the tissue. Quantitative information of small drug molecules in imaging MS was improved by using a normalization factor called the tissue extinction coefficient. The tissue extinction coefficient took account of the ion suppression effect due to endogenous.[18] Some commonly used normalization techniques for MALDI and DESI imaging MS have been discussed elsewhere.[19–20] Additional discussion on normalization is also provided in the chapter on quantitative imaging MS.

Common and open-source imaging mass spectrometry data analysis

Commercial imaging mass spectrometer systems generally come with a vendor software compatible with their proprietary data formats. There are vendor-neutral commercial software, such as LipostarMSI, that can perform imaging MS data analysis, image visualization, coregistration, univariate/multivariate analysis, automated molecular identification, segmentation.[21–22] Generally, a single proprietary imaging MS software package may not accommodate every functionality required by researchers, has a cost associated with the use, incompatible with multivendor imaging workflows, or are difficult for open sharing of data due to proprietary format or licensing requirements.

imzML

The imzML is a standardized data format compatible with imaging MS data from several imaging MS modalities and instruments. The imzML imaging MS data are divided into two separate files linked by a universally unique identifier—a binary file containing the mass spectral data and an XML file with metadata such as instrumental parameters, sample details. The XML file, based on the mzML developed by HUPO-PSI, include the spatial information such as *x–y* coordinate positions for each mass spectrum and pixel size, needed to render imaging MS. The XML and binary files are a similar size to the original RAW file. The imzML data format is used on a wide array of data analysis software. Detailed documentation on the data format and its implementation can be found at http://www.imzML.org URL.[23–24] Open-source imaging MS file converter, such as imzMLConverter, can convert mzML format data to imzML format. The proprietary RAW data need to be converted to either continuous or processed mzML format first by software, such as msConvert.[25–26] A validation tool for imzML, called imzMLValidator, has been developed in the graphical user interface and command-line versions to identify any deviations from the published imzML standard that could cause issues in processing data or to perform conditional validation on the metadata.[27] The Computis European project facilitated of several software tools to process imaging MS datasets using imzML standard format.[28]

The proposed standard for storage and management of imaging MS data, imzML format, does not support the ion mobility dimension. Additionally, the current format may not be efficient at standard data access patterns, such as fast orthogonal needed for accessing individual mass spectra and rendering ion images. HDF5 is a widespread data format without vendor lock-in for storing and accessing *n*-dimensional large scientific datasets in a self-describing file format capable of the cross-platform software platform with a high read/write performance.[29] An HDF5 file format, mzMLb, based on mzML format, has been proposed for optimal read/write speed and storage of the MS data.[30] OpenMSI is an HDF5 format web-based imaging MS software platform that supports the storage of raw data, metadata in a single. OpenMSI supports a diverse computational platform such as Matlab and R, and programming languages, such as C++, Fortran, and Python.[31–32]

Open-source imaging MS software

BioMap is one of the first image-analysis software developed under interactive data language for imaging MS and magnetic resonance imaging data.[33] The imzML file format serves as a default file format in several open-source imaging MS software. MSIReader is a popular free MATLAB platform software packages that can read imzML file and visualize MS images.[34–35] MSIReader is capable of an image overlay, quality assurance feature to gauge mass accuracy across the image, ROI, concentration calculator, and more. A screen capture of MSIReader rendering DESI image acquired in quadrupole time-of-flight mass spectrometer is shown in Fig. 10.3. MALDI Imaging Team Imaging Computing System (MITICS) is another open-source imaging MS software. MITICS is a new software compatible with many instrument platforms. MITICS software is divided into two parts; control for data acquisition and image data processing and reconstruction.[36] MassImager software includes three parts; solution, visualization, and intelligence. Some features include multiple ion visualization, multiple channel superposition, image normalization, visual resolution enhancement, image filter, ROI analysis, overlaying an optical image, automatic pattern recognition via supervised or

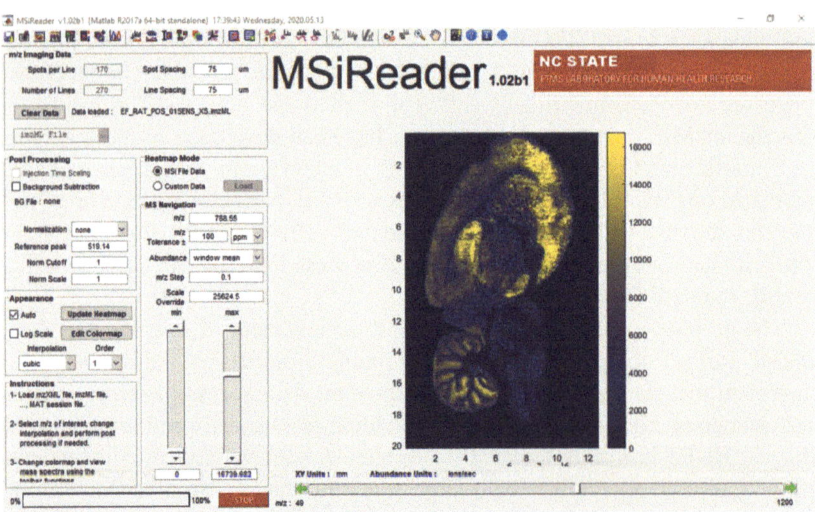

FIG. 10.3 Screen capture of open-source MSIReader software rendering DESI image of the sagittal mouse brain section acquired in quadrupole time-of-flight (QTof) mass spectrometer.

unsupervised multivariate statistical modeling.[37] IMS convolution is imaging MS software aimed at new users and has features that could automatically extract the meaningful ion images from ROIs.[38] Datacube Explorer is another software that can visualize and analyzes imaging MS imzML data.[39–40] Handling of large imaging data using innovative data structures and visualization strategies capable of rapid data exploration and visualization of two- and three-dimensional images.[40] The python BASIS (pay basis) or BASIS is an open-source computational platform for comparative analysis of imaging MS data across tissue specimens using machine learning and pattern recognition approaches from imzML format.[41] OmniSpect is an open-source web-based software that performs exhaustive computation on a remote server instead of a local computer. OmniSpect is capable of converting data, visualization, multivariate analysis in a MATLAB environment.[42]

The "R" is an open-source software of language and environment statistical data mining, modeling, and visualizing used in many research disciplines.[43] For example, Cardinal is an R package for statistical analysis of DESI and MALDI imaging data in imzML format.[44] MSI.R is a software tool for imzML imaging MS data based on R scripts that enables calculation of the density distribution of ions, automated images for a defined scan range, overlay of MS images with another digital image. MSI.R scripts have been utilized for low-temperature plasma imaging MS.[45]. The rMSI is another free and open-source R package-based software that can visualize imaging MS data with an integrated graphical user interface.[46]

Data analysis functionality for imaging mass spectrometry

Visualization

The visualization process in imaging MS entails assigning heatmap color to intensities of detected molecules on a pixel. Two- Or three-dimensional visualization of a molecule is obtained by plotting intensities pixel-by-pixel. Visualization of mass spectra to an image is the key data analysis functionality for any imaging MS software. The selection of colormaps and scales can fundamentally change the visualization of images and researchers' interpretation of data without manipulating processing parameters of data such as normalization. An example of the difference in visualization by changing interpolation, scale, and colormap on DESI data is illustrated in Fig. 10.4. Chapter 9 has a detailed discussion of the effective imaging MS visualization.[47]

Identification of molecule

The identification of ions as a molecule is a prerequisite for understanding underlying biochemical processes within the tissue. The identification strategy depends on molecule concentration, type of molecule, identification confidence required, sample complexity, etc. The first general approach for the identifi-

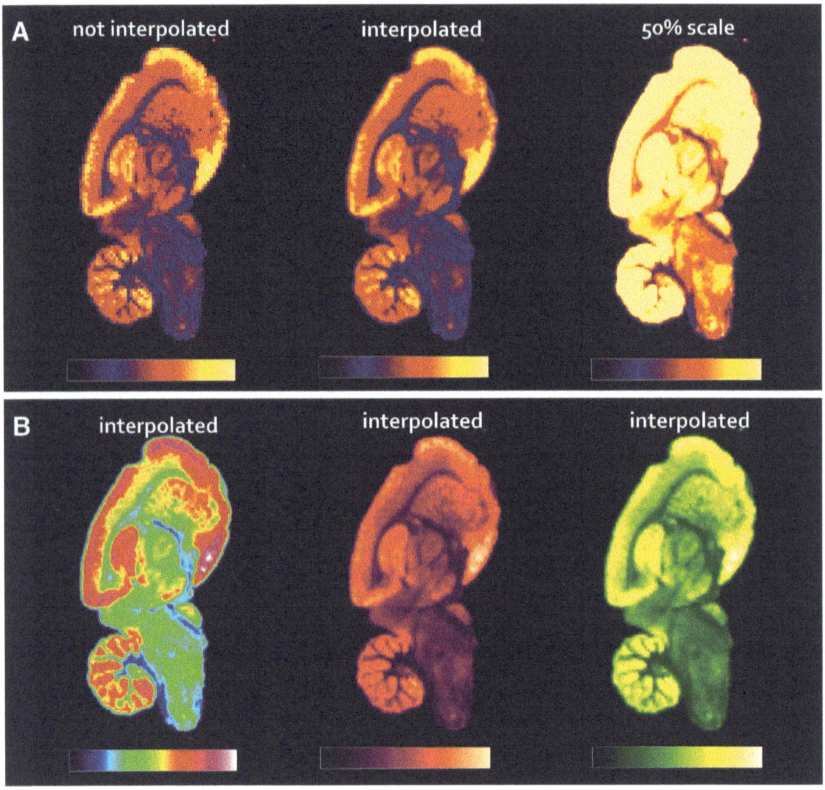

FIG. 10.4 DESI image of the sagittal rat brain section of *m/z* 782.56 lipid ion is shown. (A) The ion image visualized in the same false-color heatmap without interpolation, interpolation, and plotted to 50% scale instead of a 100% scale that is used in all other images. (B) The same ion visualized with other colormaps after interpolation and plotted to a 100% scale.

cation of molecules is by matching the accurate *m/z* value of intact ion with the known compound. The accurate *m/z* match can be aided by isotope pattern match and other orthogonal measurements such as collision cross section value. In this type of identification process, software should be able to do two tasks: (a) get accurate *m/z* value after any mass correction and (b) match those accurate *m/z* values with an internal or external database, such as publicly available curated the Human Metabolome Database.[48] METASPACE is an online metabolite annotation platform that can identify the metabolites and lipids in high mass resolution imaging experiments using an imzML file.[49] METASPACE provides metabolite annotation with a false discovery rate, which is defined as the ratio of false positives hits compared to the total annotations. The false discovery rate is generated using a decoy database consisting of the implausible adducts, such as B^+, Db^+, or Ag^+. The second approach of molecular identification utilizes fragmentation of the ions using MS/MS. In this approach, the software needs to identify the fragmentation pattern based on the fragmentation database automatically. Chapter 6 has more discussions in molecule identification strategies in imaging MS.[50]

Quantitative imaging measurements

Quantitative imaging of drugs and endogenous molecule in tissue sections require data processing steps such as normalization. The msIQuant is quantitative imaging MS software that uses vendor-neutral open access imzML data format. Besides providing many tools for MS image visualization, msIQuant can contain a quantitation feature that automatically produces calibration curves from a series of concentrations standard for a specific analyte.[51–52] iQuant2 is software for quantitative imaging using laser ablation inductively coupled plasma MS that performs semiquantitative analysis by coupling of the external correction of the relative sensitivity factor using NIST SRM610 with normalization of the corrected signal intensity.[53] MSiReader also has an MSiQuantification tool that permits users to calculate absolute concentrations from quantitative imaging MS experiments.[35] MALDIquant is an R-based software package that provides a complete and modular analysis quantitative analysis of quantitative imaging MS data across many technicals.[54] Chapter 7 has more thoughts on the quantitative imaging workflows employed in imaging MS.[55]

Multimodal

Multimodal integration of imaging MS with other imaging modalities, such as microscopy, is needed for histopathological analysis. The Pathology Interface integrates histology and imaging MS data into a web-based application to facilitate remote collaborate annotating and deliver clinical benefit.[56] An accurate association between a specific molecule and histological features can be done on imaging and staining the same tissue section.[57] In the context of imaging MS, a multimodal image can be formed by employing separate MS image acquisitions from multiple imaging MS techniques or workflow, as well as combing imaging MS image with other imaging techniques such as microscopy, magnetic resonance imaging, etc. The goal of multimodal imaging is to combine complementary information from multiple image modalities. For example, imaging MS can provide molecular distribution of chemical species such as biomarkers, while hematoxylin and eosin microscope image can show anatomical and morphological features of the tissue. The multimodal integration of the image could be done into two steps—image registration is done for spatial alignment, and image fusion is done to combine into a common fused image incorporating all imaging information.[58] For example, three-dimensional light-sheet fluorescence microscopy of the blood vessels of the brain was combined with two-dimensional MS images of a protein subunit as shown in Fig. 10.5.[59]

Most imaging MS software provides basic multimodal imaging features such as manual overlaying of digital microscopy image with MS image. For

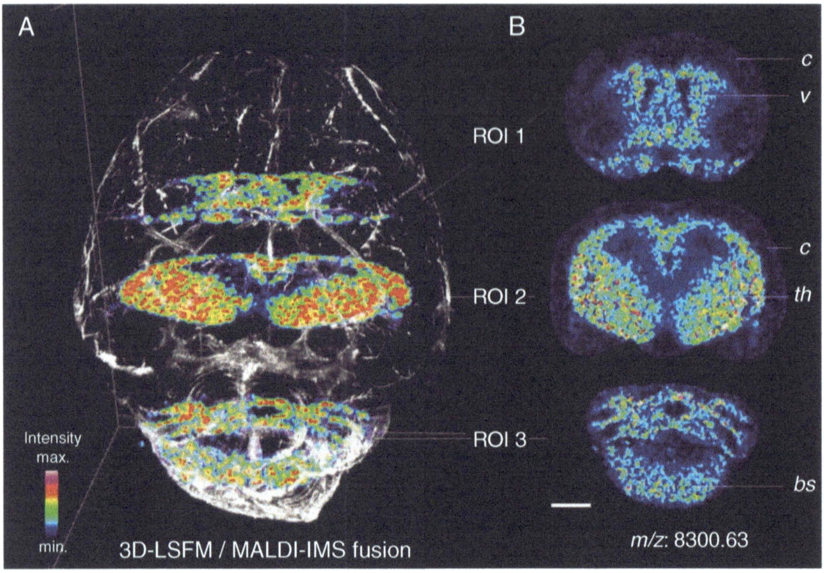

FIG. 10.5 (A) 3D light-sheet fluorescence microscopy of the blood vessels of the brain coupled with 2D MALDI images of a protein subunit (GNG3, *m/z*: 8300.63), (B) the spatial distribution of the protein subunit with brain structures annotated as cerebral cortex (*c*), ventricles (*v*), thalamus (*th*), brain stem (*bs*). Reprinted with permission from Blutke A, Sun N, Xu Z, et al. Light sheet fluorescence microscopy guided MALDI-imaging mass spectrometry of cleared tissue samples. Sci Rep. 2020;10:14461.[59] Creative Commons.

example, EasyReg2D, developed by the French Atomic Energy Commission, is multimodal imaging software that can register and fusion images. Imaging MS data can be fused with microscope histopathology images after conversion to a bidimensional array of gray intensity values (integers or floating numbers) in the American Standard Code for Information Interchange file format.[28] Chapter 11 has more details on developments in the multimodal imaging strategies used with imaging MS.[60]

Statistical analysis

Imaging MS experiment produces a large dataset with thousands of ions encompassing two-dimensional space. The dataset may be even larger due to the incorporation of orthogonal separation, such as ion mobility or the addition of another dimension of space for three-dimensional imaging. It is difficult to manually discover any meaningful information underlying biological mechanisms from these large datasets without statistical analysis. A detailed review of statistical analysis for imaging MS is done elsewhere.[2]

Exploratory statistical analysis of imaging MS data does not require a priori knowledge or tagging and can uncover trends of a large number of molecules with a handful of experiments. Statistical analysis of imaging MS data analysis is challenging due to a higher dimensionality, large file size, inherent heterogeneity of tissue, and time taken to run a large cohort of samples. Statistical analysis can be utilized to identify distinct regions of heterogeneity correlated to biomolecular MS profiles and can be used for annotating intratumor heterogeneity. The delineation of region may change depending on the multivariate method used, such as *k*-means clustering, principal component analysis, maximum autocorrelation factorization, non-negative matrix factorization, fuzzy *c*-means clustering, probabilistic latent semantic analysis.[61] A nonlinear visualization of the data using spatially mapped t-distributed stochastic neighbor embedding statistically uncovers biomolecular heterogeneity of gastric and cancer tissue in an unbiased manner.[62] These multivariate statistical tools can be used for data reduction by automated feature detection in addition to showing regions with distinct MS profiles.

Unsupervised machine learning for exploratory data analysis across the various imaging modalities, including MALDI, DESI, and SIMS, has been reviewed in the referenced publication.[3] SpectralAnalysis open-source software that can process RAW MS data (e.g., smoothing, baseline correction, normalization) and also performs multivariate analysis such as principal component analysis, non-negative matrix factorization, probabilistic latent semantic analysis, maximum autocorrelation factor, etc.[63]

Segmentation and clustering

Segmentation is the imaging data processing step where the image is demarcated into multiple segments into meaningful and easily understood compartments. Segmentation of MS image is utilized for defining a specific region boundary of disease, tissue or cell type, or other spatially differential characteristics. In imaging MS, a ROI can be manually defined, assigned based on secondary microscope images, or computed by an automatic segmentation algorithm. Segmentation of images into ROIs discriminating normal sections of tissue against diseased lesion is a crucial step before consequent informatics analyses.

Segmentation of an imaging MS data can be done by clustering spectra or pixels. A single ion image may not identify spatial correlations between analytes. Principal component and discriminant analyses are used as clustering methods to classify pixels by spectral similarity into regions using multiple correlated peaks.[5] Hierarchical clustering has been used with MALDI imaging MS to cluster and classify groups of images. Ions are clustered by in a dendrogram based on spatial correlation. The degree of similarity increases with the level of the node, with each level of nodes in the dendrogram, which can be treated as class or segment.[64] Fig. 10.6 shows a hierarchical clustering analysis of rat sagittal brain sections using *k*-Means. Each node in the cluster is represented by composite images of all of the ions in the respective nodes. Spatial segmentation of pixels by clustering mass spectra as well as by incorporating spatial relations

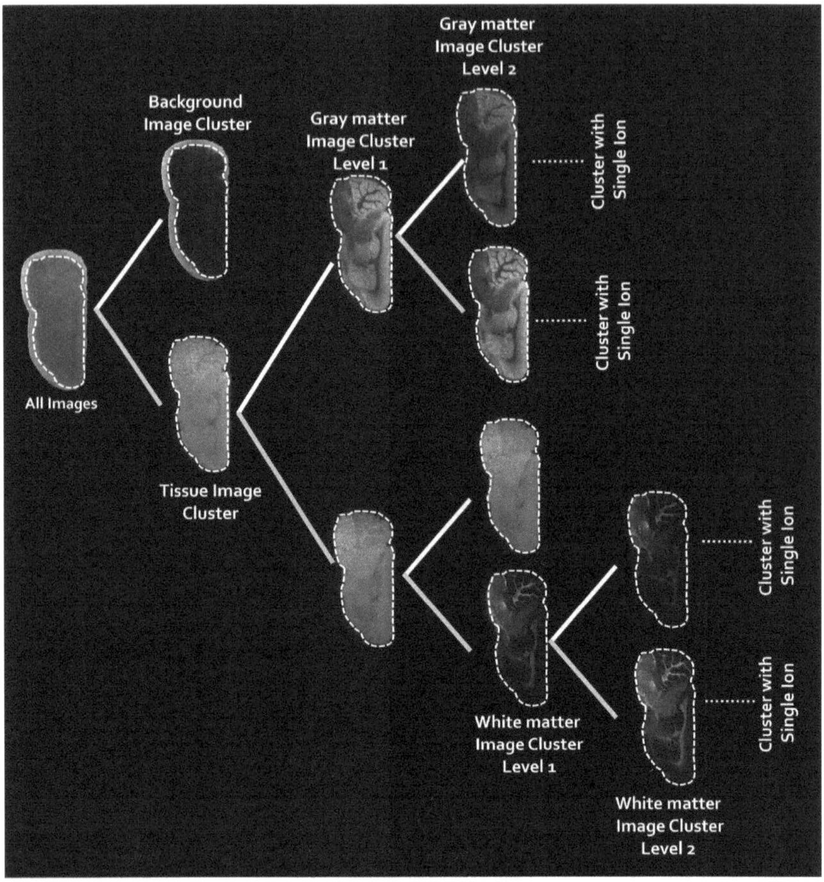

FIG. 10.6 Hierarchical clustering analysis of rat sagittal brain section MALDI imaging experiment. The hierarchical clustering analysis was created using *k*-means. Each note is represented by composite images of all of the ions in that node. The hierarchical clustering dendrogram begins on the left with a composite image of all of the analyzed ions. The first bifurcated nodes is denoted by two composite images for background ions and tissue sample. The tissue node is further divided into clusters with defined local distributions, such as nodes with ions localized in the gray matter or white matter of the brain until it terminates into a cluster containing a single image.

between the pixels so that neighboring pixels are clustered together. Spatially aware (SA) clustering utilizes distance-based clustering, and spatially aware structurally adaptive (SASA) segmentation adapts the weights assigned to the pixels in the neighborhood.[65] Spatial segmentation of imaging MS data can be improved with edge-preserving denoising of images corresponding to specific masses that reduces pixel-to-pixel.[66]

Cardinal is an R-based data analysis package that can be used for the segmentation of MALDI or DESI data. Cardinal can perform segmentation into regions based on ion intensity and image classification.[44] Cardinal implements many existing methods for segmentation, such as principle component analysis, SA, SASA segmentation, and has introduced a new unsupervised image segmentation method, called spatial shrunken centroids. This new segmentation method combines spatial aware method with nearest shrunken centroid classification for statistical regularization to prevent overfitting of data.[67] An example of segmentation comparison between *k*-means clustering of spectra, *k*-means clustering applied to the first five principal components, SA clustering, SASA clustering, spatial shrunken centroids with SA distance, and spatial shrunken centroids with SASA distance on pig fetus cross section MS image is shown in Fig. 10.7. The evaluation of commercial SCiLS with the open-source Cardinal software packages on MALDI time-of-flight imaging MS data of a Gram-negative pathogen showed both packages could perform unsupervised segmentation with similar performance with a few notable differences in identification of statistically significant features.[68]

SpectViewer is another software that can provide automatic segmentation of imaging MS data without binning. A beta version is available for academic labs by request.[28] Spatial segmentation can be performed by unsupervised hierarchical clustering of spectra, using bisecting *k*-means, into groups according to their similarities as done in the SCiLS software segmentation pipeline.[69–70] Clustering is done using a bisecting *k*-means algorithm, where the full data are split into two clusters, and at the next iteration, each of the clusters is split again until each cluster has only one spectrum. The spectrum-to-spectrum variation was reduced using edge-preserving spatial denoising prior to the segmentation.

Conclusions

Data analysis is a critical aspect of any imaging MS experiment. Research continues in computational strategies of many aspects of imaging MS data such as spectral processing, segmentation, molecule identification, multimodal integration, standardized data format, multivariate analysis, machine learning, dimensionality reduction, etc. Currently, the imzML file format has received considerable uptake in the open-source imaging software. However, an update of the format, such as mzMLb based on the HDF5 file format, may be required to utilize the benefits of high read/write speeds, optimal storage of the large raw imaging MS data. The continual development of user-friendly imaging MS software can be anticipated to encourage more biomedical researchers to utilize the imaging MS techniques. Further development of software platforms would be required before imaging MS can be used as a widespread clinical tool.

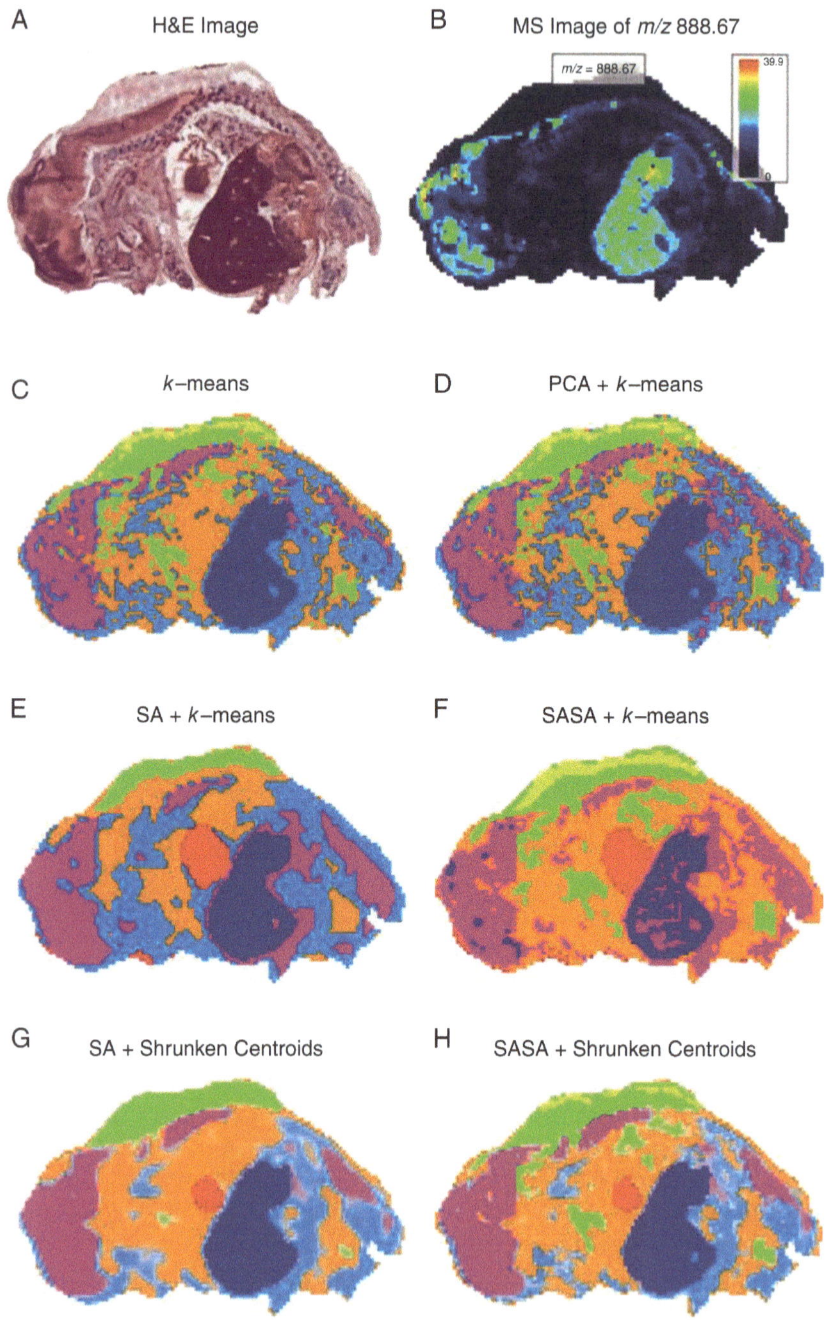

FIG. 10.7 Pig fetus cross section: (A) optical image of H&E-stained showing morphology of the brain (left), heart (center), and liver (dark region below the heart), (B) a single ion images of *m/z* 888.67 showing the brain and liver. Segmentation comparison. (C) *k*-means clustering of spectra, (D) *k*-means clustering applied to the first five principal components, (E) spatially aware (SA) clustering, (F) spatially aware structurally adaptive (SASA) clustering, (G) spatial shrunken centroids with SA distance, and spatial shrunken centroids with SASA distance. Adapted and reprinted with permission from Bemis KD, Harry A, Eberlin LS, et al. Probabilistic segmentation of mass spectrometry (MS) images helps select important ions and characterize confidence in the resulting segments. Mol Cell Proteomics. 2016;15:1761–1772.[67]

References

1. Alexandrov T. MALDI imaging mass spectrometry: statistical data analysis and current computational challenges. *BMC Bioinformatics*. 2012;13:S11.
2. Jones EA, Deininger S-O, Hogendoorn PCW, Deelder AM, McDonnell LA. Imaging mass spectrometry statistical analysis. *J Proteomics*. 2012;75:4962–4989.
3. Verbeeck N, Caprioli RM, Van de Plas R. Unsupervised machine learning for exploratory data analysis in imaging mass spectrometry. *Mass Spectrom Rev*. 2020;39:245–291.
4. Ràfols P, Vilalta D, Brezmes J, et al. Signal preprocessing, multivariate analysis and software tools for MA(LDI)-TOF mass spectrometry imaging for biological applications. *Mass Spectrom Rev*. 2018;37:281–306.
5. McCombie G, Staab D, Stoeckli M, Knochenmuss R. Spatial and spectral correlations in MALDI mass spectrometry images by clustering and multivariate analysis. *Anal Chem*. 2005;77:6118–6124.
6. Pevzner PA, Dančík V, Tang CL. Mutation-tolerant protein identification by mass spectrometry. *J Comput Biol*. 2000;7:777–787.
7. Frank AM, Pesavento JJ, Mizzen CA, Kelleher NL, Pevzner PA. Interpreting top-down mass spectra using spectral alignment. *Anal Chem*. 2008;80:2499–2505.
8. Pevzner PA, Mulyukov Z, Dancik V, Tang CL. Efficiency of database search for identification of mutated and modified proteins via mass spectrometry. *Genome Res*. 2001;11:290–299.
9. Emwas A-H, Saccenti E, Gao X, et al. Recommended strategies for spectral processing and post-processing of 1D 1H-NMR data of biofluids with a particular focus on urine. *Metabolomics*. 2018;14:31.
10. Sauve AC, Speed TP. Normalization, baseline correction and alignment of high-throughput mass spectrometry data. *Proc Gen*. 2004.
11. Yang C, He Z, Yu W. Comparison of public peak detection algorithms for MALDI mass spectrometry data analysis. *BMC Bioinformatics*. 2009;10:4.
12. Zhu P, Ding W, Tong W, Ghosal A, Alton K, Chowdhury S. A retention-time-shift-tolerant background subtraction and noise reduction algorithm (BgS-NoRA) for extraction of drug metabolites in liquid chromatography/mass spectrometry data from biological matrices. *Rapid Commun Mass Spectrom*. 2009;23:1563–1572.
13. Shin H, Mutlu M, Koomen JM, Markey MK. Parametric power spectral density analysis of noise from instrumentation in MALDI TOF mass spectrometry. *Cancer Inform*. 2007;3:117693510700300019.
14. Vrkoslav V, Muck A, Cvačka J, Svatoš A. MALDI imaging of neutral cuticular lipids in insects and plants. *J Am Soc Mass Spectrom*. 2010;21:220–231.
15. Russell Mortishire-Smith KR, Denny R, and Hughes C. Intelligent data capture: real-time noise reduction for high resolution mass spectrometry 2019.
16. Platt SJ, Richardson KG, Williams DD, Denny R, Mass spectrometer. Google Patents: 2016.
17. Deininger S-O, Cornett DS, Paape R, et al. Normalization in MALDI-TOF imaging datasets of proteins: practical considerations. *Anal Bioanal Chem*. 2011;401:167–181.
18. Hamm G, Bonnel D, Legouffe R, et al. Quantitative mass spectrometry imaging of propranolol and olanzapine using tissue extinction calculation as normalization factor. *J Proteomics*. 2012;75:4952–4961.
19. Fonville JM, Carter C, Cloarec O, et al. Robust data processing and normalization strategy for MALDI mass spectrometric imaging. *Anal Chem*. 2012;84:1310–1319.
20. Taylor AJ, Dexter A, Bunch J. Exploring ion suppression in mass spectrometry imaging of a heterogeneous tissue. *Anal Chem*. 2018;90:5637–5645.

21. Goracci L, Tortorella S, Tiberi P, et al. Lipostar, a comprehensive platform-neutral cheminformatics tool for lipidomics. *Anal Chem*. 2017;89:6257–6264.
22. Tortorella S, Tiberi P, Bowman AP, et al. LipostarMSI: comprehensive, vendor-neutral software for visualization, data analysis, and automated molecular identification in mass spectrometry imaging. *J Am Soc Mass Spectrom*. 2020;31:155–163.
23. Römpp A, Schramm T, Hester A, et al. imzML: imaging mass spectrometry markup language: a common data format for mass spectrometry imaging*Data Mining in Proteomics*: Springer; 2011:205–224.
24. Schramm T, Hester Z, Klinkert I, et al. imzML — a common data format for the flexible exchange and processing of mass spectrometry imaging data. *J Proteomics*. 2012;75:5106–5110.
25. Race AM, Styles IB, Bunch J. Inclusive sharing of mass spectrometry imaging data requires a converter for all. *J Proteomics*. 2012;75:5111–5112.
26. Adusumilli R, Mallick P. Data conversion with ProteoWizard msConvert*Proteomics*: Springer; 2017:339–368.
27. Race AM, Römpp A. Error-free data visualization and processing through imzML and mzML validation. *Anal Chem*. 2018;90:13378–13384.
28. Robbe M-F, Both J-P, Prideaux B, et al. Software tools of the Computis European project to process mass spectrometry images. *Eur J Mass Spectrom*. 2014;20:351–360.
29. The HDF Group. Available at: https://www.hdfgroup.org/ (Accessed September 5, 2020).
30. Bhamber RS, Jankevics A, Deutsch EW, Jones AR, Dowsey AW. mzMLb: a future-proof raw mass spectrometry data format based on standards-compliant mzML and optimized for speed and storage requirements. *J Proteome Res*. 2020.
31. Rübel O, Greiner A, Cholia S, et al. OpenMSI: a high-performance web-based platform for mass spectrometry imaging. *Anal Chem*. 2013;85:10354–10361.
32. de Raad M, de Rond T, Rübel O, Keasling JD, Northen TR, Bowen BP. OpenMSI arrayed analysis toolkit: Analyzing spatially defined samples using mass spectrometry imaging. *Anal Chem*. 2017;89:5818–5823.
33. Hosokawa N, Sugiura Y, Setou M. Ion image reconstruction using biomap software*Imaging Mass Spectrometry*: Springer; 2010:113–126.
34. Robichaud G, Garrard KP, Barry JA, Muddiman DC. MSiReader: an open-source interface to view and analyze high resolving power MS imaging files on Matlab platform. *J Am Soc Mass Spectrom*. 2013;24:718–721.
35. Bokhart MT, Nazari M, Garrard KP, Muddiman DC. MSiReader v1.0: evolving open-source mass spectrometry imaging software for targeted and untargeted analyses. *J Am Soc Mass Spectrom*. 2018;29:8–16.
36. Jardin-Mathé O, Bonnel D, Franck J, et al. MITICS (MALDI imaging team imaging computing system): a new open source mass spectrometry imaging software. *J Proteomics*. 2008;71:332–345.
37. He J, Huang L, Tian R, et al. MassImager: a software for interactive and in-depth analysis of mass spectrometry imaging data. *Anal Chim Acta*. 2018;1015:50–57.
38. Hayasaka T, Goto-Inoue N, Ushijima M, et al. Development of imaging mass spectrometry (IMS) dataset extractor software, IMS convolution. *Anal Bioanal Chem*. 2011;401:183–193.
39. Smith DF, Kharchenko A, Konijnenburg M, Klinkert I, Paša-Tolić L, Heeren RM. Advanced mass calibration and visualization for FT-ICR mass spectrometry imaging. *J Am Soc Mass Spectrom*. 2012;23:1865–1872.
40. Klinkert I, Chughtai K, Ellis SR, Heeren RMA. Methods for full resolution data exploration and visualization for large 2D and 3D mass spectrometry imaging datasets. *Int J Mass Spectrom*. 2014;362:40–47.

41. Veselkov K, Sleeman J, Claude E, et al. BASIS: high-performance bioinformatics platform for processing of large-scale mass spectrometry imaging data in chemically augmented histology. *Sci Rep*. 2018;8:4053.
42. Parry RM, Galhena AS, Gamage CM, Bennett RV, Wang MD, Fernández FM. OmniSpect: an open MATLAB-based tool for visualization and analysis of matrix-assisted laser desorption/ionization and desorption electrospray ionization mass spectrometry images. *J Am Soc Mass Spectrom*. 2013;24:646–649.
43. R Core Team R. foundation for Statistical Computing, 3. Austria: Vienna; 2013 0.
44. Bemis KD, Harry A, Eberlin LS, et al. Cardinal: an R package for statistical analysis of mass spectrometry-based imaging experiments. *Bioinformatics*. 2015;31:2418–2420.
45. Gamboa-Becerra R, Ramírez-Chávez E, Molina-Torres J, Winkler R. MSI. R scripts reveal volatile and semi-volatile features in low-temperature plasma mass spectrometry imaging (LTP-MSI) of chilli (Capsicum annuum). *Anal Bioanal Chem*. 2015;407:5673–5684.
46. Ràfols P, Torres S, Ramírez N, et al. rMSI: an R package for MS imaging data handling and visualization. *Bioinformatics*. 2017;33:2427–2428.
47. Shrestha B. Visualization in imaging mass spectrometry*Introduction to Spatial Mapping of Biomolecules by Imaging Mass Spectrometry*: Elsevier Science; 2021.
48. Wishart DS, Feunang YD, Marcu A, et al. HMDB 4.0: the human metabolome database for 2018. *Nucleic Acids Res*. 2018;46:D608–D617.
49. Palmer A, Phapale P, Chernyavsky I, et al. FDR-controlled metabolite annotation for high-resolution imaging mass spectrometry. *Nat Methods*. 2017;14:57–60.
50. Shrestha B. Molecule identification approaches in imaging mass spectrometry*Introduction to Spatial Mapping of Biomolecules by Imaging Mass Spectrometry*: Elsevier Science; 2021.
51. Källback P, Shariatgorji M, Nilsson A, Andrén PE. Novel mass spectrometry imaging software assisting labeled normalization and quantitation of drugs and neuropeptides directly in tissue sections. *J Proteomics*. 2012;75:4941–4951.
52. Källback P, Nilsson A, Shariatgorji M, Andrén PE. MsIQuant–quantitation software for mass spectrometry imaging enabling fast access, visualization, and analysis of large data sets. *Anal Chem*. 2016;88:4346–4353.
53. Suzuki T, Sakata S, Makino Y, et al. iQuant2: software for rapid and quantitative imaging using laser ablation-ICP mass spectrometry. *Mass Spectrom*. 2018;7:A0065–A0065.
54. Gibb S, Strimmer K. MALDIquant: a versatile R package for the analysis of mass spectrometry data. *Bioinformatics*. 2012;28:2270–2271.
55. Shrestha B. Strategies for quantitative imaging mass spectrometry*Introduction to Spatial Mapping of Biomolecules by Imaging Mass Spectrometry*: Elsevier Science; 2021.
56. Norris JL, Tsui T, Gutierrez DB, Caprioli RM. Pathology interface for the molecular analysis of tissue by mass spectrometry. *J Pathol Inform*. 2016;7:13–15.
57. Chaurand P, Schwartz SA, Billheimer D, Xu BJ, Crecelius A, Caprioli RM. Integrating histology and imaging mass spectrometry. *Anal Chem*. 2004;76:1145–1155.
58. Maintz JB, Viergever MA. A survey of medical image registration. *Med Image Anal*. 1998;2:1–36.
59. Blutke A, Sun N, Xu Z, et al. Light sheet fluorescence microscopy guided MALDI-imaging mass spectrometry of cleared tissue samples. *Sci Rep*. 2020;10:14461.
60. Shrestha B. Multimodal imaging mass spectrometry*Introduction to Spatial Mapping of Biomolecules by Imaging Mass Spectrometry*: Elsevier Science; 2021.
61. Jones EA, van Remoortere A, van Zeijl RJM, et al. Multiple statistical analysis techniques corroborate intratumor heterogeneity in imaging mass spectrometry datasets of myxofibrosarcoma. *PLOS One*. 2011;6:e24913.

62. Abdelmoula WM, Balluff B, Englert S, et al. Data-driven identification of prognostic tumor subpopulations using spatially mapped t-SNE of mass spectrometry imaging data. *Proc Natl Acad Sci*. 2016;113:12244–12249.
63. Race AM, Palmer AD, Dexter A, Steven RT, Styles IB, Bunch J. SpectralAnalysis: software for the masses. *Anal Chem*. 2016;88:9451–9458.
64. Deininger S-O, Ebert MP, Fütterer A, Gerhard M, Röcken C. MALDI imaging combined with hierarchical clustering as a new tool for the interpretation of complex human cancers. *J Proteome Res*. 2008;7:5230–5236.
65. Alexandrov T, Kobarg JH. Efficient spatial segmentation of large imaging mass spectrometry datasets with spatially aware clustering. *Bioinformatics*. 2011;27:i230–i238.
66. Alexandrov T, Becker M, Deininger S-O, et al. Spatial segmentation of imaging mass spectrometry data with edge-preserving image denoising and clustering. *J Proteome Res*. 2010;9:6535–6546.
67. Bemis KD, Harry A, Eberlin LS, et al. Probabilistic segmentation of mass spectrometry (MS) images helps select important ions and characterize confidence in the resulting segments. *Mol Cellular Proteomics*. 2016;15:1761–1772.
68. Luu GT, Condren AR, Kahl LJ, Dietrich LEP, Sanchez LM. Evaluation of data analysis platforms and compatibility with MALDI-TOF imaging mass spectrometry data sets. *J Am Soc Mass Spectrom*. 2020;31:1313–1320.
69. Krasny L, Hoffmann F, Ernst G, et al. Spatial segmentation of MALDI FT-ICR MSI data: a powerful tool to explore the head and neck tumor in situ lipidome. *J Am Soc Mass Spectrom*. 2015;26:36–43.
70. Trede D, Schiffler S, Becker M, et al. Exploring three-dimensional matrix-assisted laser desorption/ionization imaging mass spectrometry data: three-dimensional spatial segmentation of mouse kidney. *Anal Chem*. 2012;84:6079–6087.

Chapter 11

Multimodal imaging mass spectrometry

Chapter Outline

The visualization to infer the relationship between anatomical features and biochemical functions often requires several imaging modalities. Biomedical imaging systems can be classified according to the physical mechanism (e.g., X-rays, photons, acoustic waves, and ions), spatial resolution (macroscopic or microscopic), according to information obtained (anatomical, physiological, cellular, or molecular), or by types of samples analyzed or application performed.[1] A combination of imaging techniques, often referred to as multimodal imaging, gives an improved insight by overcoming the limitations of the independent modalities. Ideally, a multimodal image should provide anatomical, functional, and biomolecular information.

Multimodal imaging can be created using separate image acquisitions from multiple instruments or from one single instrument platform that combines multiple techniques. Various imaging techniques produce different types of images based on physical interactions (e.g., radiation, magnetic, acoustic, ionic, etc.) used for its creation, providing complementary information. Recently, many multimodal imaging systems combine multiple imaging techniques have been developed, such as micropositron emission tomography (μPET) and microcomputed tomography (μCT), where information with the low resolution/high sensitivity of the μPET is improved by coupling with the high resolution of μCT.[2] Multimodal imaging based on the combination of optical imaging with other clinical imaging modalities such as magnetic resonance imaging (MRI), CT, and ultrasound is reviewed herein.[3]

Imaging techniques such as imaging mass spectrometry (MS) provide molecular distribution of chemicals such as biomarkers. However, the lack of

Introduction to Spatial Mapping of Biomolecules by Imaging Mass Spectrometry. DOI: 10.1016/C2018-0-03962-6

morphological features may present uncertainty in localizing the disease or defining is structural manifestation. Complementary information gained from multiple image modalities, such as anatomical features from hematoxylin and eosin (H&E) image and molecular distribution from an imaging MS, requires a proper integration to obtain useful data. The integration of the image can be divided into two phases. The first phase is image registration, where images are brought in a spatial alignment, and the second phase is image fusion, where images are combined into a common fused image encompassing all imaging data.[4] Conceptual multimodal imaging encompassing two complementary modalities of imaging MS, matrix-assisted laser desorption/ionization (MALDI) and imaging mass cytometry, histological imaging using microscopy, and in vivo clinical imaging (MRI, PET, CT) is shown in Fig. 11.1. Each imaging modality brings unique information to the users and spans from the organ to the subcellular compartments in terms of spatial dimension.

Image registration

Image registration is the method of aligning different sets of imaging data into a common spatial coordinate.[5–6] The multiple imaging data may be microscope images with multiple stains, images using different analytical sensors, images collected at various time points, or using different depth or viewpoints.[7] Image registration is necessary to compare and integrate the information from different imaging modalities. In biomedical science, image registration is used for aligning images from multiple imaging modalities by transforming imaging data due to the differences in the acquisition.

Image registration can be achieved by the use of either extrinsic or intrinsic markers. Image registration with extrinsic markers can be done by introducing objects or markers into the imaged space. Depending on the sample, extrinsic markers could include mold, frame (e.g., stereotactic frame, dental adapter), and fiducial markers. Intrinsic markers are based on generated image information by the sample and are based on an identifiable landmark (e.g., the edge of tissue), geometrical shape, relying on segmentation, or based on voxel or pixel. Image registration can also happen using a nonimage-based registration, such as calibrating two image modality externally or bringing the sample to the exact location using robotics or computer-controlled translational stages.

In order to register, an image may require transformation. The domain of transformation is usually global, applying to the entire image. Seldom transformation will be local if a section of the image has its own transformations. The nature of image transformation can be rigid, affine, projective, or curved. Rigid transformation is when an image is transformed using only translations and rotations. Affine transformation is one of the transformations that retain parallel relationships within the image and projective if the transformation is skewed and does not preserve the parallel relationships. Many in vivo biomedical image registrations often additionally require nonrigid elastic registration to

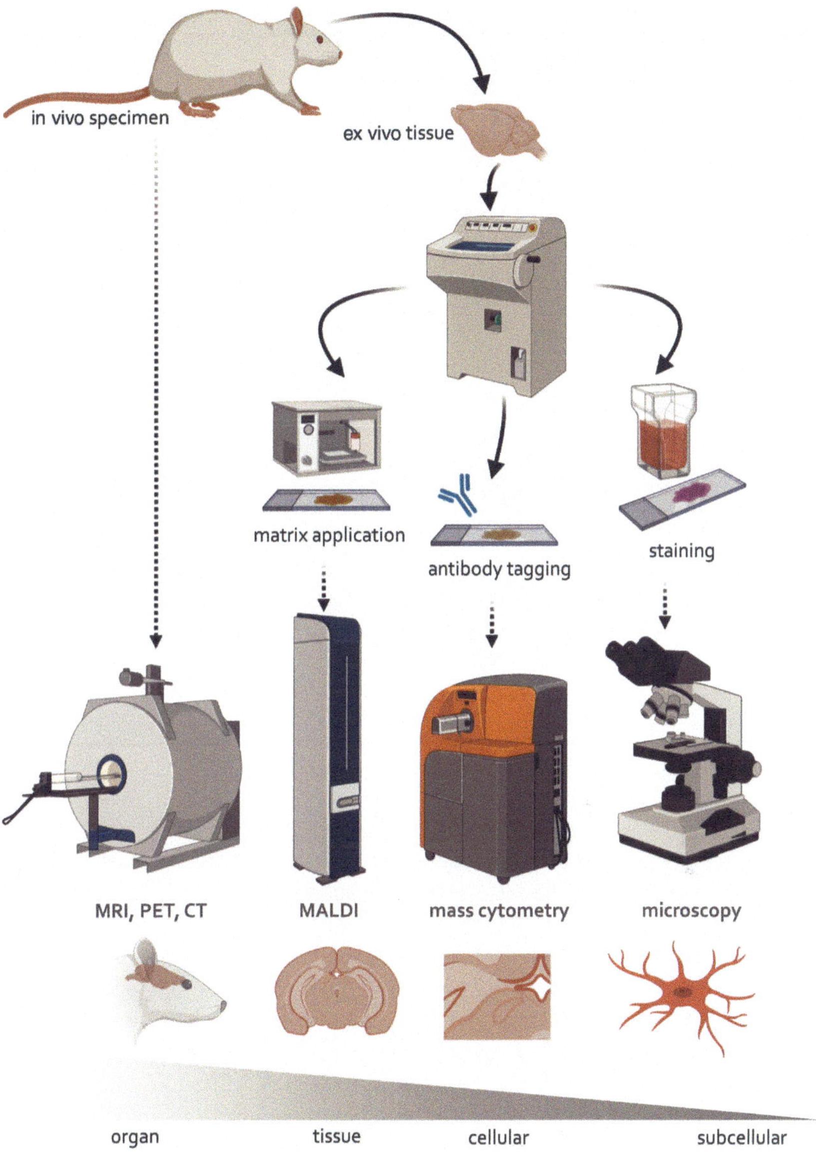

FIG. 11.1 Multimodal imaging encompassing two modalities of imaging mass spectrometry, microscopy, and in vivo clinical imaging modalities.

correct the deformation of the subject (e.g., due to breathing or other anatomical changes) or time (e.g., measure change detection of tumor size in the same subject). Many nonrigid image registration can use an established anatomical atlas as a reference or anchor for elastic registration. An example of a pictorial depiction of imaging transformation is shown in Fig. 11.2.

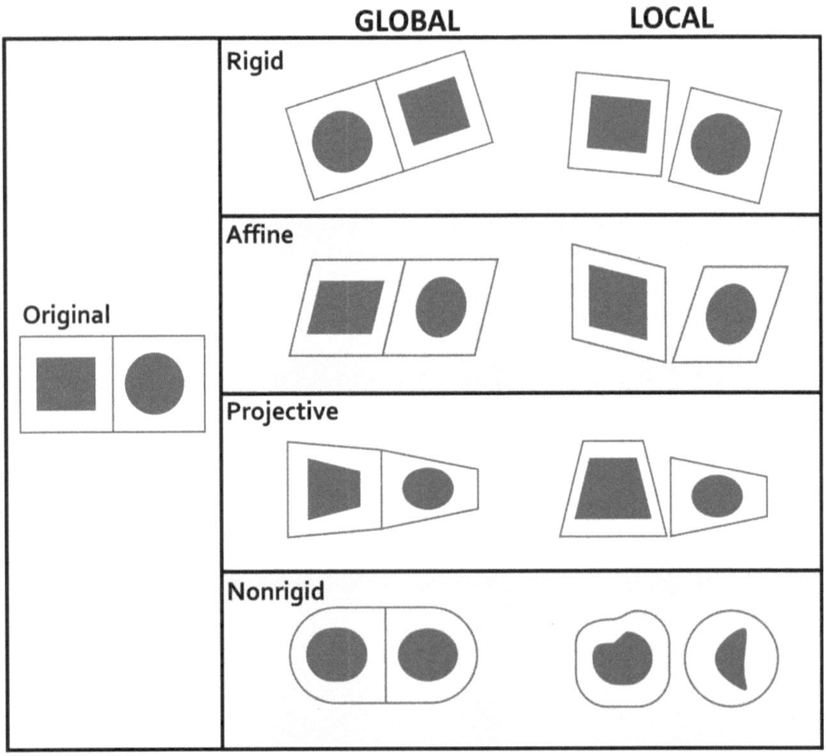

FIG. 11.2 Examples of types of transformation for image registration.

Image registration can be manual or automatic or a combination of both. In manual methods, alignments are done manually by the user. In semiautomatic methods, certain key operations are initialized but still relying on the user to guidance or final verification. In automatic image registration, image registration steps are automatically performed based on a computer algorithm. In clinical imaging, the accuracy of coregistration is critical while performing functional imaging (e.g., PET, single-photon emission computed tomography [SPECT]) with anatomical imaging (e.g., CT, MRI) at different times.[8] Many methods have been used for coregistration, common among them is the use of mutual information, regions of a similar feature in one image would correspond to regions in the other image.[9]

Multimodal imaging techniques in imaging MS

Multimodal imaging MS provides colocalization of an individual molecule with each other, as well as correlation with morphological, anatomical, or functional features. A complementary utility of several imaging modalities can be seen in an example of multimodal imaging of hallucinogenic drugs on blotter paper.

The doped paper was imaged by micro X-ray fluorescence, laser ablation (LA)-inductively coupled plasma (ICP)-optical emission spectroscopy, and MALDI with additional molecular analysis by liquid chromatography (LC) MS after extraction. The micro X-ray fluorescence was a fast, cheap, and most straightforward method for imaging chlorine and iodine elements in the drugs, and the LA-ICP optical emission spectroscopy can generate three-dimensional maps of the same elements. MALDI showed the molecular distribution of intact drugs, and LC–MS analyses confirmed the identity and aided in quantitative analysis.[10] Imaging MS modes, such as MALDI, secondary ion mass spectrometry (SIMS), and desorption electrospray ionization (DESI), have been integrated with other analytical modalities, such as microscopy, transcriptomics, spectroscopy, etc. Some of the other applications and benefits of integrating imaging MS techniques with other complementary imaging modalities for a better molecular insight of biomedical research are discussed in the cited references.[11–14]

Sample preparation and workflow strategies may need to be changed to support multimodal imaging. For example, a novel tissue embedding process based on the low-temperature hydrogel matrix of hydroxypropyl methylcellulose and polyvinylpyrrolidone is compatible with several histological, imaging MS, and molecular biology MALDI, DESI, reverse transcription polymerase chain reaction.[15]

Broadly, multimodal imaging using MS can be broadly classified into the following categories; (a) multiple ionization source or techniques on the same or serial tissue section(s), (b) imaging MS projected on another microscopy (e.g., H&E optical image) image on the same or serial tissue section(s), (c) imaging MS of a tissue section coupled with other imaging (e.g., MRI, X-ray) of bulk uncut tissues or live animal. Multimodal integration of molecular imaging technology, such as imaging MS, with other imaging modalities that provide morphological or anatomical information, would expand the capabilities to study complex biological systems in mammalian tissues, plants, and microbial systems. For example, multimodal imaging workflow with confocal microscopy and MALDI was used to study the spatial organization of the microbial communities and their metabolites and microbiome within the soil.[16] In another example, classification based on MS signal obtained from a hand-held picosecond infrared laser probe and video captured using an optical surgical tracking system was utilized for in situ augmented reality visualization of tissue classifiers.[17] A major challenge in multimodal imaging MS is coupling datasets from different ion source instruments, often using dissimilar spatial resolutions, of ions together into a coregistered image. Imaging MS is inherently a multimodal imaging techniques since hundreds, if not thousands of molecular ions are detected at once in most common imaging MS workflows. Imaging MS workflow routinely combines and overlays the image of multiple ions from the same dataset. The power of multiple ion images can be seen in an example of the MS image of the retina of the eye, which contains cells that are organized in layers, as shown in Fig. 11.3A. The overlaid ion image of lipids shows the spatial relationship between those

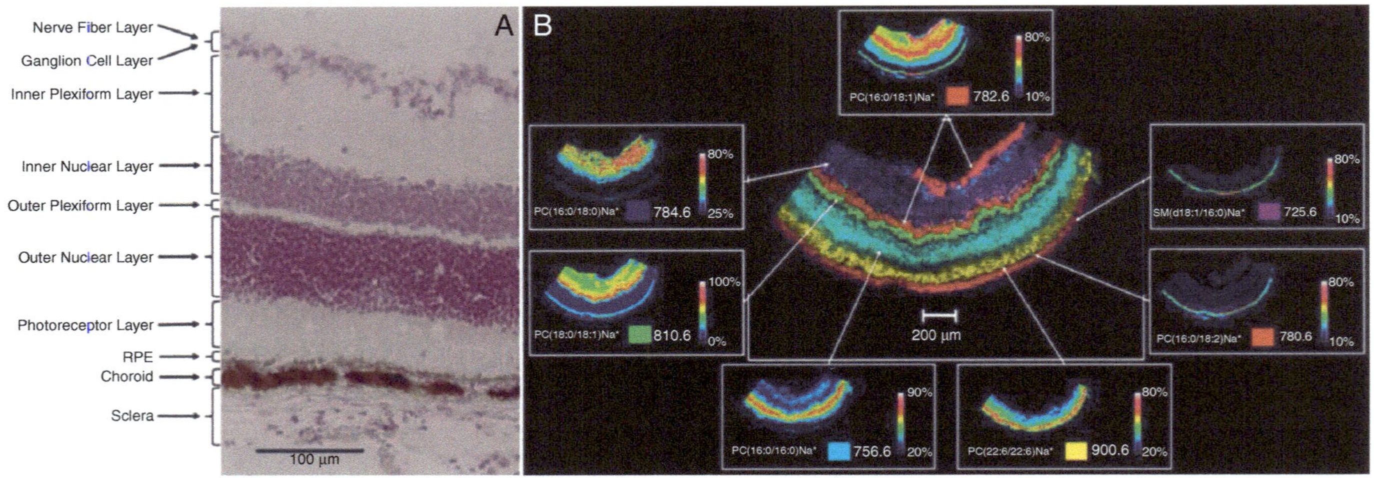

FIG. 11.3 Overlay of multiple ion images can show morphological relationship between those molecules; (A) H&E stained image of mouse retinal tissue displaying the multilayered structure; (B) overlaid (center) MALDI images of eight lipid species in the retina tissue showing the eight layers seen in H&E image with individual ion images. Adapted with permission from Anderson DMG, Ablonczy Z, Koutalos Y, et al. High resolution MALDI imaging mass spectrometry of retinal tissue lipids. J Am Soc Mass Spectrom. 2014;25:1394–1403. Copyright 2014 American Chemical Society.[18]

biomolecules within many layers of the cells, as shown in Fig. 11.3B.[18] Such coregistered images from imaging MS, and optical microscopy can be fused to extrapolate a predictive ion image with detailed features.[19]

Multimodal imaging MS with multiple ion sources

Multimodal imaging of two or more ion sources or sample preparation workflow can give a complementary molecular profile. For example, multimodal imaging of three modes of MALDI, the dual polarity of lipid was combined with protein imaging within the same tissue, allowed interpretation the lipid and protein molecule at the same myocardial infarcted region in a mouse model.[20] DESI and MALDI have been combined to improve molecular coverage of metabolites and lipids. For example, the same mouse brain tissue section was imaged using DESI, then MALDI, and in turn stained with H&E staining and imaged by optical microscopy. This multimodal imaging workflow allowed for correlating distribution of a lipid sulfatide (24:1) detected by DESI imaging with the protein MBP isoform 8 detected using MALDI in the corpus callosum substructure as shown in the H&E image as shown in Fig. 11.4.[21] Spatial distributions of metabolites in three-dimensional cell culture (e.g., lactate, glutamine, and citrate) mapped by DESI was overlaid with single-cell phenotyping obtained by imaging mass cytometry technique using LA-ICP.[22] Typically, separate mass spectrometers are used for each ion source. Sometimes more than one imaging ion source are integrated with the same mass spectrometer, such as DESI and MALDI are available with a single quadruple time-of-flight mass spectrometer.[23] In other instances, researchers modify an existing infrastructure to add another imaging modality, such as adding homebuilt picosecond infrared laser imaging technique that utilized hardware and data analysis pipelines in commercial DESI imaging source[24] or modifying electrospray ion source to install homebuilt interchangeable atmosphere pressure MALDI and laser electrospray[25] ion sources on the same mass spectrometer.

High spatial resolution SIMS was used to image the elemental and small molecules and MALDI to image phospholipids, peptides, proteins in bone, muscle, or skin from the limb of a mouse model.[26] Multimodal imaging consists of MALDI and SIMS was employed to identify molecular markers of pulmonary arterial hypertension in human lung tissue.[27] Multimodal molecular imaging of N-glycans and tryptically digested peptides from the same formalin-fixed paraffin-embedded tissue section, such as epithelial and mesenchymal tumor, was done by sequential analysis of deparaffinized tissue. First, peptide-N-glycosidase F digestion and matrix application (α-Cyano-4-hydroxycinnamic acid or CHCA) was used for N-glycan imaging and second, matrix removal, and heat-induced antigen retrieval, trypsin digestion, and matrix application (2,5-dihydroxybenzoic acid or DHB) for the MALDI imaging of digested peptides.[28]

MALDI has been coupled to other pseudoimaging MS modalities, such as liquid microjunction surface sampling coupled with LC–MS to map the

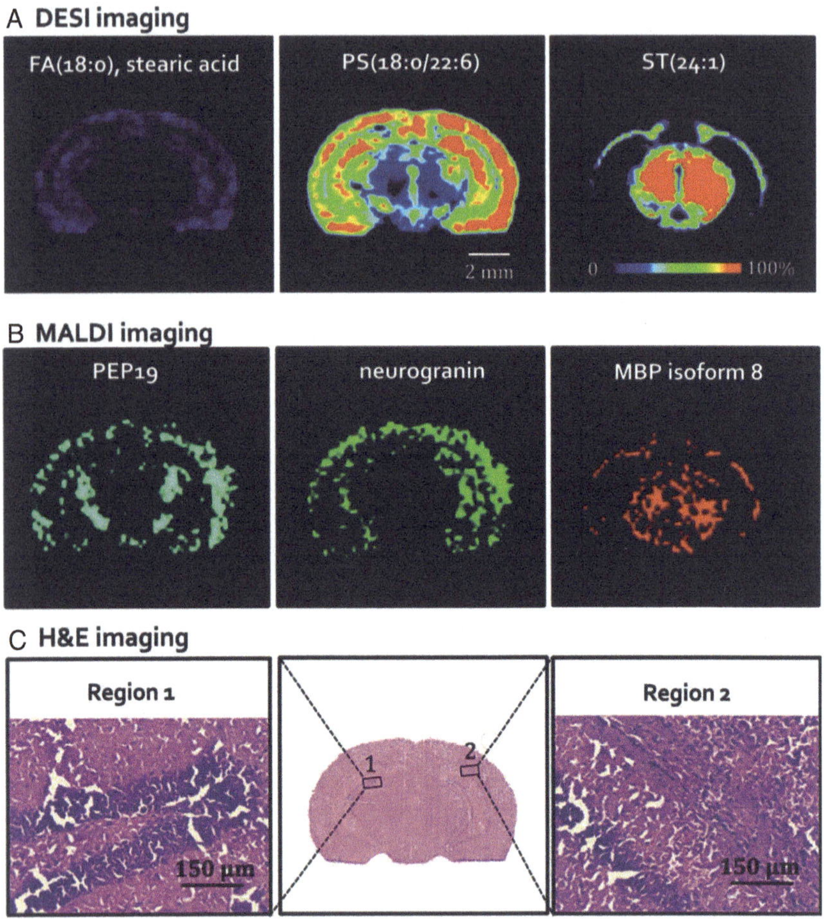

FIG. 11.4 Multimodal imaging of a mouse brain coronal section (A) acquired first by DESI, (B) MALDI imaging of the same tissue section after the DESI imaging, (C) H&E image of the same tissue after removal of the MALDI matrix. Adapted with permission from Eberlin LS, Liu X, Ferreira CR, Santagata S, Agar NY, Cooks RG. Desorption electrospray ionization then MALDI mass spectrometry imaging of lipid and protein distributions in single tissue sections. Anal Chem. 2011;83:8366–8371. Copyright 2011 American Chemical Society.[21]

distribution of tuberculosis drug in rabbit lung tissues. MALDI imaging at higher spatial resolution provided a more granular map of drug distribution, while liquid microjunction surface sampling technique confirmed the identification of the drug and provided more quantitative information.[29]

Multimodal imaging MS with optical microscopy

Microscopy imaging is a logical imaging modality to combine with imaging MS would be optical microscopy due to similarity in the sample preparation and complementary nature of structural and molecular information acquired. For example, imaging MS can locally map lipid molecules such as phosphatidylcholines or sphingomyelin onto heterogeneous tumor microenvironment images obtained from the bright field or fluorescence microscopy.[30] Optical and MS images have been coregistered by using external fiducial markers and tumor boundaries detected in both imaging modalities. A dye molecule (e.g., cresyl violet, ponceau S, or bromophenol blue) are an excellent fiducial marker for coregistration because dyes can be easily visualized optically and detected by a mass spectrometer.[31]

H&E stained slides have been used to confirm the pathological demarcation suggested by imaging MS data by correlating the MS images with the histological features of the same section by a board-certified pathologist.[32] Other applications include analysis of pressure ulcers to examine cutaneous wound healing using MALDI and H&E images,[33] a comparison of HER-2 breast cancer classification using MALDI MS image with H&E and immunohistochemistry images,[34] correlation of necrosis in breast cancer with DESI and H&E.[35]

Fluorescent protein labeling, such as the gene encoding green fluorescent protein, is commonly used in the molecular biology lab. MALDI imaging MS was employed to visualize a fluorescent protein, tandem dimer tomato red fluorescent proteins, solely present in the hypoxic area of a breast tumor xenograft mice tissue. This multimodal approach can be used to colocate other molecules that are also exclusively located or absent from the hypoxic regions.[36] MALDI imaging MS and fluorescence imaging have also been used to obtain molecular distribution and monitoring specific gene expression by using transcriptional reporters, respectively. This multimodal approach was successful at providing information on cell heterogeneity and function not available when using individual imaging techniques.[37] Multimodal combinations of light-sheet fluorescence microscopy with MALDI were utilized to guide MALDI image of brain tissue samples for detection and quantification of proteins in three dimensions.[38] A laser capture microdissection system was coupled to a LA atmospheric pressure chemical ionization was to obtain multimodal optical and chemical imaging capability. This platform was capable of multimodal imaging and coregistration of bright field, fluorescence, and MS images of mouse brain tissue.[39]

Tissues that undergo imaging MS methods such as MALDI and DESI can be compatible with histology, even maintaining the subcellular structure if the MS acquisition and sample preparation are done appropriately. The challenges in histological imaging after the imaging MS are the conservation of morphological the integrity of tissue from the analytical beam (laser or spray) during analysis and circumventing any delocalization during sample preparation or aquistion.[40]

In histology-directed strategy, imaging is performed only at regions of interest as prescribed by histological examination of a serial section by experts. Histology-directed imaging can improve the throughput of imaging MS since one does not need to image the whole tissue.[41] Histology-directed MALDI has been used to classify primary liver cancers with cross validation from p53 immunostaining.[42] A histology-directed platform with autofluorescence microscopy allows the use of the same slide for registration of microscopy and imaging MS. This approach improves the coregistration accuracy due to the use of a common tissue section allowing the acquisition of sparsely distributed features in tissue such as malaria-infected mouse hepatocytes and glomeruli in the human kidney occupying less than one-tenth of 1% tissue area.[43] This novel histology-directed platform with autofluorescence workflow and an example of coregistration of H&E image and MS image of individual Purkinje brain cells is shown in Fig. 11.5.

Conventionally, imaging MS image and histological images, such as H&E, are manually coregistered and overlaid in image processing software using external or internal fiducial markers. Automated workflow for coregistering imaging MS data to the histological image has been proposed, as shown in Fig. 11.6A. In order to automatically coregister imaging MS and histological images, a computational algorithm must align and transform the imaging data to correspond to the histological image. First, the imaging MS data are reduced using a nonlinear technique called t-distributed stochastic neighbor embedding (tSNE), which maps high-dimensional data by providing each datapoint a location in a two- or three-dimensional distribution.[44] The image generated by the tSNE map is utilized to guide automatic transformation and spatial alignment of imaging MS data with the histology. Fig. 11.6B shows a histological coronal mouse brain section, Fig. 11.6C shows an MS image of a selected ion from such

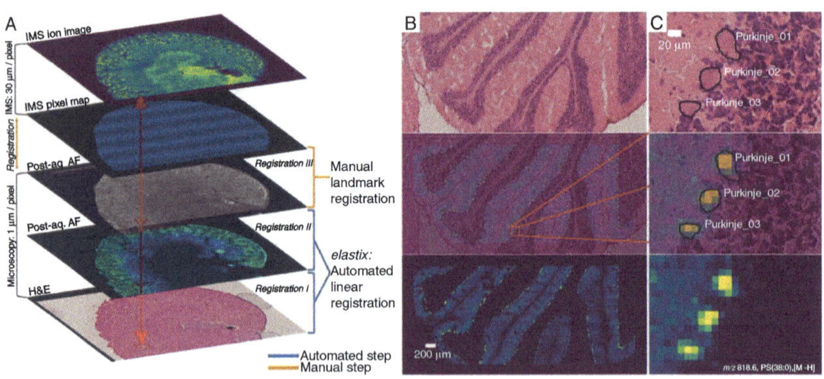

FIG. 11.5 (A) Image registration workflow for microscopy and imaging MS image shown using a rat kidney example, registration I is the with the registration of preacquisition autofluorescence and postacquisition H&E stained image, registration II is the registration between preacquisition autofluorescence and postacquisition autofluorescence (grayscale), and finally, registration III is a manual registration of laser ablation spots between postacquisition autofluorescence and imaging MS (IMS) pixel map; (B) example of image registration between MS image of Purkinje cells in the rat brain and H&E image. Adapted with permission from Patterson NH, Tuck M, Lewis A, et al. Next generation histology-directed imaging mass spectrometry driven by autofluorescence microscopy. Anal Chem. 2018;90:12404–12413. Copyright 2018 American Chemical Society.[43]

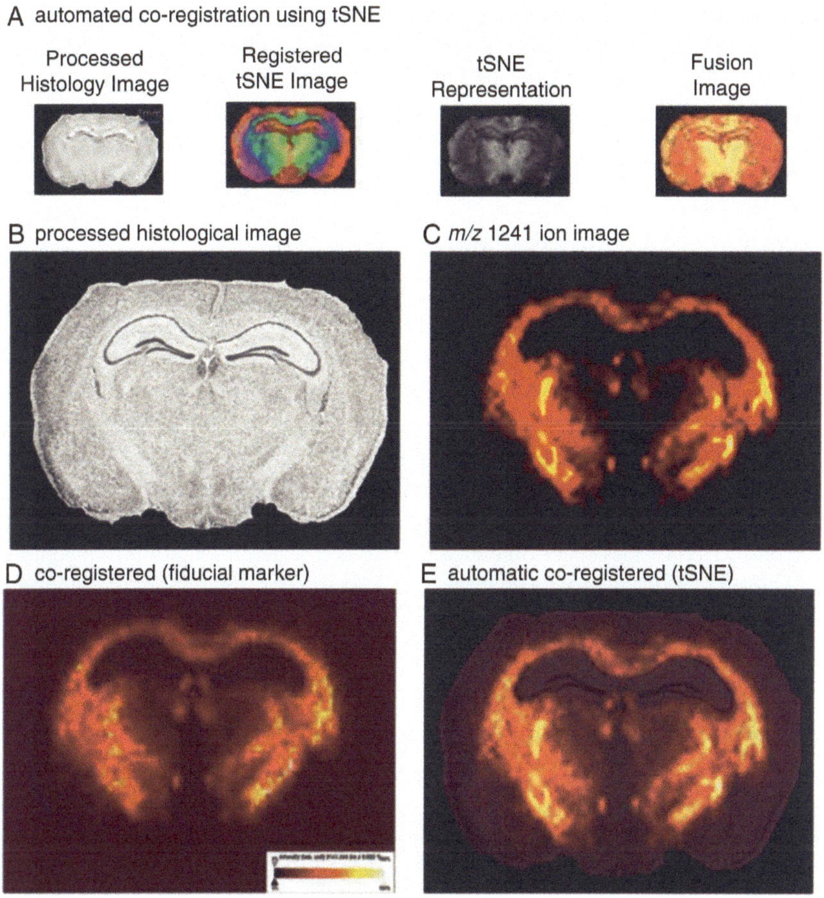

FIG. 11.6 Automatic coregistration of imaging MS data with histology image is shown in a coronal mouse brain section; (A) registration workflow showing the preprocessed histological image, low-dimensional registered tSNE image, tSNE representation, fusion image, (B) histological image; (C) ion image from MALDI; (D) coregistered image using fiducial markers; (E) registered image using automated tSNE workflow. Adapted with permission from Abdelmoula WM, Škrášková K, Balluff B, et al. Automatic generic registration of mass spectrometry imaging data to histology using nonlinear stochastic embedding. Anal Chem. 2014;86:9204–9211. Copyright 2014 American Chemical Society.[45]

section, Fig. 11.6D the coregistered the same mass spec and histological images based on manually selected fiducial markers, and Fig. 11.6E shows automatic coregistered image based on tSNE algorithm.[45] Imaging MS data can be automatically registered to a curated brain image, such as Allen brain atlas, to correlate the ion image with correct brain anatomy. In order to perform an automatic correlation with a library image, the tissue was stained with cresyl violet (Nissl stain) after MS imaging. In the workflow, the image of the tissue section was preprocessed into a binary mask isolating the tissue from the background using morphological operations (opening, closing, region-filling), and the anatomical features were correlated and fitted using gross features such as hippocampus chord length. Finally, image registration was done by automatically selecting an image from the Allen brain atlas close to the MS image sample and performing affine and nonlinear registration.[46]

Multimodal imaging MS with clinical imaging

MRI is based on nuclear magnetic resonance and was introduced in 197os when the nuclear magnetic resonance image was reconstructed by introducing gradients in the magnetic field. MRI can obtain anatomical images of all soft tissues that contain hydrogen without using ionizing radiofrequency (RF) radiation.[47] MRI uses paramagnetic, superparamagnetic, and ferromagnetic substances as contrast agents such as chelates of gadolinium. DESI imaging MS has been used to image gadolinium-based MRI contrast agents to expose tumor heterogeneity present in the absence of known mass spectrometric signals.[48] Quantitative capability of the mass spectrometer was exploited to quantify gadolinium agents after in vivo MRI using correlation from quantitative imaging of gadolinium agents in the tissue section by MALDI.[49]

The integration of the imaging MS with MRI combines molecular information of ex vivo imaging analysis with anatomical information provided by an established in vivo clinical imaging modality. For example, the integration of the DESI-MS image of oncometabolite 2-hydroxyglutarate obtained during surgical resection of gliomas onto MRI of tumors provided molecular and anatomical information.[50] Three-dimensional imaging MS data obtained by reconstruction of a series of two-dimensional images[51] has been integrated with *MRI data* collected aforehand. The integration of reconstrued three-dimensional MS imaging and MRI has been used for correlating proteomic profiles,[52] monitors molecular inflammatory response to infection,[53] with in vivo anatomical features. MRI and volume localized magnetic resonance spectroscopy was used with DESI-MS imaging to study the associated change in the alkaloid contents of regions of the nut during the growth.[54]

SPECT uses a reconstruction of X-ray computed tomography with radioactive isotopes and measure gamma (γ) rays and PET is based on the principle of detecting photon pairs from positron-emitting atoms. Both nuclear medicine imaging techniques utilize administrated radioactive isotopes as tracers for functional imaging.[55] The key difference between SPECT and PET is the type of radiotracers used. SPECT uses metastable γ-emitting radionuclides such as technetium, while one of the most common PET tracer, ^{18}F fluorine isotope in fluorodeoxyglucose measures glucose metabolism.[47] A comparison between PET imaging collected using radiolabel [18F] fluoro-6-thia-heptadecanoic acid tracer administrated to zebrafish and nonradiolabel [18F] fluoro-6-thia-heptadecanoic acid imaged by DESI showed spatial correlation.[56] In a similar study, the distributions of PET tracers in rat brain tissue sections were imaged by MALDI can also be applied during the development of novel PET radioligands to confirm its distribution.[57] Untargeted MALDI imaging MS was used to provide high spatial resolution ex vivo brain imaging complementing in vivo PET imaging to study long-term recovery after stroke.[58]

X-ray computed tomography or CT is a series of two-dimensional cross-sectional (tomographic) images computationally processed into a three-dimensional image similar to the strategy used in imaging MS to produce a three-dimensional image. A thin beam repeatedly scans the subject at many angles enabling reconstruction of any slice is like workflow used in MRI, PET, and SPECT.[47] Due to technical similarity with MRI, the registration techniques used in MRI can also be integrated with one of in vivo radiological techniques.

Other clinical imaging method, ultrasound imaging, created by propagating acoustic waves that reflects at the interface of different tissues and radiographic images, commonly known as X-ray images, can also be directly or indirectly combined with imaging MS to obtain more relevant biomedical information depending on the application.

Automatic coregistration between an MS image and in vivo three-dimensional imaging is challenging due to the sparsity of MS image data at each pixel, lack of common spatial coordinate between imaging modalities, nonlinear deformation of live tissues, and high dimensionality of data.[59] Automatic coregistration is critical in image-guided surgery, where preoperative pathological images taken before surgery correspond to anatomical features of the intraoperative image taken during surgery physical anatomy.[60] Automatic registration of MS imaging data and MRI has been demonstrated for a patient-derived, xenograft mouse brain model of glioblastoma after administration drug, erlotinib.[59]

Multimodal imaging MS with other imaging techniques

There are many more biomedical imaging modalities that can benefit from multimodal combination with imaging MS. MALDI imaging was coupled with confocal Raman microspectroscopy to characterize and relate the distribution of biomolecules in biofilms and planktonic cells of the bacterium *Pseudomonas aeruginosa*.[61] In another example, confocal Raman microscopy imaging and SIMS imaging, laser desorption–ionization MS imaging was used to study the chemical and structural maps of the plant with the help of manual coregistration by optical fiducial markers.[62]

Atomic force microscopy (AFM) can perform topography and viscoelastic surface analysis by measuring the deflection of a probe tip mounted on a microcantilever. A hybrid atmospheric pressure AFM coupled with MS imaging system based on atmospheric pressure chemical ionization was able to correlate topographic surface images (pixel size = 0.2 μm × 0.8 μm) obtained from AFM with the chemical image (pixel size = 2.5 μm × 2.0 μm) acquired by the mass spectrometer on a living bacterial colony on an agar gel.[63]

Conclusions

Multimodal imaging MS can overcome the limitations of the individual imaging MS technique—providing broader information than just each independent imaging modality. Depending on the application, imaging MS may not be sensitive as a label-based imaging techniques, or not have enough throughput, or limited by spatial resolution, or not consist of an adequate field of view or ability to perform repetitive temporal in situ analysis on a live subject. The complementary visual information provided by multimodal imaging can obtain the relationship between spatial features and biomolecular, making painting a broader picture of biological processes within the tissue. The correlation of spatial distribution of molecular entities (e.g., drugs, metabolites, lipids, and protein) with morphological, anatomical, or functional features enhances the utility of imaging in biomedical research and clinical applications. Research in multimodal imaging MS is still at very early stages. The early proof-of-concept studies in image registration for seamless integration of imaging MS with other imaging modality are encouraging. Innovation in mass spectrometer instrumentation, autoregistration, and workflow compatible with multimodal imaging MS is needed to expand the multimodal imaging MS into routine biomedical and clinical applications.

References

1. Weissleder R, Nahrendorf M. Advancing biomedical imaging. *Proc Natl Acad Sci*. 2015;112:14424–14428.
2. van der Horst G, Buijs JT, van der Pluijm G. Chapter 46 - Pre-clinical molecular imaging of "the seed and the soil" in bone metastasis. In: Heymann D, ed. *Bone Cancer*. 2nd ed San Diego: Academic Press; 2015:557–570.
3. Venugopal V, Fang Q, Intes X. Multimodal diffuse optical imaging for biomedical applications*Biophotonics for Medical Applications*: Elsevier; 2015:3–24.
4. Maintz JA, Viergever MA. A survey of medical image registration. *Med Image Anal*. 1998;2:1–36.
5. Hajnal JV, Hill DL. *Medical Image Registration*. 2001.
6. Zitova B, Flusser J. Image registration methods: a survey. *Image Vis Comput*. 2003;21:977–1000.
7. Brown LG. A survey of image registration techniques. *ACM Comput Surv (CSUR)*. 1992;24:325–376.
8. Farncombe TH. Multimodality imaging with MR/PET and MR/SPECT. *Medical Imaging: Technology AND Applications*. 2017:359–402.
9. Pluim JP, Maintz JA, Viergever MA. Mutual-information-based registration of medical images: a survey. *IEEE Transact Med Imaging*. 2003;22:986–1004.
10. Lützen E, Holtkamp M, Stamme I, et al. Multimodal imaging of hallucinogens 25C- and 25I-NBOMe on blotter papers. *Drug Test Anal*. 2020;12:465–471.
11. Porta Siegel T, Hamm G, Bunch J, Cappell J, Fletcher JS, Schwamborn K. Mass spectrometry imaging and integration with other imaging modalities for greater molecular understanding of biological tissues. *Mol Imaging Biol*. 2018;20:888–901.
12. Ho YN, Shu LJ, Yang YL. Imaging mass spectrometry for metabolites: technical progress, multimodal imaging, and biological interactions. *Wiley Interdiscipl Rev Syst Biol Med*. 2017;9:e1387.

13. Buchberger AR, DeLaney K, Johnson J, Li L. Mass spectrometry imaging: a review of emerging advancements and future insights. *Anal Chem*. 2018;90:240–265.
14. Neumann EK, Djambazova KV, Caprioli RM, Spraggins JM. Multimodal imaging mass spectrometry: next generation molecular mapping in biology and medicine. *J Am Soc Mass Spectrom*. 2020.
15. Dannhorn A, Kazanc E, Ling S, et al. Universal sample preparation unlocking multimodal molecular tissue imaging. *Anal Chem*. 2020;92:11080–11088.
16. Bhattacharjee A, Velickovic D, Wietsma TW, et al. Visualizing microbial community dynamics via a controllable soil environment. *mSystems*. 2020;5(1):e00645 –19.
17. Woolman M, Qiu J, Fischer C, et al. In situ tissue pathology from spatially encoded mass spectrometry classifiers visualized in real time through augmented reality. *Chem Sci*. 2020.
18. Anderson DMG, Ablonczy Z, Koutalos Y, et al. High resolution MALDI imaging mass spectrometry of retinal tissue lipids. *J Am Soc Mass Spectrom*. 2014;25:1394–1403.
19. Van de Plas R, Yang J, Spraggins J, Caprioli RM. Image fusion of mass spectrometry and microscopy: a multimodality paradigm for molecular tissue mapping. *Nat Methods*. 2015;12:366–372.
20. Kaya I, Samfors S, Levin M, Boren J, Fletcher JS. Multimodal MALDI imaging mass spectrometry reveals spatially correlated lipid and protein changes in mouse heart with acute myocardial infarction. *J Am Soc Mass Spectrom*. 2020.
21. Eberlin LS, Liu X, Ferreira CR, Santagata S, Agar NY, Cooks RG. Desorption electrospray ionization then MALDI mass spectrometry imaging of lipid and protein distributions in single tissue sections. *Anal Chem*. 2011;83:8366–8371.
22. Flint LE, Hamm GR, Ready JD, et al. The characterization of an aggregated three-dimensional cell culture model by multimodal mass spectrometry imaging. *Anal Chem*. 2020.
23. Škrášková K, Claude E, Jones EA, Towers M, Ellis SR, Heeren RM. Enhanced capabilities for imaging gangliosides in murine brain with matrix-assisted laser desorption/ionization and desorption electrospray ionization mass spectrometry coupled to ion mobility separation. *Methods*. 2016;104:69–78.
24. Katz L, Woolman M, Talbot F, et al. Dual laser and desorption electrospray ionization mass spectrometry imaging using the same interface. *Anal Chem*. 2020;92:6349–6357.
25. Shrestha B, Nemes P, Nazarian J, Hathout Y, Hoffman EP, Vertes A. Direct analysis of lipids and small metabolites in mouse brain tissue by AP IR-MALDI and reactive LAESI mass spectrometry. *Analyst*. 2010;135:751–758.
26. Chughtai S, Chughtai K, Cillero-Pastor B, et al. A multimodal mass spectrometry imaging approach for the study of musculoskeletal tissues. *Int J Mass Spectrom*. 2012;325:150–160.
27. Van Nuffel S, Quatredeniers M, Pirkl A, et al. Multimodal imaging mass spectrometry to identify markers of pulmonary arterial hypertension in human lung tissue using MALDI-ToF, ToF-SIMS, and Hybrid SIMS. *Anal Chem*. 2020;92:12079–12087.
28. Heijs B, Holst S, Briaire-de Bruijn IH, et al. Multimodal mass spectrometry imaging of N-glycans and proteins from the same tissue section. *Anal Chem*. 2016;88:7745–7753.
29. Prideaux B, ElNaggar MS, Zimmerman M, Wiseman JM, Li X, Dartois V. Mass spectrometry imaging of levofloxacin distribution in TB-infected pulmonary lesions by MALDI-MSI and continuous liquid microjunction surface sampling. *Int J Mass Spectrom*. 2015;377:699–708.
30. Chughtai K, Jiang L, Greenwood TR, Glunde K, Heeren RM. Mass spectrometry images acylcarnitines, phosphatidylcholines, and sphingomyelin in MDA-MB-231 breast tumor models. *J Lipid Res*. 2013;54:333–344.
31. Chughtai K, Jiang L, Greenwood TR, et al. Fiducial markers for combined 3-dimensional mass spectrometric and optical tissue imaging. *Anal Chem*. 2012;84:1817–1823.

32. Schwamborn K, Krieg RC, Reska M, Jakse G, Knuechel R, Wellmann A. Identifying prostate carcinoma by MALDI-Imaging. *Int J Mol Med*. 2007;20:155–159.
33. Taverna D, Pollins AC, Sindona G, Caprioli RM, Nanney LB. Imaging mass spectrometry for assessing cutaneous wound healing: analysis of pressure ulcers. *J Proteome Res*. 2015;14:986–996.
34. Rauser S, Marquardt C, Balluff B, et al. Classification of HER2 receptor status in breast cancer tissues by MALDI imaging mass spectrometry. *J Proteome Res*. 2010;9:1854–1863.
35. Tata A, Woolman M, Ventura M, et al. Rapid detection of necrosis in breast cancer with desorption electrospray ionization mass spectrometry. *Sci Rep*. 2016;6:35374.
36. Chughtai K, Jiang L, Post H, et al. Mass spectrometric imaging of red fluorescent protein in breast tumor xenografts. *J Am Soc Mass Spectrom*. 2013;24:711–717.
37. Si T, Li B, Zhang K, Xu Y, Zhao H, Sweedler JV. Characterization of bacillus subtilis colony biofilms via mass spectrometry and fluorescence imaging. *J Proteome Res*. 2016;15:1955–1962.
38. Blutke A, Sun N, Xu Z, et al. Light sheet fluorescence microscopy guided MALDI-imaging mass spectrometry of cleared tissue samples. *Sci Rep*. 2020;10:14461.
39. Lorenz M, Ovchinnikova OS, Kertesz V, Van Berkel GJ. Laser microdissection and atmospheric pressure chemical ionization mass spectrometry coupled for multimodal imaging. *Rapid Commun Mass Spectrom*. 2013;27:1429–1436.
40. Agar NYR, Yang HW, Carroll RS, Black PM, Agar JN. Matrix solution fixation: histology-compatible tissue preparation for MALDI mass spectrometry imaging. *Anal Chem*. 2007;79:7416–7423.
41. Cornett DS, Mobley JA, Dias EC, et al. A novel histology-directed strategy for MALDI-MS tissue profiling that improves throughput and cellular specificity in human breast cancer. *Mol Cell Proteomics*. 2006;5:1975–1983.
42. Jeon YE, Lee SC, Paik SS, et al. Histology-directed matrix-assisted laser desorption/ionization analysis reveals tissue origin and p53 status of primary liver cancers. *Pathol Int*. 2011;61:449–455.
43. Patterson NH, Tuck M, Lewis A, et al. Next generation histology-directed imaging mass spectrometry driven by autofluorescence microscopy. *Anal Chem*. 2018;90:12404–12413.
44. Maaten L, Hinton G. Visualizing data using t-SNE. *J Mach Learn Res*. 2008;9:2579–2605.
45. Abdelmoula WM, Škrášková K, Balluff B, et al. Automatic generic registration of mass spectrometry imaging data to histology using nonlinear stochastic embedding. *Anal Chem*. 2014;86:9204–9211.
46. Abdelmoula WM, Carreira RJ, Shyti R, et al. Automatic registration of mass spectrometry imaging data sets to the allen brain atlas. *Anal Chem*. 2014;86:3947–3954.
47. Suetens P. *Fundamentals of Medical Imaging*. 2017;1–218.
48. Tata A, Zheng J, Ginsberg HJ, Jaffray DA, Ifa DR, Zarrine-Afsar A. Contrast agent mass spectrometry imaging reveals tumor heterogeneity. *Anal Chem*. 2015;87:7683–7689.
49. Aichler M, Huber K, Schilling F, et al. Spatially resolved quantification of gadolinium(III)-based magnetic resonance agents in tissue by MALDI imaging mass spectrometry after in vivo MRI. *Angew Chem Int Ed Engl*. 2015;54:4279–4283.
50. Santagata S, Eberlin LS, Norton I, et al. Intraoperative mass spectrometry mapping of an onco-metabolite to guide brain tumor surgery. *Proc Natl Acad Sci*. 2014;111:11121–11126.
51. Thiele H, Heldmann S, Trede D, et al. 2D and 3D MALDI-imaging: conceptual strategies for visualization and data mining. *Biochim Biophys Acta Proteins Proteomics*. 2014;1844:117–137 (1, Part A).
52. Sinha TK, Khatib-Shahidi S, Yankeelov TE, et al. Integrating spatially resolved three-dimensional MALDI IMS with in vivo magnetic resonance imaging. *Nat Methods*. 2008;5:57–59.

53. Attia AS, Schroeder KA, Seeley EH, et al. Monitoring the inflammatory response to infection through the integration of MALDI IMS and MRI. *Cell Host Microbe*. 2012;11:664–673.
54. Srimany A, George C, Naik HR, Pinto DG, Chandrakumar N, Pradeep T. Developmental patterning and segregation of alkaloids in areca nut (seed of Areca catechu) revealed by magnetic resonance and mass spectrometry imaging. *Phytochemistry*. 2016;125:35–42.
55. Fleckenstein P, Tranum-Jensen J. *Anatomy in Diagnostic Imaging*: John Wiley & Sons; 2014.
56. Jones E, Henderson F, Midey A, et al. Comparison of lipid imaging in a zebrafish melanoma model by positron emission tomography (PET) and desorption electrospray ionization-mass spectrometry (DESI-MS). *Drug Metab Pharmacokinet*. 2017;32:S42–S43 1, Supplement.
57. Goodwin RJA, Mackay CL, Nilsson A, et al. Qualitative and quantitative MALDI imaging of the positron emission tomography ligands raclopride (a D2 dopamine antagonist) and SCH 23390 (a D1 dopamine antagonist) in rat brain tissue sections using a solvent-free dry matrix application method. *Anal Chem*. 2011;83:9694–9701.
58. Henderson F, Hart PJ, Pradillo JM, et al. Multimodal imaging of long-term recovery post-stroke by positron emission tomography and matrix-assisted laser desorption/ionisation mass spectrometry. *Rapid Commun Mass Spectrom*. 2018;32:721–729.
59. Abdelmoula WM, Regan MS, Lopez BGC, et al. Automatic 3D nonlinear registration of mass spectrometry imaging and magnetic resonance imaging data. *Anal Chem*. 2019;91:6206–6216.
60. Alam F, Rahman SU, Ullah S, Gulati K. Medical image registration in image guided surgery: Issues, challenges and research opportunities. *Biocybern Biomed Eng*. 2018;38:71–89.
61. Masyuko RN, Lanni EJ, Driscoll CM, Shrout JD, Sweedler JV, Bohn PW. Spatial organization of *Pseudomonas aeruginosa* biofilms probed by combined matrix-assisted laser desorption ionization mass spectrometry and confocal Raman microscopy. *Analyst*. 2014;139:5700–5708.
62. Li Z, Chu L-Q, Sweedler JV, Bohn PW. Spatial correlation of confocal Raman scattering and secondary ion mass spectrometric molecular images of lignocellulosic materials. *Anal Chem*. 2010;82:2608–2611.
63. Ovchinnikova OS, Kjoller K, Hurst GB, Pelletier DA, Van Berkel GJ. Atomic force microscope controlled topographical imaging and proximal probe thermal desorption/ionization mass spectrometry imaging. *Anal Chem*. 2014;86:1083–1090.

Chapter 12

Validation and standardization of imaging mass spectrometry

Chapter Outline

Imaging mass spectrometry (MS) is a budding analytical field with a constant influx of novel research tools, software, and workflows. Due to innovative and fast-paced evolution of workflow, it is often perceived as a complex analytical workflow. On the other hand, the adoption of routine imaging MS workflows in pharmaceutical research and clinical realm with more regulations necessitates analytical validation.

Validation is performed to ensure the analytical measurements using the outlined workflow closely represent the true value of the analyte on the sample and fit for the intended purpose.[1] Method validation is indispensable for obtaining reliable results. Validation is a set of procedures for showing the reliability of the analytical methods. The intricacy of validation depends on the complexity of the technique, the steps involved in the workflow, the degree of reliability needed, the regulatory scrutiny, etc.

Disparities in instrumentation, software, and workflows can introduce inconsistencies in imaging MS analysis and can slow down its adoption in translational and clinical research. Validation shows if the analytical methods are performing as expected under intended use and accept confidence. Imaging MS validation should be fit-for-purpose. Before beginning validation, it is essential to establish acceptability criteria for the analytical validation considering the institutional or regulatory requirements, industry guidance, biological variation, required scientific information, etc.

Introduction to Spatial Mapping of Biomolecules by Imaging Mass Spectrometry. DOI: 10.1016/C2018-0-03962-6

Being an evolving and still a niche analytical technique, imaging MS does not have extensive validation guidelines like other established routine techniques, such as liquid chromatography (LC)–MS—e.g., C62–LC–MS methods guideline from Clinical and Laboratory Standard Institute.[2] LC–MS validation is described in many tutorial reviews such as one referenced.[3-4]

Analytical validation of imaging MS involves the determination of the reliable feasibility of analyte measurement, as well as accuracy and precision of amplitude of the measurement. If an analyte cannot be reliably and somewhat quantitatively measured from the tissue-of-interest, it has little or no value in biomedical applications. The validation process for imaging MS may include determining limits of detection, limits of quantification, and reference typical values of the molecule-of-interest in tissue-of-interest. Precision involves both repeatability and reproducibility. Repeatability is defined as the agreement between successive measurements made under the same conditions over a short period, and reproducibility is the agreement between measurements made at different times, locations, and/or multiple operators.

Imaging MS, in particular quantitative imaging of pharmaceuticals drugs, will need a set of validation method, such as ones used for validation of high-performance liquid chromatography methods as prescribed by the US Food and Drug Administration (FDA), United States Pharmacopeia, and the International Council for Harmonisation of Technical Requirements for Pharmaceuticals for Human Use.[5] A general bioanalytical method validation guideline for new drug applications is already provided by the US FDA.[6] The guiding principle remains defining and validating the bioanalytical method, operating conditions, optimization steps, limitations, and suitability. Method development and assessing suitability for validation include the optimization of reference standards, critical reagents, calibration curve, quality control samples, selectivity and specificity, sensitivity, accuracy, precision, and stability in the matrix.

Validation of imaging MS

The analytical workflow starts from the definition of requirement leading to the development of the method and its validation encompassing basic research, translational science, and clinical practice. All analytical methods have errors, random or systematic. The goal of validation is to define the nature and magnitude of these errors to determine if the imaging MS methods meet the requirements for its intended purpose. Validation includes documenting specific variables over some time by multiple operators and evaluation of how the method performs in real-life samples, preferably in the clinical lab. A flowchart of an analytical assay from method validation to clinical practice is shown in Fig. 12.1. Within here, some conceptual steps of validation are discussed for academic discourse. The specific terminology used, not used, or differences may depend on field, industry, geographic location, or regulatory definition. Any discussion here should not be taken as guidance or a prescription for clinical assays.

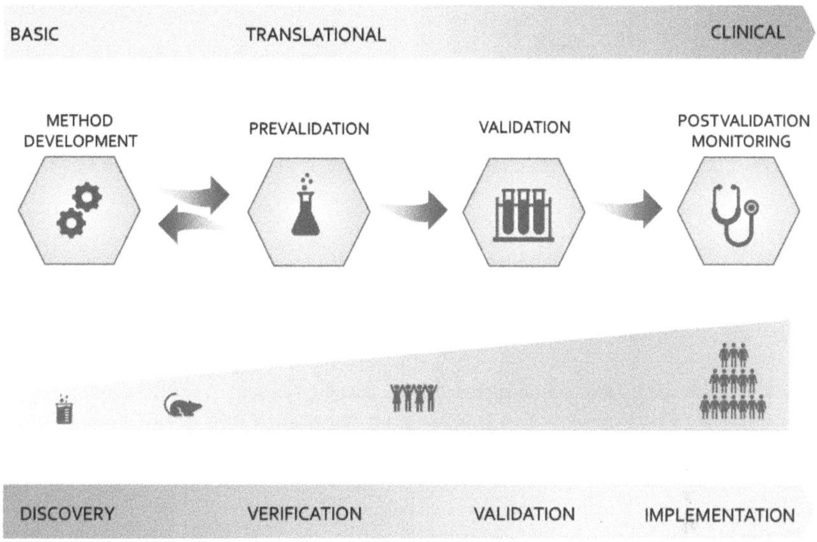

FIG. 12.1 A conceptual flowchart of an analytical assay from the method validation to clinical implementation is shown.

Before extensive validation, often, the prevalidation step is done, which is sometimes referred to as method qualification or screening. Method qualification assesses the suitability of the method for its intended purpose. The prevalidation method qualification is done to evaluate the quality of the analytical method and to check if it is ready for validation. The prevalidation studies are smaller scale than the field validation studies and carried out to determine the general accuracy and precision of the method and to evaluate if any modification is needed to the method. Like validation studies, the parameters examined may include specificity, sensitivity, matrix effects, linearity, robustness, etc. Prevalidation often overlaps with the method development.

A proposed holistic approach to validation can be divided into four sections, as shown in Fig. 12.2—(a) applicability and acceptability of the assay; (b) specificity; (c) sensitivity with calibration profile, dynamic range, the limit of detection; and (d) accuracy. During the laboratory work, the following items may be evaluated for validation. The exact definitions and usage of the terms, as well as experimental parameters to determine them or express them, differ between literature sources or specific guidelines. Here are some conceptual frameworks of those concepts.

Specificity

Specificity is the degree of selectivity that a method can accurately measure an analyte in the presence of other interferences. In LC–MS, an acceptable criterion for interference signal is considered less than 15%–20% of limit

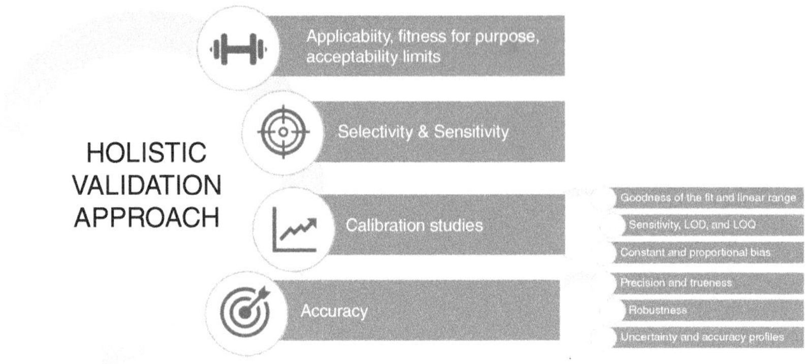

FIG. 12.2 Scheme for the holistic approach to the analytic validation. Adapted with permission from González AG, Herrador MÁ. A practical guide to analytical method validation, including measurement uncertainty and accuracy profiles. *Trends Anal Chem.* 2007;26:227–238.

of quantitation of the ion-of-interest. Identity confirmation is also needed to demonstrate that analyte is not falsely identified. There are no guidelines for imaging MS identity confirmation. For comparison, the Association of Official Analytical Chemist guideline for LC–MS calls for identification of three or four characteristic fragments or the use of multiple MS analysis (MS_n).[7]

Robustness

Robustness or ruggedness can be defined as the competence of the analytical technique to stay unaffected or remain within the normal performance range due to minor variations in operating parameters. The variations may be influenced due to sample preparation, such as handling, tissue thawing time, a slight change in solvents, etc., or environmental factors such as humidity, room temperature, etc. Robustness can be conveyed as a relative standard deviation (RSD) percentage, variance, or the coefficient of variation (CV) obtained with the deliberately introduced minor variation to the method with respect to a reference value obtained before.[3] For LC–MS or high-performance liquid chromatography analysis, robustness is not universally included in validation guidelines.

Sensitivity

The sensitivity is often defined as limit of quantitation (LOQ), which is the lowest concentration that can be reproducibility measured between the runs with less than 20% CV. Matrix effect is the influence of other components in the sample that affects ionization efficiency of the ion-of-interest and, therefore, skews measurement by ion suppression or ion enhancement. If possible, the magnitude of ion suppression or ion enhancement should be assessed.

Linearity and dynamic range

Linearity is the ability of the MS method to provide the ion intensity proportional to the concentration of the analyte. Linearity for matrix-specific, i.e., depends on matrix and hard to achieve in the heterogeneous sample. The linearity of ion-of-interest should be measured at the sample matrix. Dynamic range is a difference in intensity between the smallest and largest signal. Linear dynamic range can be thought of as the range of that linearity. The linearity of analytical measurement is critical for quantitative imaging MS.

Stability

Stability measurement due to sample preparation steps is tricky to accomplish due to the variation in storage conditions and the time required to undertake such studies. Analyte stability during sample preparation and imaging experiments run should be measured for a prescribed workflow. Any point of the deviation should be appropriately noted. For example, if thawing the tissue more than a specific time changes the stability of the analyte, it should be noted and included in the standard operating procedure. An acceptable criterion for stability in other mass spectrometric assays is less than 15% compared against the freshly obtained sample. For LC–MS validation, analyte stability may not be specifically because accuracy and imprecision may account for a change instability.

Accuracy and imprecision

Accuracy and imprecision are related to a degree of agreement between the replicate measurements and the true value. Imprecision is the evaluation of agreement between repeated analysis of the same sample and can be measured within-run or between-run. Generally, an imprecision of 15% or less between the control material with a known concentration and its measurement is acceptable. In LC–MS, "within-run imprecision" is performed on subsequent analysis, and "between-run imprecision" is performed over several days apart. Imprecision may also be expressed as the RSD, variance, or CV, like robustness. The main difference in robustness is measured by deliberately introducing changes in the method, while measurement of imprecision is intended to measure unintentional changes during the regular operation.

Postvalidation

Monitoring of the analytical method is required after the validation and may include checking instrument performance using system suitability tests, analysis of quality control material, sample analysis performance overtime. Postvalidation may be required after major instrument maintenance.

Standardization of imaging MS

Standardization is a communal consensus-driven process openly created by researchers in the field. Although the adoption and compliance of standardization are voluntary, but after standardization it is well adopted, it can affect the field and drive the regulatory policy. Standardization brings compatibility and interoperability, benefiting users, streamline manufacturing, and government regulation. Standardization benefits the research field by supporting the dissemination of the results and broader adoption.[8]

As imaging MS is moving toward wider biomedical and clinical usage, standardization will be required to obtain comparative results. Variation in sample preparation, instrumentation, data processing, and reporting can result in disparate results that cannot be justifiably compared. Broader implementation of biomedical research requires an accurate comparison of various methods and tools that can be eventually developed into standardized protocols and guidelines for more extensive and accessible implementation of imaging MS in clinical and pharmaceutical assays.[9]

Standardization

Imaging MS is a dynamic field with frequent evolution in workflows on various instrument configurations and data processing architecture. For instance, literature has multiple workflows for analyzing peptides in formalin-fixed paraffin-embedded tissue with variation in dewaxing, antigen retrieval, and digestion. In addition to the availability of diverse workflows, fundamentally different instrument configurations and a variety of data processing may conflate the harmonization and standardization. Standardization of any workflows may require a comparison of many objective criteria, such as imprecision, sensitivity, robustness during their development and assessment for clinical usage. Reproducibility of imaging results can depend on multiple parameters but can also be affected by one key factor. For example, reproducibility of on-tissue trypsin-digested peptides in formalin-fixed paraffin-embedded samples can be mainly achieved by standardizing trypsin application and digestion workflows with similar instrumentation.[10] Statistical scores can provide objective metrics, such as one used for standardization of on-tissue digestion of tissue independent of any user bias.[11]

Standardized data format—imzML

A common standardized data format can facilitate an exchange of information and a comparison of data produced by dissimilar workflows and instruments.[12] The imzML is a standardized eXtensible Markup Language data format based on mzML from the HUPO Proteomics Standards Initiative.[13-14] Exchanging and processing of imaging MS data from several instruments in a variety of software platforms is only possible due to the standardized and open data format like imzML. Additional background information and detail technical description can be found at www.imzml.org URL.

Reporting standards for imaging MS

One of the critical aspects of reproducible imaging data is a sufficient and accurate reporting of experimental and processing details. There are several proposals for reporting imaging MS data referenced here.[15-16] The experiments in imaging MS often have different goals and diverse applications. Thus, defining a monolithic criterion of reporting for all imaging MS data will be a difficult task. The initial guidelines provided by the referenced publication intend to describe a general template applicable for different applications.[15]

Minimum information about a mass spectrometry imaging experiment or MSICheck is proposed to increase uptake of reporting less exhaustive but critical details of imaging MS experiments.[16] Minimum information about a mass spectrometry imaging experiment can be divided into five categories—general, sample preparation, imaging MS data, processing, and others. The general category has experimental details such as aim, hypothesis, sample type, and can be used as a conduit between Imaging MS and other scientific fields, while the other category has more technical information specific to a particular experiment such as MS/MS identification, quantitation method. A robust reporting of an experimental result is often the first step for a reproducible experiment.

Reporting guidelines for imagining MS for biomedical applications can consist of details on (a) tissue sample—origins, physiology, sampling, storage; (b) sample preparation—mounting, washing, on-tissue digestion or derivatization, matrix application; (c) data acquisition—mass analyzer parameters, ion source, pixel size; (d) data processing—software, normalization, scale, visualization parameters; (e) compound identification workflow; (f) segmentation, clustering or other data analysis, (g) other analytical information—microscope image, clinical information, LC–MS data. The suggested reporting categories are also tabulated in Table 12.1. The minimal reporting details may depend on the type and aim of the imaging MS experiment. For example, accurate laser fluence may be a critical parameter in a fundamental matrix-assisted laser desorption/

TABLE 12.1 Imaging MS experiment reporting categories.

Tissue sample	Origins, physiology, sampling, storage
Sample preparation	Mounting, washing, on-tissue digestion or derivatization, matrix application
Data acquisition	Mass analyzer parameters, ion source, pixel size
Data processing	Software, normalization, scale, visualization parameters
Compound identification workflow	Mass accuracy, collision cross section (CCS), MS/MS
Segmentation, clustering, or other data analysis	
Other analytical information	Microscope image, clinical assay, LC–MS

ionization (MALDI) experiment, but not in other biomedical studies. However, good practice should be reporting beyond the minimal reporting standard.

Efforts to develop a standardization with predefined sample preparation protocols, similar acquisition tools, and data processing parameters are needed to assess reproducibility in multicenter studies.

Multicenter validation of imaging MS

The reproducibility of an imaging MS assay can be demonstrated by multicenter or multisite imaging experiments, where multiple laboratories analyze identical samples when adhering to standardized sample preparation, acquisition parameters, and data analysis workflow. Standardization is generally done by analyzing serial-sections from the same tissue stored as bulk or sections in −80 °C freezer. However, long-term storage can result in repeated freeze-thaw cycle or dehydration and oxidation due to freezer burn. The variation in measurement can be improved by addressing the repeatability, which is the deviation during successive analysis using the same tool, and the reproducibility, which is the variation when the analysis was done, preferably at a different laboratory setting. The variation in reproducible results can be attributed to the sample preparation, environmental factors, instrumentation used, or data analysis.

A few publications, listed in Table 12.2, have examined the multicenter cross-validation imaging experiment done at several laboratories across geography. The table intends to give a glimpse of the breadth and type of multicenter studies as well as the type of sample used. Detail discussions of each of the multicenter studies are in the respective publications of those studies. A detailed comparison of some of those studies is in the referenced review.[17] A first reported multicenter comparison of MALDI imaging was focused on breast cancer in two centers using independent tissue banks, infrastructure, methods, and practitioners.[18] As shown in Table 12.2, predominantly multicenter studies have utilized the most common ion source, MALDI, except for three, which used secondary ion mass spectrometry and desorption electrospray ionization. Besides one, all of the multicenter studies have been focused on the qualitative assay. The quantitative imaging MS studies across two sites and using three analysts was performed using a set protocol for sample preparation, acquisition, and data analysis—achieving low teens RSDs and about 80% accurate when compared to LC–MS/MS quantification.[19]

Conclusions

Despite distinct technical advantages for the analysis of molecular distribution on tissue, the inconsistent results are restricting imaging MS technologies broader usage in translational and clinical research. Imaging MS has massive potential in clinical application to satisfy an unmet diagnostic need in the area

TABLE 12.2 Summary of multicenter studies in imaging MS. Based on data in the referenced publication.[17]

Site number	Sample size	Ion source	Analyte class	Tissue type	Tissue condition	Validation/QC	Year[Ref]
2	30	MALDI	Proteins	Human breast cancer	Fresh frozen	Immunohistochemistry	2014[18]
2	16	DESI	Lipids	Human esophageal cancer	Fresh frozen	Principal component analysis and hierarchical cluster analysis	2014[20]
5		MALDI SIMS	Lipids proteins	Mouse brain	Fresh frozen	Comparison to brain atlas[21]	2015[22]
1 (22)	102	MALDI	Peptides	Human lesion	FFPE	Fisher's exact test	2016[23]
4	20(×8)	MALDI	Peptides metabolites	Multiorgan (8) mouse tissue microarray	FFPE	Univariate/multivariate classified validation set	2018[24]
2	103	DESI	Lipids	Human breast cancer	Fresh frozen	Intralaboratory validation set	2018[25]
5	2×10, 3, 3×6	MALDI	Peptides	Mouse intestine, human ovarian teratoma, six tissue microarray	FFPE	Spatial segmentation, PCA, number of the peak above S/N 3	2018[10]
2	6	MALDI	Drugs	Rat liver	Fresh frozen	LC–MS quantitation	2019[19]

where a commercial diagnostic test has not been developed, or the test is not accurate or adequate. To my knowledge, none of the imaging MS assays are approved as in vitro diagnostic product or IVD by the US FDA. Standardization and harmonization with authoritative guidance documents will be needed to increase the utilization of imaging MS in the pharmaceutical and clinical laboratory. A realistic implementation of imaging MS in a clinical assay is impossible without a fit-for-purpose standardization and validation—perhaps one tailored to a specific clinical application.

References

1. González AG, Herrador MÁ. A practical guide to analytical method validation, including measurement uncertainty and accuracy profiles. *Trends Anal Chem*. 2007;26:227–238.
2. (CLSI), C. L. S. I., C62-liquid chromatography-mass spectrometry methods; approved guideline, CLSI document C62-A. Clinical Laboratory Standards Institute Wayne (PA): 2014.
3. Kruve A, Rebane R, Kipper K, et al. Tutorial review on validation of liquid chromatography–mass spectrometry methods: Part I. *Anal Chim Acta*. 2015;870:29–44.
4. Kruve A, Rebane R, Kipper K, et al. Tutorial review on validation of liquid chromatography–mass spectrometry methods: Part II. *Anal chim Acta*. 2015;870:8–28.
5. Shabir GA. Validation of high-performance liquid chromatography methods for pharmaceutical analysis: understanding the differences and similarities between validation requirements of the US Food and Drug Administration, the US Pharmacopeia and the International Conference on Harmonization. *J Chromatogr A*. 2003;987:57–66.
6. Bioanalytical method validation guidance for industry. Food and Drug Administration (FDA). 2018.
7. AOAC guidelines for single laboratory validation of chemical methods for dietary supplements and botanicals. Gaithersburg, MD, USA: AOAC International; 2002:12–19.
8. Consortium C Standardization guidelines for IST research projects interfacing with ICT standards organizations. Available at: https://www.w3.org/2004/copras/docu/D27.pdf, Accessed 23 January 2021.
9. McDonnell LA, Heeren RMA, Andrén PE, Stoeckli M, Corthals GL. Going forward: Increasing the accessibility of imaging mass spectrometry. *J Proteomics*. 2012;75:5113–5121.
10. Ly A, Longuespée R, Casadonte R, et al. Site-to-site reproducibility and spatial resolution in MALDI–MSI of peptides from formalin-fixed paraffin-embedded samples. *Proteomics Clin Appl*. 2019;13:1800029.
11. Erich K, Sammour DA, Marx A, Hopf C. Scores for standardization of on-tissue digestion of formalin-fixed paraffin-embedded tissue in MALDI-MS imaging. *Biochim Biophys Acta Proteins Proteomics*. 2017;1865:907–915.
12. Orchard S. Data standardization and sharing—the work of the HUPO-PSI. *Biochim Biophys Acta Proteins Proteomics*. 2014;1844:82–87 Part A.
13. Römpp A, Schramm T, Hester A, et al. imzML: imaging mass spectrometry markup language: a common data format for mass spectrometry imaging*Data Mining in Proteomics*: Springer; 2011:205–224.
14. Schramm T, Hester Z, Klinkert I, et al. imzML—a common data format for the flexible exchange and processing of mass spectrometry imaging data. *J Proteomics*. 2012;75:5106–5110.

15. McDonnell LA, Römpp A, Balluff B, et al. Discussion point: reporting guidelines for mass spectrometry imaging. *Anal Bioanal Chem.* 2015;407:2035–2045.
16. Gustafsson OJR, Winderbaum LJ, Condina MR, et al. Balancing sufficiency and impact in reporting standards for mass spectrometry imaging experiments. *GigaScience.* 2018;7(10):giy102. doi:10.1093/gigascience/giy102.
17. Ucal Y, Coskun A, Ozpinar A. Quality will determine the future of mass spectrometry imaging in clinical laboratories: the need for standardization. *Expert Rev Proteomics.* 2019;16:521–532.
18. Dekker TJ, Balluff BD, Jones EA, et al. Multicenter matrix-assisted laser desorption/ionization mass spectrometry imaging (MALDI MSI) identifies proteomic differences in breast-cancer-associated stroma. *J Proteome Res.* 2014;13:4730–4738.
19. Barry JA, Ait-Belkacem R, Hardesty WM, et al. Multicenter Validation Study of Quantitative Imaging Mass Spectrometry. *Anal Chem.* 2019;91:6266–6274.
20. Abbassi-Ghadi N, Jones EA, Veselkov KA, et al. Repeatability and reproducibility of desorption electrospray ionization-mass spectrometry (DESI-MS) for the imaging analysis of human cancer tissue: a gateway for clinical applications. *Anal Methods.* 2015;7:71–80.
21. Richard L. Sidman BK, Misra B Stephen Senft High Resolution Mouse Brain Atlas. Available at: http://www.hms.harvard.edu/research/brain/intro.html, Accessed 23 January 2021.
22. Römpp A, Both JP, Brunelle A, et al. Mass spectrometry imaging of biological tissue: an approach for multicenter studies. *Anal Bioanal Chem.* 2015;407:2329–2335.
23. Lazova R, Seeley EH, Kutzner H, et al. Imaging mass spectrometry assists in the classification of diagnostically challenging atypical Spitzoid neoplasms. *J Am Acad Dermatol.* 2016;75:1176–1186.
24. Buck A, Heijs B, Beine B, et al. Round robin study of formalin-fixed paraffin-embedded tissues in mass spectrometry imaging. *Anal Bioanal Chem.* 2018;410:5969–5980.
25. Porcari AM, Zhang J, Garza KY, et al. Multicenter study using desorption-electrospray-ionization-mass-spectrometry imaging for breast-cancer diagnosis. *Anal Chem.* 2018;90:11324–11332.

Chapter 13

Toward clinical imaging mass spectrometry

Chapter Outline

Since its inception, the research and development in imaging mass spectrometry (MS) have been driven by aspiration for clinical applications. Imaging MS can map simultaneously hundreds of metabolites, lipids, drugs, and proteins on diverse types of tissues with minor variations in sample preparation. Imaging MS can detect molecular distributions in tissues that can identify earlier changes that have not manifested in morphological changes visible under a microscope.

In clinical laboratories, matrix-assisted laser desorption/ionization (MALDI) mass spectrometers are already used for obtaining improved taxonomical classification of bacteria, yeasts, and fungi.[1] The rapid, high-throughput, low-cost, and user-friendly microbial diagnosis obtained by MALDI is used for improved treatment regiments such as prescription of more suitable antibiotics.[2]

The selection of the imaging modality depends on the scientific or clinical needs, as shown in Fig. 13.1.[3] For instance, faster imaging systems that provide macroscopic anatomical visualization are prevalent in clinical and preclinical space, while those that give detailed microscopic features are more common in basic research.

Biomedical visualization in clinic

In the clinic, biomedical visualization is generally divided into two fields—radiologists producing medical images from live subjects and pathologists who work on excised tissue samples using histologic techniques. For sample

Introduction to Spatial Mapping of Biomolecules by Imaging Mass Spectrometry. DOI: 10.1016/C2018-0-03962-6

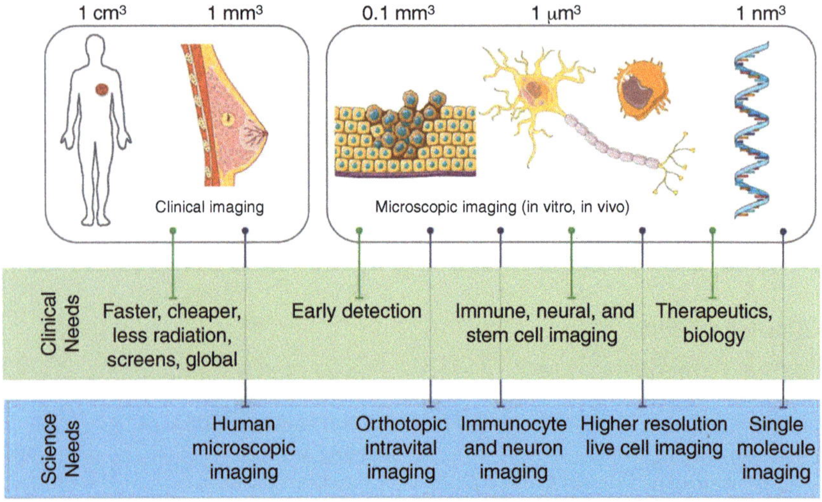

FIG. 13.1 Overview of clinical and basic science imaging needs. Reprinted with permission from Weissleder R, Nahrendorf M. Advancing biomedical imaging. *Proc Natl Acad Sci*. 2015;112:14424–14428.[3]

workflow, imaging MS is very similar to histopathology, while the instrumentation and computer-aided data analysis aspects are more akin to radiology. In clinical practice, pathology provides a definite diagnostic of the disease. In contrast, radiology provides impressions and recommendations.[4] Regardless of techniques used, the clinical imaging aims to provide clinicians with a source of information to advance the diagnosis of disease progression and efficacy of treatment in the body or organ. There are many types of biomedical imaging used in clinics, some of which are discussed briefly below.

Histopathology

Histopathology is the study of cells and tissues with disease using a microscope. Pathologists assess the morphology characteristics of cells and tissues to evaluate pathological changes providing diagnosis and classification based on the histological patterns and cell counting.[5] In histopathology workflow, tissues are collected from patients, sent to a pathology lab where it is processed, and sectioned into glass slides for analysis under a light microscope by a pathologist. A simple hematoxylin and eosin staining of sections may be often adequate and suitable for morphological diagnosis of pathological tissue lesions from biopsy and surgical specimens. Immunohistochemistry (IHC) staining can be a secondary tool in those cases where a simple morphological examination of tissue is inconclusive. IHC staining is a histological assay that integrates immunological and biochemical changes by using antigen and antibody reaction.[6] Immunohistochemical examination of tissue can detect tissue antigens such as amino acids, proteins and provides spatial discrimination among cell populations providing the diagnosis of tumors.[7–9] Imaging MS can provide a similar type of additional information to the pathologist like IHC stained tissues.[10]

Digital pathology

Optical microscopy is still considered a gold standard in pathology and widely used. The hospital often archives the glass slide or tissue block that can be examined again in the future.[11] Digital pathology, where pathological images are captured and transmitted digitally, is still not widely implemented due to a substantial cost of establishing digital infrastructure, purchasing equipment, performing validations, and changing workflows. Digital pathology holds the potential of supporting pathologists in routine tasks and accelerating scientific evaluation and exchange, thus potentially improving the diagnostic accuracy.[12]

Computer-generated tumor grading cannot and should not replace a trained pathologist. Currently, artificial intelligence cannot state why and how a specific diagnosis is made and is incapable of human judgment calls or recommendations.[13] However, computer algorithms can aid pathologists in screening more samples with efficient and accurate outcomes.[14] Computer-assisted diagnosis has been implemented in clinical practices for various imaging techniques, such as radiography, magnetic resonance imaging (MRI), mammography, and computed tomography (CT). Computational development and lessons learned during the implementation of techniques such as artificial intelligence in clinical, histopathological will be invaluable for future implementation of such methods in imaging MS workflow.

In vivo imaging

Radiography or X-rays were in clinical practice a few months after the demonstration of bone structures of hand by Wilhelm Röntgen at the end of the 19th century. Indisputably, imaging MS will take longer than X-rays to get into clinics from basic research.[15] Imaging MS is an ex vivo technique and also does not require radiation such as X-ray or radioactive sample preparation. The X-ray can damage cells and tissues, even at minimal doses. Thus, the International Commission on Radiological Protection recommends keeping the magnitude of doses "as low as reasonably achievable" by using necessary protective measures—time, distance, and shielding.[16] X-ray CT or CT uses X-ray, introduced in the late 1960s, produces a computationally processed three-dimensional image from two-dimensional images similar to the strategy used in most three-dimensional imaging MS. X-ray radiation doses are relatively higher in CT in the order of one to two orders of magnitude higher than a radiographic image. MRI is introduced in late 197os uses paramagnetic, superparamagnetic, and ferromagnetic material as contrast agents is generally safe than X-ray-based CT or positron emission tomography scans.[15] Safety precautions should be taken to avoid ferromagnetic objects with iron during the examination, but nonferromagnetic metals such as aluminum are considered safe. Ultrasound imaging is nondestructive, minimally invasive, comparatively affordable, portable, and has an excellent temporal resolution first introduced in the mid-20th century.

Labeling in clinical imaging

Nuclear medicine imaging utilizes radioactive isotopes as tracers for clinical imaging. Depending on radiotracer and detector used, the techniques can be called single-photon emission computed tomography detecting gamma (γ) rays and positron emission tomography detecting photon pairs from positron emission.[17] Autoradiography is a classic in vitro histochemical counterpart used for imaging tissue sections. In autoradiography, specific binding of a radioligand to its biological target is visualized.[18] Depending on type and concentration, tracer molecules are not specific and can accumulate in other organs, remain in the body, or exert unintentional contamination. Radiolabeling techniques, such as whole-body autoradiography, used radiolabeled drugs to enhance the signal and improve the quantification of molecular imaging.[19] However, radiolabel autoradiography imaging can paint a misleading map since it does not distinguish between a drug and its potential metabolites.[20] A nonradiolabeling approach can also be used in imaging MS to enhance the detection of hard to detect molecules. For instance, imaging mass cytometry, and multiplexed ion beam imaging uses metal-tagged antibodies to simultaneous image dozens of protein markers.[21–22]

Clinical application of imaging MS

In basic clinical research, imaging MS is often used to understand the tissue localization of differentially expressed endogenous disease biomarkers such as metabolites, lipids, proteins or to interrogate the distribution of exogenous treatment agents such as drugs. Perspectives and applications of imaging MS in clinics have been discussed in some other publications.[10,23–27] Some of the potential applications of imaging MS in the clinic are discussed below.

Pathology

The sample collection and preparation workflow of imaging MS is very similar to the pathological workflow in the clinic. Molecular images of tissue biopsies generated by imaging MS can be a helpful tool in pathology like other additional assays such as IHC. Many imaging MS modalities, specifically MALDI and desorption electrospray ionization (DESI), have been employed on finding molecular markers of disease in cancer research.[27] MALDI has been used to examine HER2 status directly from breast cancer tissues status,[28] identify protein patterns associated with histological grades of human glial neoplasms in the brain,[29] differentiate benign Spitz nevus mole from spitzoid malignant melanoma cancer,[30] distinguish pathologic stage 1a from more advanced stage 1b or higher gastric cancer tumors,[31] endobronchial lesions that predictive of lung cancer,[32] distinguish between classical Hodgkin lymphoma and lymphadenitis,[33] discriminate hepatocellular carcinomas with peritumoral cirrhosis,[34] delineate tumor margins in renal carcinoma,[35] identify preneoplastic pancreatic lesions and early pancreatic ductal adenocarcinoma,[36] stratify cervical Pap smears, classify high-grade osteosarcoma, leiomyosarcoma, myxofibrosarcoma from undifferentiated pleomorphic sarcoma,[37] identify ovarian cancer biomarker,[38] and more. An example of MALDI imaging to distinguish HER2 status in breast cancer tissue section and its validation by IHC assay is shown in Fig. 13.2. Oncology research studies have shown the advantages of using small metabolites and lipids mapped by DESI imaging MS such as for delineating breast cancer margins,[39] diagnosing of prostate cancer,[40] molecular subtypes classification of breast cancers,[41] phospholipidomes of esophageal adenocarcinoma,[42] and many others. DESI imaging MS has been used to examine metabolic and lipid signature from normal kidney, benign renal oncocytoma, and renal cell carcinomas. DESI imaging MS can discriminate between noncancerous renal oncocytoma and renal cell carcinomas that often have similar morphological

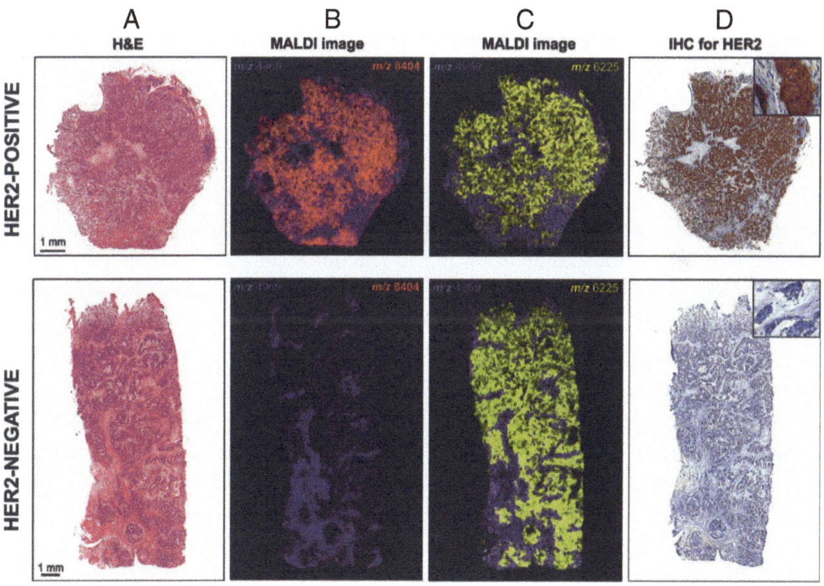

FIG. 13.2 (A) Optical H&E microscopic images of a HER2-positive and HER2-negative, (B) *m/z* 8404 ion image (shown in red) showing highly upregulated mass spectral feature in HER2-positive tissue, (C) *m/z* 6225 (shown in yellow) ion images associated with both HER2-positive and HER2-negative cancer cells and *m/z* 4969 (shown in blue) ion image of tumor stroma, (D) immunohistochemistry for HER2 shows strong immunoreactivity (score 3+) of the positive case. Adapted with permission from Rauser S, Marquardt C, Balluff B, et al. Classification of HER2 receptor status in breast cancer tissues by MALDI imaging mass spectrometry. *J Proteome Res*. 2010;9:1854–1863. Copyright 2010 American Chemical Society.[28]

features and hard to differentiate using just histology.[43] In another example, DESI imaging was successful at differentiating morphologically similar malignant and benign thyroid nodules obtained during preoperative fine-needle aspiration biopsy.[44]

Intraoperative histopathology

Several research teams have demonstrated clinical workflow and power of imaging MS for intraoperative histopathology during surgery.[45–46] Ambient MS methods that do not require any sample preparation, such as DESI, have been utilized to map oncometabolites to guide and assess tumor margin during glioma brain surgery.[47–48] Other slower methods, such as MALDI with time-consuming steps (e.g., matrix application, data acquisition, and data processing), have been adapted to fit intraoperative tissue imaging. For example, hour-long conventional MALDI imaging MS was reduced to a 5-minute long workflow by using precoated slides and rapid high-dimensional data visualization tools, expanding its clinical utility for intraoperative diagnostics.[49]

Tissue microarrays

Tissue microarray (TMA) provides an avenue for high-throughput evaluation of a large number of tissue biopsies. Tumor or benign tissue control can be harvested using a needle biopsy gun (e.g., 18-gauge). Multiple cores, typically a few millimeters in diameter, are excised from tissue, embedded as a grid, and mounted as an array in a glass slide. MALDI imaging has been demonstrated for analysis of formalin-fixed paraffin-embedded (FFPE) TMA consisting of hundreds of needle core lung-tumor biopsies distinguishing between adenocarcinoma from squamous cell carcinoma.[50] The on-tissue tryptic-digested peptide images from the lung biopsies were correlated with a hematoxylin and eosin stained serial sections. MALDI imaging MS has also been used to differentiate between adenocarcinomas and healthy tissues in gastric carcinoma using trypsin-digested peptide signal from FFPE TMA.[51]

MALDI imaging has also been used for a large-scale TMA from 789 patients showing significant spectral differences between papillary and clear cell renal cell cancer in the kidney, as well as the discovery of one spectral feature associated with shortened survival.[52] A large-scale MALDI TMA studied FFPE bladder cancer issue from 697 patients discovering statistically relevant mass spectral signals for tumor aggressiveness (stage, grade, or nodal status), solid or papillary growth patterns, cell proliferation, tumor recurrence, decreased survival.[53] Fig. 13.3 shows an overview of TMA studies in bladder cancer. Another research studied large-scale FFPE esophageal cancer TMA from 300 adenocarcinomas and 177 squamous cell carcinomas, revealing mass spectral signals associated uniquely with squamous cell carcinomas or adenocarcinomas, as well as indicative of tumor stage, histological grade, and lymph node metastasis.[54] N-linked glycans were analyzed from the FFPE

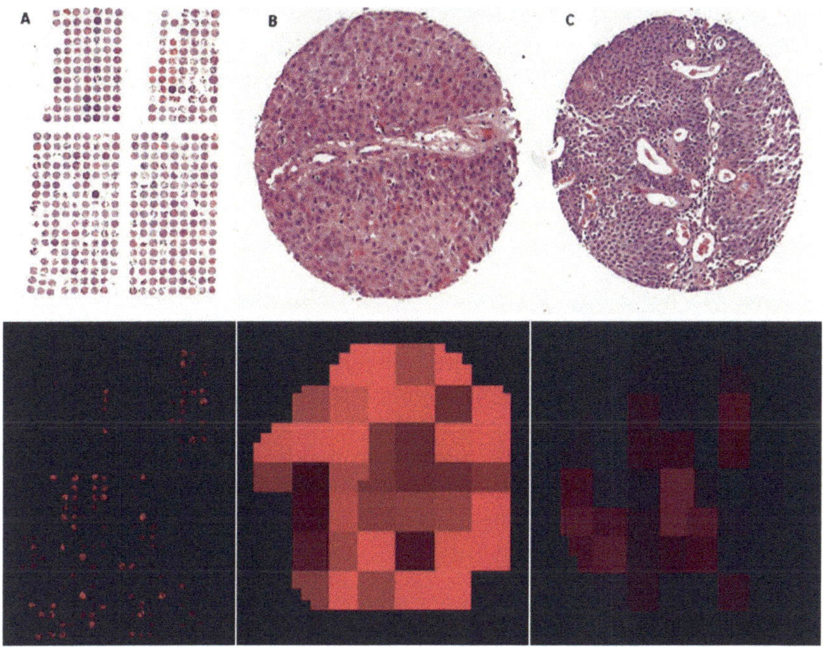

FIG. 13.3 H&E images are shown in top and corresponding MALDI images of ion *m/z* 1406.9 are shown in bottom. (A) Tissue microarray slide, (B) a representative bladder cancer spot (diameter 0.6 mm), (C) cancer-negative tissue spot. Reprinted with permission from Steurer S, Singer JM, Rink M, et al. MALDI imaging-based identification of prognostically relevant signals in bladder cancer using large-scale tissue microarrays. *Urol Oncol Semin Orig Invest.* 2014;32:1225–1233.[53]

TMA section from human pancreatic cancer, prostate cancers, and human hepatocellular carcinoma to distinguish between cancerous and noncancerous tissues.[55] DESI imaging has been used for the identification of metabolomic signatures associated with prostate cancer using imaging MS analysis of in needle core biopsies represented as a 0.6 mm TMA.[56]

Treatment response

The response of therapeutic treatment in patients can be done by monitoring treatment agents (drugs), its metabolites, or any other prognostic biomarkers. In one study, molecular markers of paclitaxel and radiation therapies response during pretreatment were imaged by MALDI and validated by IHC. Patients who achieved a complete pathological response had higher α-defensins peptide expression image by MALDI than those with residual disease.[57] In another study, esophageal adenocarcinoma biopsies were analyzed by MALDI imaging to evaluate molecular changes associated with response to chemotherapy of cisplatin with 5-fluorouracil. Biopsies were obtained from patients before undergoing chemotherapy and again after 3–4 weeks after its completion. Tumor regression in response to the chemotherapy was histopathologically appraised based on the remaining tumor size in comparison to the original. Residual tumor less than 10% of the original was classified as grade 1 and responders, while grade 2 (10% to 50%) and grade 3 (more than 50%) were classified as nonresponders. The proteomic signatures of the responder in comparison to nonresponders can be insightful in predicting the chemotherapy response.[58]

Clinical implementation of imaging MS

Imaging MS has a huge opportunity to fulfill an unmet diagnostic need in the clinical space, such as histopathology, without a reliable test. A conceptual framework of intraoperative histopathology with imaging MS is shown in Fig. 13.4. The implementation of imaging MS in a clinical setting cannot take place without standardization and validation designed for a specific clinical application. Some work has been started in multicenter validation and standardization, yet many works remain creating standardize workflow for a specific assay. A detailed discussion of validation and standardization is given in another chapter of this book.

Medical device

Governmental regulatory bodies will evaluate the approval of imaging MS in medical intervention as any other medical device and regulate it based on benefits against any potential risks. The regulatory framework differs significantly according to countries. In general, medical device regulation encompasses regulatory mandate, premarket evaluation, adverse event reporting, quality system, postapproval, and postmarket surveillance.[59] Studies comparing international regulatory approaches to medical devices are referenced here.[59–61] The Food and Drug Administration (FDA) in the United States has the mandate to provide reasonable assurance of the safety and efficacy for medical devices since the Medical Device Amendments to the Federal Food, Drug, and Cosmetic Act in 1976. The act established three classes of medical devices based on their

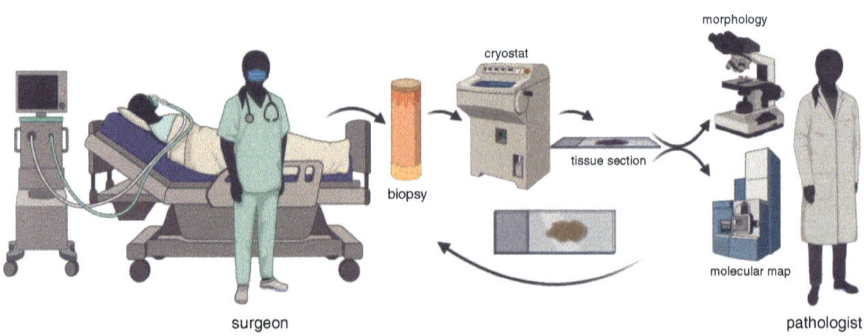

FIG. 13.4 Conceptual framework of intraoperative histopathology that also includes molecular imaging modalities such as imaging mass spectrometry is shown here.

risk and regulatory requirement—class I (low risk, e.g., wheelchairs), class II (moderate risk, e.g., needles), class III (high risk, e.g., pacemakers). It also established a regulatory pathway for new medical devices for using premarket approval and premarket notification (510(k)). The 1997 Food and Drug Administration Modernization Act created "least burdensome" provisions for premarket review and established the De Novo program.[62] Bruker Biotyper and bioMérieux Vitek MALDI systems for clinical use for the identification of microorganisms are classified as class II devices.[63] In the European Union, medical devices can be marketed after obtaining a Conformité Européenne mark from a Notified Body, an organization designated by a member state. Regulation of medical devices in emerging economies such as China is established recently. Many developing countries lack a regulatory framework for medical devices. Thus the World Health Organization has provided guidance on the global model regulatory framework for medical devices.[61] At present, I am unaware of any imaging MS system that is approved to be used as a medical device. Any potential future medical device based on imaging MS would have to comply with all these requirements depending on the locality.

Laboratory-developed test

Clinical usage of imaging MS assay may start as a laboratory-developed test (LDT). An LDT, as defined by the US FDA, is an in-house developed in vitro diagnostic (IVD) test that is used within a single Clinical Laboratory Improvement Amendments (CLIA) certified laboratory.[64] The CLIA regulates the laboratories offering LDTs to ensure accuracy and reliability of patient specimens test. The LDTs are considered "devices" similar to other IVD tests and subjected to regulatory oversight by the FDA for their safety and efficacy. In contrast to other devices, LDTs generally do not undergo premarket review by the FDA. The review assures the validity of the analytical and clinical validity of the IVD tests. LDTs still need to establish analytical validity.[65] A CLIA requirement for LDTs may include accuracy, imprecision, analytical sensitivity, analytical specificity, reportable range, reference intervals, system calibration, documentation of control procedures, etc. Now defunct, New River Labs aimed to develop and commercialize LDTs using proteomic spectral fingerprint from histology-guided imaging MS for diagnostic and prognostic assays for tissue-based diseases such as melanoma.

Outlook

Since its introduction about 20 years ago, imaging MS has made massive progress in terms of acquisition tools, sample preparation, biomedical and clinical workflows, and software. The availability of commercial imaging mass spectrometer, sample preparation tools, and processing software has contributed to its adoption in biomedical and clinical research. As imaging MS moves toward the clinical arena, the standard to communicate medical information should be considered. Digital Imaging and Communications in Medicine, an international standard for medical images and associated metadata, is implemented in many clinical imaging modalities such as X-ray, CT, MRI, ultrasound.[66–67] However, Digital Imaging and Communications in Medicine has seen low adoption in digital pathology, although the format and communication protocol for pathology have been defined.[68]

The Gartner's hype cycle, an example shown in Fig. 13.5, depicts a development arc of emerging technologies with the five phases—innovative trigger, the peak of inflated expectations in the tool, the trough of disillusionment by results, the slope of enlightenment, and finally the plateau of productivity.[69] It is evident that imaging MS for clinical analysis is not yet at the plateau of productivity. Depending on the specific disease, application, and technology, imaging MS may be anywhere in the first four phases. A more in-depth insight into the biomolecular process of disease, such as cancer, obtained by imaging studies, will have a transformative effect on healthcare. For example, molecular imaging in cancer can improve fundamental understanding of cancer, as well as usher in earlier detection, stratification of treatments, evaluation of therapies. Thus far, no imaging mass spectrometric assays have gone beyond exploratory preclinical research and influenced the treatment regiments of patients.

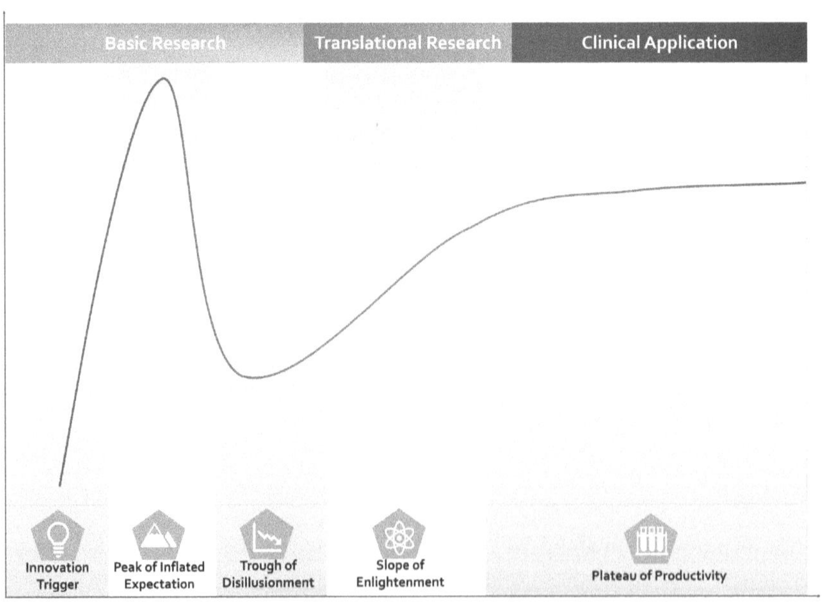

FIG. 13.5 Hype cycle, a graphical representation of a new technology, such as imaging MS, from innovation to adoption is depicted by five stages of development encompassing basic research to clinical application.

The translation of imaging MS into the clinic has been slower than was initially anticipated by some of us. The reasons may be a lack of standardization, availability of commercial tools, regulatory hurdles, market dynamics such as reimbursement strategies. Despite these hurdles, imaging MS assays are poised to enter the clinical application. For instance, imaging MS has a conceivable application in clinical pathology but has not yet been implemented in clinical settings. We need to facilitate continual advancement toward the clinic and hope one day; imaging MS will be routinely used by pathologists to guide them to identify the molecular changes leading to pathological diagnosis and monitor prognosis of therapy.

References

1. Croxatto A, Prod'hom G, Greub G. Applications of MALDI-TOF mass spectrometry in clinical diagnostic microbiology. *FEMS Microb Rev*. 2012;36:380–407.
2. Rodríguez-Sánchez B, Cercenado E, Coste AT, Greub G. Review of the impact of MALDI-TOF MS in public health and hospital hygiene, 2018. *Eurosurveillance*. 2019;24:1800193.
3. Weissleder R, Nahrendorf M. Advancing biomedical imaging. *Proc Natl Acad Sci*. 2015;112:14424–14428.
4. Hipp JD, Fernandez A, Compton CC, Balis UJ. Why a pathology image should not be considered as a radiology image. *J Pathol Inform*. 2011;2:26.
5. Burns ER. Clinical histology. *Clin Anat*. 2006;19:156–163.
6. Dey P. Immunocytochemistry in histology and cytology*Basic and Advanced Laboratory Techniques in Histopathology and Cytology*: Springer; 2018:149–169.
7. Brandtzaeg P. The increasing power of immunohistochemistry and immunocytochemistry. *J Immunol Methods*. 1998;216:49–67.
8. Buchwalow IB, Böcker W. Diagnostic immunohistochemistry. *Immunohistochemistry: Basics and Methods*. 2010:109–127.
9. Hornick JL. Novel uses of immunohistochemistry in the diagnosis and classification of soft tissue tumors. *Mod Pathol*. 2014;27:S47–S63.
10. Schwamborn K, Kriegsmann M, Weichert W. MALDI imaging mass spectrometry — from bench to bedside. *Biochim Biophys Acta Proteins Proteomics*. 2017;1865:776–783.
11. Sandy Mullay TRN, Jeffrey Prichard SJE. Evolving workflow drives anatomic pathology design. *MedicalLab Manag*. 2017;6(10):8.
12. Al-Janabi S, Huisman A, Van Diest PJ. Digital pathology: current status and future perspectives. *Histopathology*. 2012;61:1–9.
13. Rodriguez-Antolin C. Artificial intelligence will not replace health professionals, but the proper use of artificial intelligence will make health professionals better. *Cancer Res*. 2020;80:1912–1913.
14. Tizhoosh HR, Pantanowitz L. Artificial intelligence and digital pathology: challenges and opportunities. *J Pathol Inform*. 2018;9:38.
15. Suetens P. *Fundamentals of Medical Imaging*: Cambridge university press; 2017:1–218.
16. Miller DL, Schauer D. The ALARA principle in medical imaging. *Philosophy*. 1983;44:595–600.
17. Fleckenstein P, Tranum-Jensen J. *Anatomy in Diagnostic Imaging*: John Wiley & Sons; 2014:3–45.
18. Zanzonico P. Principles of nuclear medicine imaging: planar, SPECT, PET, multi-modality, and autoradiography systems. *Radiat Res*. 2012;177:349–364.

19. Solon EG. Autoradiography techniques and quantification of drug distribution. *Cell Tissue Res*. 2015;360:87–107.
20. Solon EG, Schweitzer A, Stoeckli M, Prideaux B, Autoradiography, MALDI-MS, and SIMS-MS imaging in pharmaceutical discovery and development. AAPS J 2010;12:11–26.
21. Chang Q, Ornatsky OI, Siddiqui I, Loboda A, Baranov VI, Hedley DW. Imaging mass cytometry. *Cytom Part A*. 2017;91:160–169.
22. Levenson RM, Borowsky AD, Angelo M. Immunohistochemistry and mass spectrometry for highly multiplexed cellular molecular imaging. *Lab Invest*. 2015;95:397–405.
23. Seeley EH, Caprioli RM. MALDI imaging mass spectrometry of human tissue: method challenges and clinical perspectives. *Trends Biotechnol*. 2011;29:136–143.
24. Spraggins JM, Schwamborn K, Heeren RM, Eberlin LS. The importance of clinical tissue imaging. *Clin Mass Spectrom*. 2019;12:47.
25. Ucal Y, Durer ZA, Atak H, et al. Clinical applications of MALDI imaging technologies in cancer and neurodegenerative diseases. *Biochim Biophys Acta Proteins Proteomics*. 2017;1865:795–816.
26. Leung F, Eberlin LS, Schwamborn K, Heeren RMA, Winograd N, Cooks RG. Mass spectrometry-based tissue imaging: the next frontier in clinical diagnostics?. *Clin Chem*. 2019;65:510–513.
27. Cole LM, Clench MR. Mass spectrometry imaging tools in oncology. *Biomark Med*. 2015;9:863–868.
28. Rauser S, Marquardt C, Balluff B, et al. Classification of HER2 receptor status in breast cancer tissues by MALDI imaging mass spectrometry. *J Proteome Res*. 2010;9:1854–1863.
29. Cornett DS, Mobley JA, Dias EC, et al. A novel histology-directed strategy for MALDI-MS tissue profiling that improves throughput and cellular specificity in human breast cancer. *Mol Cell Proteomics*. 2006;5:1975–1983.
30. Lazova R, Seeley EH, Keenan M, Gueorguieva R, Caprioli RM. Imaging mass spectrometry–a new and promising method to differentiate Spitz nevi from Spitzoid malignant melanomas. *Am J Dermatopathol*. 2012;34:82–90.
31. Kim HK, Reyzer ML, Choi IJ, et al. Gastric cancer-specific protein profile identified using endoscopic biopsy samples via MALDI mass spectrometry. *J Proteome Res*. 2010;9:4123–4130.
32. Rahman SMJ, Gonzalez AL, Li M, et al. Lung cancer diagnosis from proteomic analysis of preinvasive lesions. *Cancer Res*. 2011;71:3009–3017.
33. Schwamborn K, Krieg RC, Jirak P, et al. Application of MALDI imaging for the diagnosis of classical Hodgkin lymphoma. *J Cancer Res Clin Oncol*. 2010;136:1651–1655.
34. Le Faouder J, Laouirem S, Chapelle M, et al. Imaging mass spectrometry provides fingerprints for distinguishing hepatocellular carcinoma from cirrhosis. *J Proteome Res*. 2011;10:3755–3765.
35. Oppenheimer SR, Mi D, Sanders ME, Caprioli RM. Molecular analysis of tumor margins by MALDI mass spectrometry in renal carcinoma. *J Proteome Res*. 2010;9:2182–2190.
36. Grüner BM, Hahne H, Mazur PK, et al. MALDI imaging mass spectrometry for in situ proteomic analysis of preneoplastic lesions in pancreatic cancer. *PLOS One*. 2012;7:e39424.
37. Lou S, Balluff B, de Graaff MA, et al. High-grade sarcoma diagnosis and prognosis: Biomarker discovery by mass spectrometry imaging. *Proteomics*. 2016;16:1802–1813.
38. Lemaire R, Ait Menguellet S, Stauber J, et al. Specific MALDI imaging and profiling for biomarker hunting and validation: fragment of the 11S proteasome activator complex, reg alpha fragment, is a new potential ovary cancer biomarker. *J Proteome Res*. 2007;6:4127–4134.

39. Calligaris D, Caragacianu D, Liu X, et al. Application of desorption electrospray ionization mass spectrometry imaging in breast cancer margin analysis. *Proc Natl Acad Sci USA*. 2014;111:15184–15189.
40. Banerjee S, Zare RN, Tibshirani RJ, et al. Diagnosis of prostate cancer by desorption electrospray ionization mass spectrometric imaging of small metabolites and lipids. *Proc Natl Acad Sci*. 2017;114:3334–3339.
41. Santoro AL, Drummond RD, da Silva IT, et al. In situ DESI - MSI lipidomic profiles of breast cancer molecular subtypes and precursor lesions. *Cancer Res*. 2020;80:1246–1257.
42. Abbassi-Ghadi N, Antonowicz SS, McKenzie JS, et al. De novo lipogenesis alters the phospholipidome of esophageal adenocarcinoma. *Cancer Res*. 2020;80:2764–2774.
43. Zhang J, Li SQ, Lin JQ, Yu W, Eberlin LS. Mass spectrometry imaging enables discrimination of renal oncocytoma from renal cell cancer subtypes and normal kidney tissues. *Cancer Res*. 2020;80:689–698.
44. DeHoog RJ, Zhang J, Alore E, et al. Preoperative metabolic classification of thyroid nodules using mass spectrometry imaging of fine-needle aspiration biopsies. *Proc Natl Acad Sci*. 2019;116:21401–21408.
45. Calligaris D, Norton I, Feldman DR, et al. Mass spectrometry imaging as a tool for surgical decision-making. *J Mass Spectrom*. 2013;48:1178–1187.
46. Eberlin LS, Norton I, Dill AL, et al. Classifying human brain tumors by lipid imaging with mass spectrometry. *Cancer Res*. 2012;72:645–654.
47. Santagata S, Eberlin LS, Norton I, et al. Intraoperative mass spectrometry mapping of an onco-metabolite to guide brain tumor surgery. *Proc Natl Acad Sci*. 2014;111:11121–11126.
48. Pirro V, Alfaro CM, Jarmusch AK, Hattab EM, Cohen-Gadol AA, Cooks RG. Intraoperative assessment of tumor margins during glioma resection by desorption electrospray ionization-mass spectrometry. *Proc Natl Acad Sci*. 2017;114:6700–6705.
49. Basu SS, Regan MS, Randall EC, et al. Rapid MALDI mass spectrometry imaging for surgical pathology. *NPJ Precis Oncol*. 2019;3:17.
50. Groseclose MR, Massion PP, Chaurand P, Caprioli RM. High-throughput proteomic analysis of formalin-fixed paraffin-embedded tissue microarrays using MALDI imaging mass spectrometry. *Proteomics*. 2008;8:3715–3724.
51. Morita Y, Ikegami K, Goto-Inoue N, et al. Imaging mass spectrometry of gastric carcinoma in formalin-fixed paraffin-embedded tissue microarray. *Cancer Sci*. 2010;101:267–273.
52. Steurer S, Seddiqi AS, Singer JM, et al. MALDI imaging on tissue microarrays identifies molecular features associated with renal cell cancer phenotype. *Anticancer Res*. 2014;34:2255–2261.
53. Steurer S, Singer JM, Rink M, et al. MALDI imaging–based identification of prognostically relevant signals in bladder cancer using large-scale tissue microarrays1These authors contributed equally to this work. *Urol Oncol Semin Orig Invest*. 2014;32:1225–1233.
54. Quaas A, Bahar AS, von Loga K, et al. MALDI imaging on large-scale tissue microarrays identifies molecular features associated with tumour phenotype in oesophageal cancer. *Histopathology*. 2013;63:455–462.
55. Powers TW, Neely BA, Shao Y, et al. MALDI imaging mass spectrometry profiling of n-glycans in formalin-fixed paraffin embedded clinical tissue blocks and tissue microarrays. *PLOS One*. 2014;9:e106255.
56. Morse N, Jamaspishvili T, Simon D, et al. Reliable identification of prostate cancer using mass spectrometry metabolomic imaging in needle core biopsies. *Lab Invest*. 2019;99:1561–1571.
57. Bauer JA, Chakravarthy AB, Rosenbluth JM, et al. Identification of markers of taxane sensitivity using proteomic and genomic analyses of breast tumors from patients receiving neoadjuvant paclitaxel and radiation. *Clin Cancer Res*. 2010;16:681–690.

58. Aichler M, Elsner M, Ludyga N, et al. Clinical response to chemotherapy in oesophageal adenocarcinoma patients is linked to defects in mitochondria. *J Pathol.* 2013;230:410–419.
59. Kramer DB, Tan YT, Sato C, Kesselheim AS. Ensuring medical device effectiveness and safety: a cross–national comparison of approaches to regulation. *Food Drug Law J.* 2014; 69(1):1–23, i.
60. Kramer DB, Tan YT, Sato C, Kesselheim AS. Postmarket surveillance of medical devices: a comparison of strategies in the US, EU, Japan, and China. *PLoS Med.* 2013;10(9):e1001519.
61. Organization WH, WHO global model regulatory framework for medical devices including in vitro diagnostic medical devices. 2017.
62. FDA, A History of medical device regulation & oversight in the United States. Available at: https://www.fda.gov/medical-devices/overview-device-regulation/history-medical-device-regulation-oversight-united-states (Accessed July 20, 2020).
63. FDA, Device classification under section 513(f)(2)(de novo). Available at: https://www.accessdata.fda.gov/scripts/cdrh/cfdocs/cfpmn/denovo.cfm (Accessed July 20, 2020).
64. U.S. Department of Health and Human Services, Office of In Vitro Diagnostics and Radiological Health, Center for Biologics Evaluation and Research. Framework for regulatory oversight of laboratory developed tests (LDTs). 2014.
65. FDA, Discussion paper on laboratory developed tests (LDTs) 2017.
66. Pianykh OS. *Digital Imaging and Communications in Medicine (DICOM): A Practical Introduction and Survival Guide*: Springer Science & Business Media; 2009:1–389.
67. Herrmann M, Clunie D, Fedorov A, et al. Implementing the DICOM standard for digital pathology. *J Pathol Inform.* 2018;9:37 37.
68. Singh R, Chubb L, Pantanowitz L, Parwani A. Standardization in digital pathology: supplement 145 of the DICOM standards. *J Pathol Inform.* 2011;2:23.
69. Gartner's 2015 hype cycle for emerging technologies identifies the computing innovations that organizations should monitor. Available at: https://www.gartner.com/en/research/methodologies/gartner-hype-cycle. Accessed on January 23, 2021.

Chapter 14

Imaging mass spectrometry: endogenous mammalian metabolites

Chapter Outline

Metabolites are small molecules involved in metabolism operating vital cell functions (e.g., pluripotency, energy management, apoptosis), ecological functions (pigmentation, antibiotic), epigenetic mechanism, regulation of post-translational modifications, drug interactions, etc.[1-2] Metabolite molecules can be endogenously produced by the body or host microorganisms, or exogenically by diet, drug or environmental exposure.[3-5]

Imaging mass spectrometry (MS) can accurately depict the spatial distribution of endogenous metabolites in the tissue section. More thorough discussions on imaging metabolites by MS have been reviewed in the references here.[6-13] Matrix-assisted laser desorption/ionization (MALDI) imaging MS has been extensively used to image higher molecular weight biomolecules, such as peptides and proteins.[14-16] The application of MALDI imaging MS for small molecular weight metabolites was challenging due to poor ionization of small molecule, ion suppression from more abundant lipids, and the spectral interferences in the low *m/z* range produced by endogenous molecules, organic matrices, or background ions.

Mapping of metabolite distribution on the tissue by imaging MS can be improved using five strategies encompassing each step of the experiment encompassing sample collection of tissue to detect metabolite ions. As the first step in the imaging MS experiment, the bulk tissue sample or sections can be fixed to reduce metabolic turnover or degradation. Second, the cryosectioned tissue can be processed to improve metabolite detection by several preparation methods, such as washing off MS signal interfering lipids, specific chemical

Introduction to Spatial Mapping of Biomolecules by Imaging Mass Spectrometry. DOI: 10.1016/C2018-0-03962-6

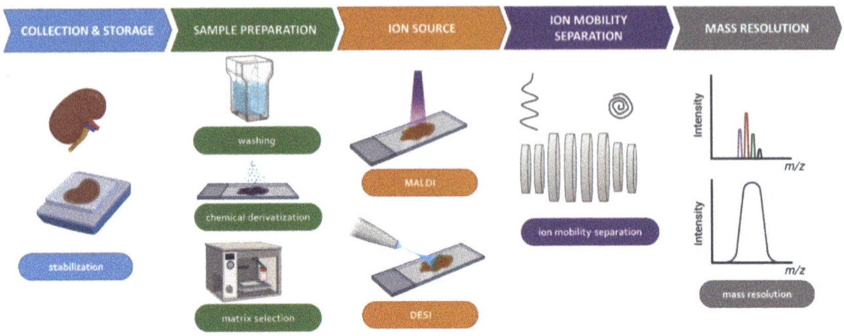

FIG. 14.1 Strategies to enhance the metabolite detection for imaging mass spectrometry from sample collection to ion detection.

derivatization to boost ionization of metabolites, or using matrix designed for small molecules imaging for MALDI. Additionally, multiple imaging MS ion sources can be utilized to improve the ionization of different molecular classes of metabolite molecules. Ion sources based on electrospray ionization (ESI), such as desorption electrospray ionization (DESI) and laser ablation electrospray ionization (LAESI), have been found to detect a complementary set of molecules in comparison to laser/desorption-based MALDI.[17-18] Finally, a gas-phase ion mobility separation or higher mass resolution during imaging MS can separate the interfering MS signal. Some of these strategies can be combined to further improve the in situ detection and imaging of metabolites from the tissue. Fig. 14.1 shows these steps for improved metabolite imaging by MS.

Improved sample preparation for metabolite imaging

Some metabolic events are too dynamic and susceptible to postmortem changes, such as high-energy phosphates molecules, including glucose intermediates and phosphonucleotides.[19] Additional sample preparation steps to fix the tissue are required to detect dynamic and labile metabolites in tissue. Imaging MS on tissue fixed with focused microwave irradiation has been used to visualize metabolic flux of glucose-derived metabolites in the mouse brain,[20] study glucose oxidation and lactate utilization of the tricarboxylic acid cycle in damaged stroke tissue in the heart.[21]

In situ detection and imaging of small molecules, such as metabolites by MALDI remains to be challenging due to ion suppression from more abundant and easily ionizable molecules in tissue such as phospholipids. Selective removal of the interfering molecule by quickly rinsing tissue section can improve the ionization of metabolites. Brain tissue section rinsed quickly with chloroform for 15 seconds at −20°C was found to removed most glycerophospholipids and glycerolipids from tissue section, improving detection of metabolites such as ATP-related metabolites, amino acids, glycolysis-related, inosine, glutamine, glutathione, taurine, spermine, etc.[22]

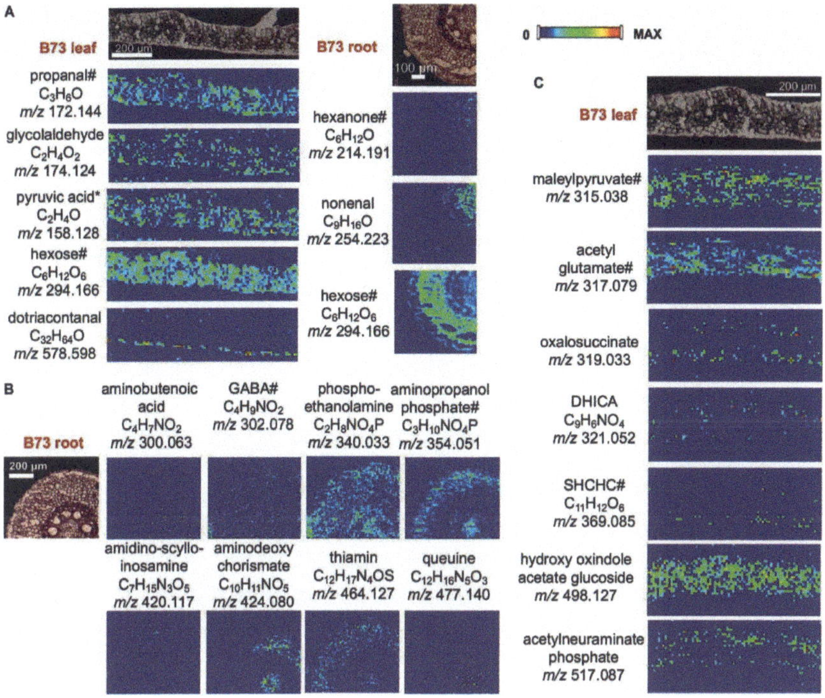

FIG. 14.2 MALDI MS images of metabolites derivatized with (A) Girard's reagent T (GT), (B) coniferyl aldehyde (CA), and (C) 2-picolylamine (2-PA). All images are acquired in positive ion mode. Reprinted with permission from Dueñas ME, Larson EA, Lee YJ. Toward mass spectrometry imaging in the metabolomics scale: increasing metabolic coverage through multiple on-tissue chemical modifications. Front Plant Sci. 2019;10:860–860.[24] Creative commons.

One of the most popular strategies to detect metabolites includes on-tissue chemical derivatization of an analyte to improve the detection sensitivity. The chemical derivatization will address isobaric chemical interferences endogenous to tissue or exogenous interferences due to the sample preparation, ion suppression effects, poor ionization efficiency, or in-source fragmentation. For example, on-tissue chemical derivatization using Amplifex Diene reagent as a derivatization reagent used to image localization of vitamin D metabolites in mouse kidney by MALDI and DESI imaging MS.[23] A comprehensive study on-tissue chemical derivatization using three common derivatization methods—coniferyl aldehyde for primary amines, Girard's reagent T for carbonyl groups, and 2-picolylamine for carboxylic acids—showed an increase in the number of metabolites for imaging MS of leaves and roots in corn tissue, as shown in Fig. 14.2.[24]

Enhancing metabolite detection in MALDI by matrix selection

There has been considerable research effort on rationally designing matrices for small molecules. An overview of the development of MALDI matrices capability for small molecules analysis is given in the review.[25] An ideal matrix for small molecule analysis has strong ultraviolet absorption around MALDI laser, gives a high sensitivity for the analyte in a complex mixture, has a low background interference in the low molecular weight region, can be conveniently applied on tissue, cost-effective, and safe.

A rationally designed matrix on Brønsted–Lowry concept is based on mixing an acidic analyte with a strong base matrix, and vice versa produced ions that are oppositely charged to those of the absorbing partner. For example, a very strong base, "proton-sponge," also known as 1,8-dimethylamino naphthalene, was used as a matrix for several acid metabolites in plant leaves, blood, and fly.[26] One of the purposefully synthesized matrices, 2,3,4,5-tetrakis(3′,4′-dihydroxylphenyl)-thiophene, was found useful in analyzing low molecular weight amines such as creatinine, glycine, alloxan, allantoin, and 3-hydroxyhippuric acid in the urine sample.[27] 1-Naphthylhydrazine hydrochloride has been proposed as a matrix to detect small molecules such as $[\text{glucose} + \text{Cl}]^-$ ion in serum.[28] A rationally designed matrix, N-(1-naphthyl) ethylenediamine dinitrate, a strong UV absorption, hydrogen binding, and nitrate anion donors, has been used to analyze small molecules metabolites in negative ion mode.[29] The 1,5-naphthalenediamine hydrochloride matrix showed improved detection of small molecules, such as amino acids, carboxylic acids, nucleotide derivatives, peptide, and lipids in the liver, brain, and kidneys tissue from mice. The matrix was able to simultaneously image 19 endogenous metabolites involved in ATP metabolism, tricarboxylic acid cycle, glutamate–glutamine cycle, and malate–aspartate shuttle. An example of 1,5-naphthalenediamine matrix for imaging metabolites involved in ATP metabolism in the brain tissue is shown in Fig. 14.3. The prepared matrix has low cost, strong ultraviolet absorption, high salt tolerance capacity, and fewer background signals, especially in the low mass range below $m/z < 500$.[30] The usability of the matrix for imaging a specific analyte or group of molecules always needs to evaluated in the context of the native interference found in the specific tissue microenvironment. A matrix may be excellent at detecting the analyte in standard or plant tissue but fail to detect the same compound at the same concentration in a specific animal tissue due to (biological) matrix effect.

Increased mass resolution for metabolite imaging

Fourier transform ion cyclotron resonance (FTICR) MS provides the highest mass resolution revealing metabolic spectral features indistinguishable in low-resolution mass spectrometers, such as time-of-flight.[31] The ultrahigh mass resolving power of FTICR mass spectrometer has been used to detect and image metabolites and drugs using MALDI.[32] The increased mass resolution helped to separate matrix ions from small molecules and also assisted in more confident identification due to an increase in mass accuracy. A 15T high field FTICR coupled with MALDI was able to image metabolites and critical metabolic proteins

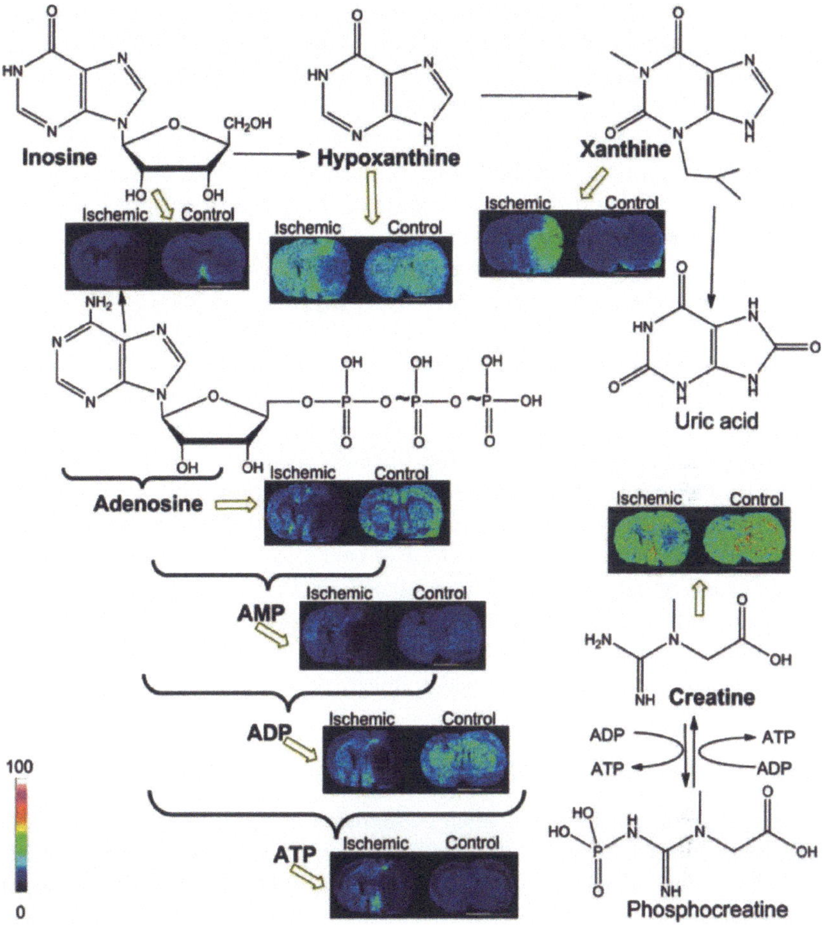

FIG. 14.3 Negative ion-mode MALDI imaging MS with 1,5-naphthalenediamine (1,5-DAN) hydrochloride matrix showed the distribution of metabolites involved in ATP metabolism in the brain between control and ischemic brain. Scale bar is 1 mm and spatial resolution 50 μm. Reprinted with permission from Liu H, Chen R, Wang J, et al. 1,5-Diaminonaphthalene hydrochloride assisted laser desorption/ionization mass spectrometry imaging of small molecules in tissues following focal cerebral ischemia. Anal Chem. 2014;86:10114–10121. Copyright 2014 American Chemical Society.[30]

in a mouse model of high-grade glioma, albeit metabolite imaging utilized different tissue sample preparation protocols and different lower molecular settings for the mass.[33] MALDI FTICR MS imaging was utilized to map the distribution of endogenous metabolites, such as ADP, AMP, ATO, glutathione, in breast cancer spheroids.[34] The ultrahigh mass resolution MALDI FTICR has also been utilized to analyze small molecules in formalin-fixed paraffin-embedded tissue samples.[35-36]

In another example, MALDI-FTICR imaging retina detected metabolites such as nucleotides, central carbon metabolism pathway intermediates,

2-oxocarboxylic acid metabolism, oxidative phosphorylation, glycerophospholipid metabolism, and cysteine and methionine metabolites.[37] Similar ultrahigh-resolution MALDI FTICR has also been used to examine the intracellular metabolism of *Escherichia coli* in whole-cell without imaging. Ultrahigh resolution helped in the identification of metabolite using high mass accuracy and elemental composition.[38] The ultrahigh mass resolution tends to be slower and require more acquisition time to image tissue. The longer imaging time also increases increasing imaging complexity due to matrix evaporation for a handful of matrices. Also, the FTICR instruments tend to be more expensive and require more technical maintenance by experienced users, making it an unpractical choice for routine clinical applications.

Besides, ultrahigh resolving FTICR, the spectral overlap of the matrix with imaged metabolites can be addressed by coupling MALDI or other imaging ion source to a high-resolution linear ion trap-Orbitrap, quadrupole time-of-flight (q-ToF), or quadrupole ion trap time-of-flight mass spectrometers.[39-41] For example, several small lipids, such as lysophosphatidylcholines (e.g., LPC 16:0, LPC 18:0, and LPC 18:1), were imaged by quadrupole traveling wave ion q-ToF mass spectrometer with MALDI in isotopically labeled atherosclerotic aorta plaques in mice.[42] MALDI quadrupole ion trap time-of-flight was utilized to visualize endogenous molecules, such as metabolites and lipids, in Japanese rice fish, medaka sections.[40] MALDI, coupled with linear ion trap Orbitrap mass spectrometer, has been utilized to image metabolites in plant tissues.[41, 43]

Imaging metabolite with ion mobility separation MS

Imaging MS of metabolites is often confounded by spectral interference from background ions, such as MALDI matrix clusters or electrospray solvent peaks, and endogenous molecules, such as phospholipids. Due to the direct nature of the sampling method in imaging MS techniques, interfering endogenous or background molecules cannot be separated in sample preparation steps. Due to the same reason, the imaging MS ion source can be coupled with the traditional liquid chromatography or gas chromatography separation, which takes place before the sample is ionized and transferred for detection by a mass spectrometer.

Instead, imaging MS techniques can utilize a gas-phase separation using ion mobility to separate the ions based on their drift against the buffer gas in the electric field. The ion mobility separation takes place after the ionization and before the detection of the metabolite ions from the tissue. The mobilities of ions are based on collision cross sections of the ions against the buffer gas and can serve as a measurement yardstick for their size and shape. Ion mobility separation can separate isobaric and isomeric ions beyond the capability of an ultrahigh resolving mass spectrometer providing a superior level of chemical selectivity. During imaging MS, ion mobility is capable of isolating low abundance isobaric metabolite species, which otherwise will be obscured by a more dominant overlapping ion signal. The ion mobility separation often separates molecules, such

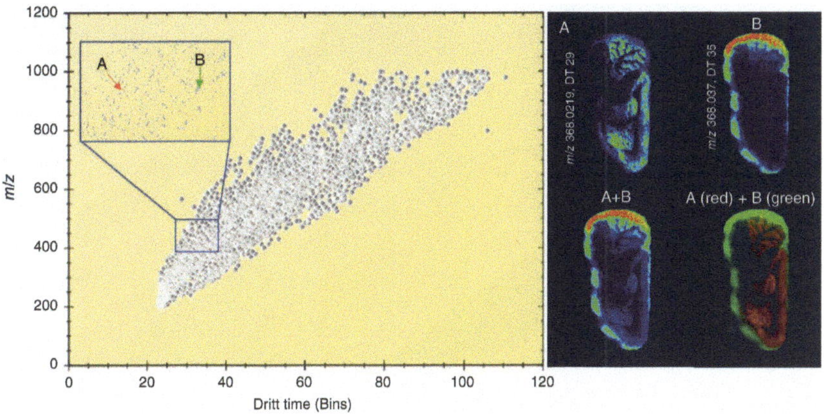

FIG. 14.4 MALDI imaging using traveling wave ion mobility spectrometry (TWIMS) quadrupole mass spectrometer. The left portion shows the mobility plot. The right portion shows two images of isobaric ions on the top and images showing overlaid combined images of the two ions in the bottom.

as lipids, by compound class and charge.[44] Many imaging ion sources, such as MALDI,[45-47] DESI,[48-49] andLAESI,[50] have been used with commercial traveling wave ion mobility or trapped ion mobility systems coupled to a q-ToF mass spectrometer. The imaging MS coupled with ion mobility separation can result in more precise and accurate distributions of the metabolite of interest due to the removal of background isobaric signal, as shown in Fig. 14.4.

Using multiple ion sources for metabolite imaging MS

MALDI and ESI ion sources have been long believed to generate a complementary set of ions due to ionization mechanism and different ion suppression.[51-52] A recent statistical analysis study of 72,000 peptides showed complementary coverage with an overlap of about 40% for MALDI and ESI.[53] Some metabolites within the tissue can be easily detected using the electrospray-based ambient ionization method compared to MALDI.

Ambient ionization source, such as DESI and LAESI, directly extract and ionize, detect and image analyte off the tissues sections directly. Fig. 14.5 shows a panel of metabolites imaged on the mouse brain section using DESI ion source on a q-ToF mass spectrometer. In both DESI and LAESI, ions are extracted and ionized in concert from a tissue section without sample preparation. In DESI, a confined area Or pixel of tissue is extracted and ionized by impinging focused electrospray droplets, while in LAESI, a midinfrared beam ablates and extracts each pixel of the sample, and the resulting ablation plume is ionized by intercepting electrospray. Another ion source, such as laser ablation atmospheric pressure photoionization (LAAPPI), utilizes the same sampling strategies as LAESI of using an infrared for extracting samples off the tissue. Instead of ESI, LAAPPI

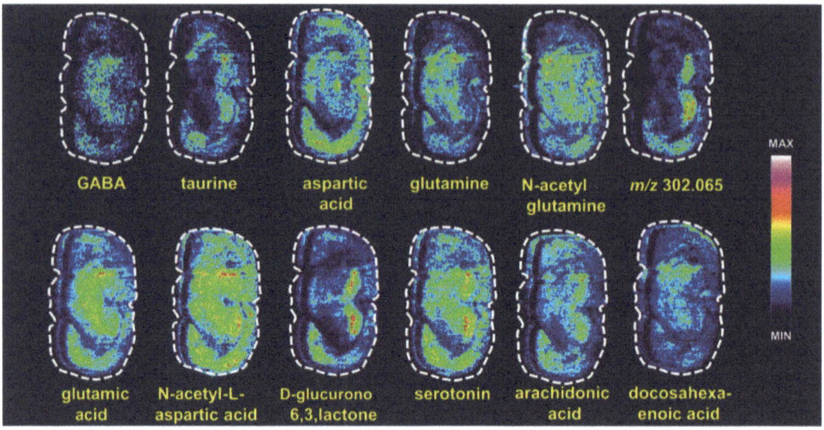

FIG. 14.5 Metabolites imaged on the mouse brain section using desorption electrospray ionization (DESI) ion source on a quadrupole time-of-flight (q-ToF) mass spectrometer. Reprinted with permission from Midey A, Olivos H, Shrestha B. Spatial mapping of cellular metabolites using DESI ion mobility mass spectrometry. In: Single Cell Metabolism. Springer; 2020:181–190.[49]

uses an orthogonal hot solvent (e.g., toluene or anisole) jet from a microfluidic device and photoionization using a 10-eV ultraviolet krypton discharge lamp for ionization. LAAPPI is able to ionize and detect different sets of molecules complementary to ESI-based ion source. For example, LAAPPI ionizes neutral and nonpolar compounds in addition to polar compounds ionized using ESI.[54]

Matrix-free laser desorption/ionization (LDI) has shown the capability of enhancing desorption and ionization biomolecules using interactions between pulsed laser radiation and nanostructured materials.[55] These LDI methods utilize the same MALDI instrumentation, but instead of a matrix, the sample is placed on a nanomaterial substrate. Endogenous metabolites in the mouse brain and kidney tissue sections were imaged using a matrix-free nanophotonic LDI on a highly uniform silicon nanopost array.[56] In another example, nanostructure initiator mass spectrometry with optionally fluorinated gold nanoparticles were used to image metabolites in biological systems.[57]

Outlook

Imaging MS is a powerful technique to map spatial distributions of several endogenous metabolites within the tissue. The spatial distribution of various metabolites can correlate their distribution within the tissue morphology—giving rise to the field of spatial metabolomics. The last decade has seen many major developments in the detection and imaging of metabolites by a mass spectrometer, such as the development of ambient ionization, several chemical derivatization workflows, incorporation of ion mobility separation, etc. In addition to development in instrumentation, a further advance in sample preparation and storage will be critical for metabolite imaging by MS. Accurate and high-throughput metabolite identification and quantitation remain a bottleneck step for imaging MS experiments. With the development of tools and workflows, we can anticipate spatial metabolomics using imaging MS being increasingly capable of uncovering many biochemical interactions not possible by bulk metabolomics.

References

1. Johnson CH, Ivanisevic J, Siuzdak G. Metabolomics: beyond biomarkers and towards mechanisms. *Nat Rev Mol Cell Biol*. 2016;17:451–459.
2. Wishart DS. Emerging applications of metabolomics in drug discovery and precision medicine. *Nat Rev Drug Discov*. 2016;15:473–484.
3. Louis P, Hold GL, Flint HJ. The gut microbiota, bacterial metabolites and colorectal cancer. *Nat Rev Microbiol*. 2014;12:661–672.
4. Walker DI, Valvi D, Rothman N, Lan Q, Miller GW, Jones DP. The metabolome: a key measure for exposome research in epidemiology. *Curr Epidemiol Rep*. 2019;6:93–103.
5. Prakash C, Shaffer CL, Nedderman A. Analytical strategies for identifying drug metabolites. *Mass Spectrom Rev*. 2007;26:340–369.
6. Ho YN, Shu LJ, Yang YL. Imaging mass spectrometry for metabolites: technical progress, multimodal imaging, and biological interactions. *Wiley Interdiscip Rev Syst Biol Med*. 2017;9:e1387.
7. Dueñas ME, Larson EA, Lee YJ. Toward mass spectrometry imaging in the metabolomics scale: increasing metabolic coverage through multiple on-tissue chemical modifications. *Front Plant Sci*. 2019;10 860.
8. Balluff B, McDonnell LA. Mass spectrometry imaging of metabolites. *Methods Mol Biol*. 2018;1730:345–357.
9. Sugiura Y, Setou M. Imaging mass spectrometry for visualization of drug and endogenous metabolite distribution: toward in situ pharmacometabolomes. *J Neuroimmune Pharmacol*. 2010;5:31–43.
10. Greer T, Sturm R, Li L. Mass spectrometry imaging for drugs and metabolites. *J Proteomics*. 2011;74:2617–2631.
11. Kaspar S, Peukert M, Svatos A, Matros A, Mock H-P. MALDI-imaging mass spectrometry – an emerging technique in plant biology. *Proteomics*. 2011;11:1840–1850.
12. Sturtevant D, Lee Y-J, Chapman KD. Matrix assisted laser desorption/ionization-mass spectrometry imaging (MALDI-MSI) for direct visualization of plant metabolites in situ. *Curr Opin Biotechnol*. 2016;37:53–60.
13. Trim PJ, Snel MF. Small molecule MALDI MS imaging: Current technologies and future challenges. *Methods*. 2016;104:127–141.
14. Cillero-Pastor B, Heeren RM. Matrix-assisted laser desorption ionization mass spectrometry imaging for peptide and protein analyses: a critical review of on-tissue digestion. *J Proteome Res*. 2014;13:325–335.
15. Stoeckli M, Chaurand P, Hallahan DE, Caprioli RM. Imaging mass spectrometry: a new technology for the analysis of protein expression in mammalian tissues. *Nat Med*. 2001;7:493–496.
16. MacAleese L, Stauber J, Heeren RM. Perspectives for imaging mass spectrometry in the proteomics landscape. *Proteomics*. 2009;9:819–834.
17. Takats Z, Wiseman JM, Cooks RG. Ambient mass spectrometry using desorption electrospray ionization (DESI): instrumentation, mechanisms and applications in forensics, chemistry, and biology. *J Mass Spectrom*. 2005;40:1261–1275.

18. Shrestha B, Nemes P, Nazarian J, Hathout Y, Hoffman EP, Vertes A. Direct analysis of lipids and small metabolites in mouse brain tissue by AP IR-MALDI and reactive LAESI mass spectrometry. *Analyst*. 2010;135:751–758.
19. Hattori K, Kajimura M, Hishiki T, et al. *Paradoxical ATP elevation in ischemic penumbra revealed by quantitative imaging mass spectrometry*. NYUSA: Mary Ann Liebert, Inc. 140 Huguenot Street, 3rd Floor New Rochelle; 2010.
20. Sugiura Y, Honda K, Kajimura M, Suematsu M. Visualization and quantification of cerebral metabolic fluxes of glucose in awake mice. *Proteomics*. 2014;14:829–838.
21. Sugiura Y, Katsumata Y, Sano M, et al. Visualization of in vivo metabolic flows reveals accelerated utilization of glucose and lactate in penumbra of ischemic heart. *Sci Rep*. 2016;6:32361.
22. Yang H, Ji W, Guan M, et al. Organic washes of tissue sections for comprehensive analysis of small molecule metabolites by MALDI MS imaging of rat brain following status epilepticus. *Metabolomics*. 2018;14:50.
23. Smith KW, Flinders B, Thompson PD, et al. Spatial localization of vitamin D metabolites in mouse kidney by mass spectrometry imaging. *ACS Omega*. 2020;5:13430–13437.
24. Dueñas ME, Larson EA, Lee YJ. Toward mass spectrometry imaging in the metabolomics scale: increasing metabolic coverage through multiple on-tissue chemical modifications. *Front Plant Sci*. 2019;10:860 860.
25. Calvano CD, Monopoli A, Cataldi TRI, Palmisano F. MALDI matrices for low molecular weight compounds: an endless story?. *Anal Bioanal Chem*. 2018;410:4015–4038.
26. Shroff R, Rulíšek L, Doubský J, Svatoš A. Acid–base-driven matrix-assisted mass spectrometry for targeted metabolomics. *Proc Natl Acad Sci*. 2009;106:10092–10096.
27. Chen S, Chen L, Wang J, et al. 2,3,4,5-tetrakis(3′,4′-dihydroxylphenyl)thiophene: a new matrix for the selective analysis of low molecular weight amines and direct determination of creatinine in urine by MALDI-TOF MS. *Anal Chem*. 2012;84:10291–10297.
28. He Q, Chen S, Wang J, et al. 1-Naphthylhydrazine hydrochloride: a new matrix for the quantification of glucose and homogentisic acid in real samples by MALDI-TOF MS. *Clin Chim Acta*. 2013;420:94–98.
29. Chen R, Chen S, Xiong C, et al. N-(1-Naphthyl) ethylenediamine dinitrate: a new matrix for negative ion MALDI-TOF MS analysis of small molecules. *J Am Soc Mass Spectrom*. 2012;23:1454–1460.
30. Liu H, Chen R, Wang J, et al. 1,5-diaminonaphthalene hydrochloride assisted laser desorption/ionization mass spectrometry imaging of small molecules in tissues following focal cerebral ischemia. *Anal Chem*. 2014;86:10114–10121.
31. Marshall AG, Hendrickson CL, Jackson GS. Fourier transform ion cyclotron resonance mass spectrometry: a primer. *Mass Spectrom Rev*. 1998;17:1–35.
32. Cornett DS, Frappier SL, Caprioli RM. MALDI-FTICR imaging mass spectrometry of drugs and metabolites in tissue. *Anal Chem*. 2008;80:5648–5653.
33. Dilillo M, Ait-Belkacem R, Esteve C, et al. Ultra-high mass resolution MALDI imaging mass spectrometry of proteins and metabolites in a mouse model of glioblastoma. *Sci Rep*. 2017;7:603.
34. Tucker LH, Hamm GR, Sargeant RJE, et al. Untargeted metabolite mapping in 3D cell culture models using high spectral resolution FT-ICR mass spectrometry imaging. *Anal Chem*. 2019;91:9522–9529.
35. Buck A, Ly A, Balluff B, et al. High-resolution MALDI-FT-ICR MS imaging for the analysis of metabolites from formalin-fixed, paraffin-embedded clinical tissue samples. *J Pathol*. 2015;237:123–132.

36. Ly A, Buck A, Balluff B, et al. High-mass-resolution MALDI mass spectrometry imaging of metabolites from formalin-fixed paraffin-embedded tissue. *Nat Protoc*. 2016;11:1428–1443.
37. Sun N, Ly A, Meding S, et al. High-resolution metabolite imaging of light and dark treated retina using MALDI-FTICR mass spectrometry. *Proteomics*. 2014;14:913–923.
38. Yukihira D, Miura D, Saito K, Takahashi K, Wariishi H. MALDI–MS-based high-throughput metabolite analysis for intracellular metabolic dynamics. *Anal Chem*. 2010;82:4278–4282.
39. Goodwin RJA, Mackay CL, Nilsson A, et al. Qualitative and quantitative MALDI imaging of the positron emission tomography ligands raclopride (a D2 dopamine antagonist) and SCH 23390 (a D1 dopamine antagonist) in rat brain tissue sections using a solvent-free dry matrix application method. *Anal Chem*. 2011;83:9694–9701.
40. Zaima N, Hayasaka T, Goto-Inoue N, Setou M. Imaging of metabolites by MALDI mass spectrometry. *J Oleo Sci*. 2009;58:415–419.
41. Korte AR, Yandeau-Nelson MD, Nikolau BJ, Lee YJ. Subcellular-level resolution MALDI-MS imaging of maize leaf metabolites by MALDI-linear ion trap-Orbitrap mass spectrometer. *Anal Bioanal Chem*. 2015;407:2301–2309.
42. Castro-Perez J, Hatcher N, Kofi Karikari N, et al. In vivo isotopically labeled atherosclerotic aorta plaques in ApoE KO mice and molecular profiling by matrix-assisted laser desorption/ionization mass spectrometric imaging. *Rapid Commun Mass Spectrom*. 2014;28:2471–2479.
43. Korte AR, Lee YJ. MALDI-MS analysis and imaging of small molecule metabolites with 1,5-diaminonaphthalene (DAN). *J Mass Spectrom*. 2014;49:737–741.
44. Paglia G, Williams JP, Menikarachchi L, et al. Ion mobility derived collision cross sections to support metabolomics applications. *Anal Chem*. 2014;86:3985–3993.
45. Oshikata M, Sugiura Y, Yokota N, Setou M. MALDI imaging with ion-mobility MS: Waters Corporation*Imaging Mass Spectrometry*: Springer; 2010:221–231.
46. Neumann EK, Migas LG, Allen JL, Caprioli RM, Van de Plas R, Spraggins JM. Spatial metabolomics of the human kidney using maldi trapped ion mobility imaging mass spectrometry. *Anal Chem*. 2020.
47. Paglia G, Kliman M, Claude E, Geromanos S, Astarita G. Applications of ion-mobility mass spectrometry for lipid analysis. *Anal Bioanal Chem*. 2015;407:4995–5007.
48. Langridge JI, Claude E. Matrix-assisted laser desorption and desorption electrospray ionization mass spectrometry coupled to ion mobility. In *Ion Mobility-Mass Spectrometry*, Springer; 2020:245–265.
49. Midey A, Olivos H, Shrestha B. Spatial mapping of cellular metabolites using DESI ion mobility mass spectrometry. In *Single Cell Metabolism*, Springer; 2020:181–190.
50. Shrestha B, Vertes A. High-throughput cell and tissue analysis with enhanced molecular coverage by laser ablation electrospray ionization mass spectrometry using ion mobility separation. *Anal Chem*. 2014;86:4308–4315.
51. Bodnar WM, Blackburn RK, Krise JM, Moseley MA. Exploiting the complementary nature of LC/MALDI/MS/MS and LC/ESI/MS/MS for increased proteome coverage. *J Am Soc Mass Spectrom*. 2003;14:971–979.
52. Tomlinson L, Fuchser J, Fütterer A, et al. Using a single, high mass resolution mass spectrometry platform to investigate ion suppression effects observed during tissue imaging. *Rapid Commun Mass Spectrom*. 2014;28:995–1003.
53. Nadler WM, Waidelich D, Kerner A, et al. MALDI versus ESI: the impact of the ion source on peptide identification. *J Proteome Res*. 2017;16:1207–1215.
54. Vaikkinen A, Shrestha B, Kauppila TJ, Vertes A, Kostiainen R. Infrared laser ablation atmospheric pressure photoionization mass spectrometry. *Anal Chem*. 2012;84:1630–1636.

55. Stolee JA, Walker BN, Zorba V, Russo RE, Vertes A. Laser–nanostructure interactions for ion production. *Phys Chem Chem Phys*. 2012;14:8453–8471.
56. Stopka SA, Rong C, Korte AR, et al. Molecular imaging of biological samples on nanophotonic laser desorption ionization platforms. *Angew Chem Int Ed Engl*. 2016;55:482–486.
57. Palermo A. Charting metabolism heterogeneity by nanostructure imaging mass spectrometry: from biological systems to subcellular functions. *J Am Soc Mass Spectrom*. 2020.

Chapter 15

Imaging mass spectrometry: Glycans

Chapter Outline

Glycans are oligosaccharides that are covalently linked to amino acid side chains of glycoproteins. Glycosylation is a common cotranslational modification during synthesis or post-translational modification after the initial synthesis.[1] Glycans, generally found on the exterior surface of cells, have major metabolic, structural, and physical roles in biological systems. The same glycan might have different functions at different locations, during a different developmental stage, or different environmental conditions.[2]

Many infectious agents or symbiotic organisms recognize glycans with a high degree of specificity due to the dense coating of diverse glycan species on all cell walls and extracellular molecules.[3] There are three major motifs of glycosylation, N-linked glycans attached to asparagine residues, O-linked glycans attached to serine or threonine residues, or glycosaminoglycan. Glycans have a fundamental role in biomolecular processes related to cancer such as cell signaling, tumor invasion, tumor angiogenesis, immune modulation metastasis, etc.[4] Specific glycans can serve as biomarkers of disease status. Changes in glycosylation regulate the progression of cancer, giving biomarker targets for potential therapeutic intervention. For example, a specific N-linked glycosylation profile was observed in the areas of tumor tissues with necrosis.[5]

The basic strategy of glycan analysis by mass spectrometry (MS) involves removing the glycan with a suitable glycosidase, performing optional separation, and using a mass spectrometer to measure the mass of the glycans. Liquid chromatography MS is a commonly used analytical technique for analyzing glycans. Ion mobility separation coupled to liquid chromatography/MS is an emerging technique to separate and characterize glycan and other polysaccharides, as well as a tool to understand the conformational change in proteins after glycan binding.[6] Matrix-assisted laser desorption/ionization (MALDI) MS has been utilized to oligosaccharides, such as N-linked glycans since 1990s.[7–8] MALDI time-of-flight was used to mass spectrometers were used to perform

Introduction to Spatial Mapping of Biomolecules by Imaging Mass Spectrometry. DOI: 10.1016/C2018-0-03962-6

glycan profiling.[9] MALDI MS/MS with traveling wave ion mobility for the structural analysis of N-linked glycans released from glycoproteins in negative ion mode using 2,4,6-trihydroxyacetophenone (THAP) doped with ammonium nitrate.[10] Detailed discussions on the utility of MS tools on glycomics and glycan analysis are reviewed here.[11–12]

Mass spectrometer can image N-linked glycans after cleaving off the molecules by an endoglycosidase enzyme, peptide N-glycosidase F (PNGase F), using a direct ionization source. PNGase F can be applied on fresh tissue directly or on an archived formalin-fixed paraffin-embedded (FFPE) tissue after washing for release of N-glycans for imaging MS.[13]

MALDI imaging of glycans

The initial workflow for mapping the spatial distribution of N-linked glycans within tissues used on-tissue application and incubation of PNGase F to release N-linked glycans and imaging performed by MALDI MS.[14] MALDI mass spectrum and images of fresh-frozen mouse brain tissue treated with or without PNGase F are shown in Fig. 15.1. Briefly, fresh-frozen tissue sections (e.g., mouse brains, human kidney) were sequentially washed twice with 70%

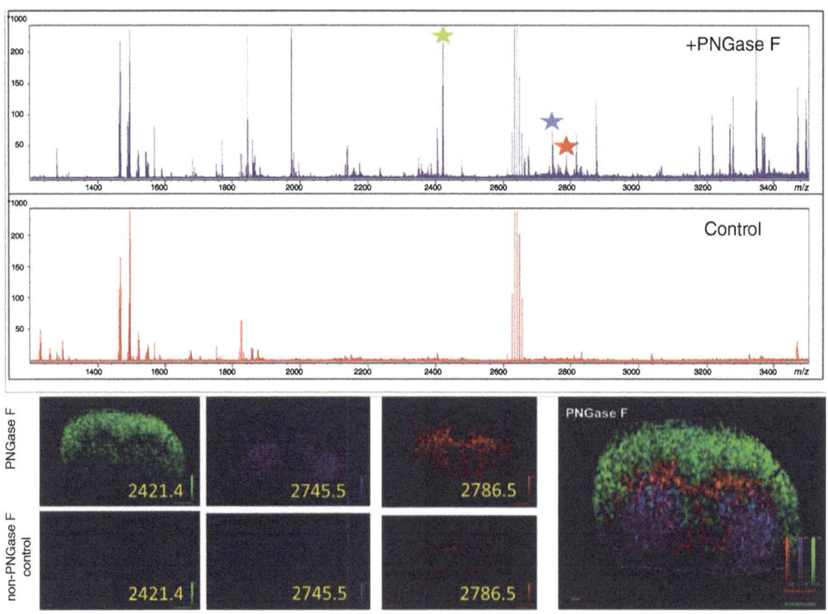

FIG. 15.1 MALDI mass spectrum of mouse brain tissue treated with on-tissue PNGase F digestion compared with one without the digestion, which also displayed regional expression distributions in the mouse brain. The bottom panel shows an image of N-glycans with or without the PNGaseF application. An overlay of the three N-glycans is shown on the bottom right. Reprinted (adapted) with permission from Powers TW, Jones EE, Betesh LR, et al. Matrix assisted laser desorption ionization imaging mass spectrometry workflow for spatial profiling analysis of N-linked glycan expression in tissues. Anal Chem. 2013;85:9799–9806. Copyright 2013 American Chemical Society.[14]

ethanol and once with 100% ethanol for a couple of minutes. PNGase F (1 mg/mL) aqueous solution was sprayed, and sprayed tissue was incubated at 37 °C for about 2 hours in a humidified chamber. The tissue section was dried. After applying 2,5-dihydroxybenzoic acid matrix in 50% methanol with 1% trifluoroacetic acid, tissue was imaged by MALDI.[14–15] The on-tissue digestion workflow with PNGase F is compatible with MALDI imaging and subsequent histopathology analysis. A detailed discussion of protocol describing MALDI imaging analysis of N-glycans imaging by MALDI MS is described in the references here.[13,16–19] MALDI analysis of N-linked glycans on FFPE murine kidney was done along with on-target MS analysis of extracts and in situ MALDI profiling to help strengthen the identification of the molecular species.[20]

N-glycans, O-glycans, and glycosphingolipid variations in colorectal cancer include an increase in β1,6-branching, higher abundance of (poly-) *N*-acetyllactosamine extensions of N-glycans, increase in truncated high-mannose type glycans, and decrease in bisected structures. Colorectal cancer is also related to O-glycan changes such as decreased core 3 and 4 glycans, increased core 1 glycans, (sialyl) T-antigen, (sialyl) Tn-antigen, and general higher density of O-glycans, as well as changes in glycosphingolipid.[21] Prostate produces many glycoproteins. Prostate-specific antigen is a frequently analyzed glycoprotein biomarker for detecting and progression of prostate cancer. Several research studies have characterized the changes in O-linked and N-linked glycosylations related to prostate cancer that are reviewed here.[22]

The signal of glycan imaging can be improved by using different ion sources or improving ionization efficiency. MALDI-2 is a new variant of MALDI that improves the ion yield by irradiating the initial MALDI plume with a second orthogonal laser pulse.[23] MALDI-2 analysis of N-glycans has reported showing about one order of improvement in positive ion mode and three orders of magnitude in the negative ion mode.[24] The fragmentation of negative ions provides rich structural information on the composition of N-glycans and can be used in direct tandem MS analysis from thin tissue sections.

On-tissue chemical derivatization using a hydrazine reagent (Girard's reagent P) has been shown to improve imaging N-glycans from tissue sections and other signal-to-noise ratios enhancement for other sugars such as from 28-fold for maltose to 230-fold for glucose.[25] Briefly, FFPE tissue sections were sequentially deparaffinized by three washes with xylene and washed by decreasing concentrations of ethanol to remove xylene. The tissue section was heated in citrate buffer (pH 6.0) at 95°C for antigen retrieval by breaking formalin protein crosslinks and to increases enzyme access. N-glycans were released by spraying PNGase F dissolved in an ammonium bicarbonate solution and incubating at 37°C in a humidified chamber. The chemical derivatization was performed by spraying GP reagent in 50% (v/v) methanol aqueous containing 10% acetic acid was sprayed onto the tissue sections and incubating in a chamber with acetic acid vapor. Finally, the MALDI matrix, such as 2,5-dihydroxybenzoic acid, was sprayed onto the tissue section.[25]

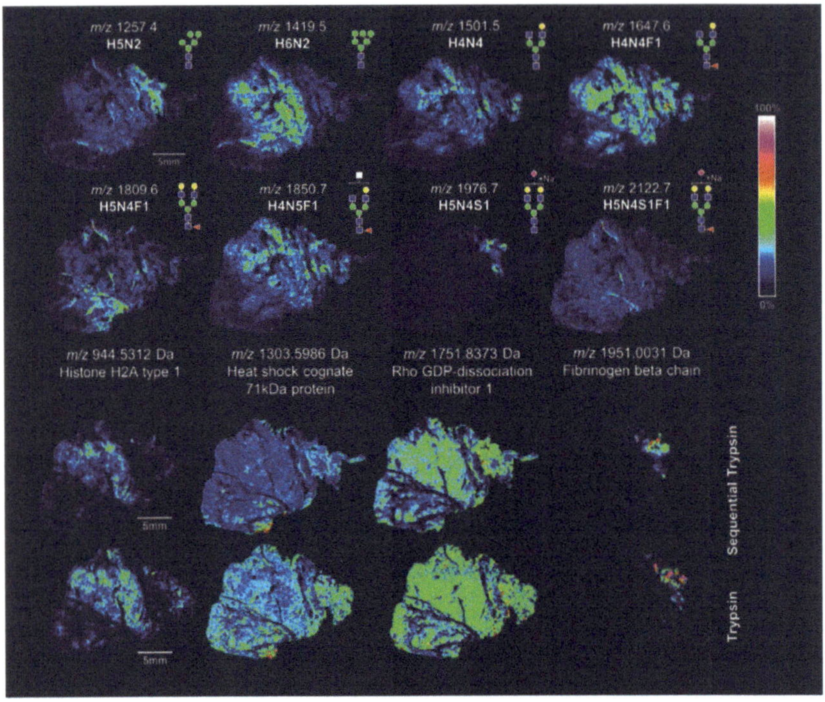

FIG. 15.2 The top two rows show MALDI MS images of N-glycans before sequential trypsin digestion. The bottom two rows show MALDI MS images of peptides after sequential trypsin digestion after PNGase F application for N-glycans and trypsin digestion by itself. Sequential trypsin digestion and control show similar peptide distributions. All of the images were obtained from RMS-normalized datasets. Reprinted (adapted) with permission from Heijs B, Holst S, Briaire-de Bruijn IH, et al. Multimodal mass spectrometry imaging of N-glycans and proteins from the same tissue section. Anal Chem. 2016;88:7745–7753. Copyright 2016 American Chemical Society.[26]

Multimodal imaging of N-linked glycans and proteins from the same FFPE tissue section was demonstrated for several tumors, such as epithelial and mesenchymal tumor types. By combining the sequential imaging and sampling workflow, N-glycan and peptides were imaged without significant lateral diffusion of the analyte molecules or change in sensitivity due to the additional sample preparation.[26] MALDI images of N-glycans before sequential trypsin digestion and peptide images obtained with or without sequential workflow are shown in Fig. 15.2.

Targeted glycan imaging

The linkage information between the monosaccharides is generally not obtained in MALDI because the sialic acid structure of glycan molecules is labile and fragmented due to in-source or postsource decay. However, the linkage may have biological significance, such as α2,3-linked sialic acids are associated with cancer, but not the α2,6-linked counterpart.[4,27] Derivatization of sialic acid using ethyl esterification and dimethylamidation can successfully stabilize sialic acids in a linkage without noticeable molecular diffusion in tissue. Imaging derivatized tissue provides an accurate linkage-specific spatial distribution of glycan on the tissue.[28] Briefly, on-tissue derivatization is performed after paraffin removal by sequential solvent washes with xylene, ethanol, and water. The derivatization was performed by incubating the tissues slides in a derivatization solution of 1-ethyl-3-(3-dimethylamino)propyl)carbodiimide as a carboxylic acid activator and 1-hydroxybenzotriazole as a catalyst with dimethylamine in dimethyl sulfoxide in a 60 °C oven for 1 hour followed by 2-hour incubation after the addition of an ammonia solution. After derivatization, tissue sections are washed by sequential ethanol and water before incubation with PNGase F and α-2,4-hydroxycinammic acid matrix. A comparison of MALDI images after PNGase F enzyme application, enzyme application with derivatization, and just derivatization without enzyme application is shown in Fig. 15.3.

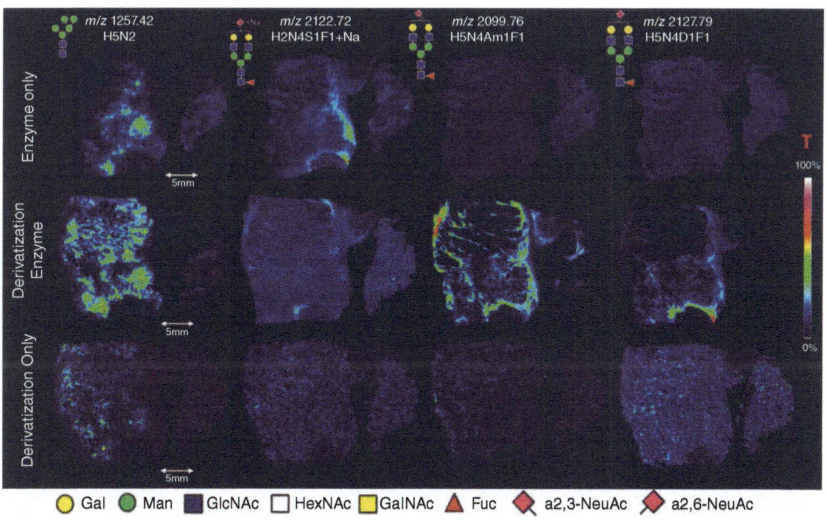

FIG. 15.3 A comparison of MALDI images of colon carcinoma FFPE tissues after PNGase F enzyme application, an enzyme with derivatization and just derivatization, is shown. The high-mannose N-glycan Hex5HexNAc2 (H5N2) similar localizations in the enzyme only and derivatized section. The sialylated species Hex5HexNAc4-NeuAc1dHex1 (H5N4S1F1) at m/z 2122.72 detected in the underivatized enzyme only section, whereas the derivatized sialylated N-glycans Hex5HexNAc4(α2,3)-NeuAc1dHex1 (H5N4Am1F1) at m/z 2099.76 and Hex5HexNAc4(α2,6)NeuAc1dHex1 (H5N4D1F1) at m/z 2127.79 each showed unique distributions. *Note*: Yellow circle = galactose, green circle = mannose, blue square = N-acetylglucosamine, white square = N-acetylhexosamine, yellow square = N-acetylgalactosamine, red triangle = fucose, purple diamond = N-acetylneuraminic acid. Reprinted (adapted) with permission from Holst S, Heijs B, de Haan N, et al. Linkage-specific in situ sialic acid derivatization for N-glycan mass spectrometry imaging of formalin-fixed paraffin-embedded tissues. Anal Chem. 2016;88:5904–5913. Copyright 2016 American Chemical Society.[28]

Specific chemical modifications in N-linked glycans, such as the addition of fucose sugar units or fucosylation, have been linked with cancers. Imaging MS workflow using recombinant endoglycosidase F3 or Endo F3 can cleave a core-fucosylated N-glycans linked to glycoproteins, can show specific anomeric linkages and the site of fucosylation.[29] Endo F3 cleaves between the two core N-acetylglucosamine residues at the protein attachment site result in a mass shift of 349 for the core-fucosylated N-glycan in comparison to PNGaseF.[29] The imaging of liver and prostate tissues showed core-fucosylated glycans at specific tissue regions of tissue.[29]

Outlook

In summary, the fundamental workflow for obtaining N-glycans' spatial distribution from various tissue samples is established using MALDI imaging MS. The sensitivity of N-glycan imaging from the tissue can be enhanced by increasing the ionization efficiency of N-glycans by chemical derivatization or using a new ionization approach such as MALDI-2. An enhancement in the sensitivity is not just allowing imaging low-abundance species, but also imaging of glycans in small pixel size such as single-cell level. The advancements in glycan imaging can anticipate to facilitate investigation of glycosylation changes in more detail, strengthening our fundamental understanding of molecular biology. The glycan imaging also shows a promise in future utility in clinical imaging for disease such as cancer.

References

1. Varki A. Biological roles of glycans. *Glycobiology*. 2016;27:3–49.
2. Varki A, Cummings RD, Esko JD, et al. *Essentials of Glycobiology*. Cold Spring Harbor Laboratory Press; 2015: Accessed 23 January 2021.
3. Esko JD, Kimata K, Lindahl U. Chapter 17 Proteoglycans and Sulfated Glycosaminoglycans. Essentials of Glycobiology. *Cold Spring Harbor Laboratory Press*; 2009.
4. Pinho SS, Reis CA. Glycosylation in cancer: mechanisms and clinical implications. *Nat Rev Cancer*. 2015;15:540–555.
5. Scott DA, Norris-Caneda K, Spruill L, et al. Specific N-linked glycosylation patterns in areas of necrosis in tumor tissues. *Int J Mass Spectrom*. 2019;437:69–76.
6. Gray CJ, Thomas B, Upton R, et al. Applications of ion mobility mass spectrometry for high throughput, high resolution glycan analysis. *Biochim Biophys Acta*. 2016;1860:1688–1709.
7. Mock KK, Davey M, Cottrell JS. The analysis of underivatised oligosaccharides by matrix-assisted laser desorption mass spectrometry. *Biochem Biophys Res Commun*. 1991;177:644–651.
8. Harvey DJ. Matrix-assisted laser desorption/ionization mass spectrometry of carbohydrates. *Mass Spectrom Rev*. 1999;18:349–450.
9. Harvey DJ, Küster B, Wheeler SF, Hunter AP, Bateman RH, Dwek RA. Matrix-assisted laser desorption/ionization mass spectrometry of N-linked carbohydrates and related compounds. In: Burlingame AL, Carr SA, Baldwin MA, eds. *Mass Spectrometry in Biology & Medicine*. Totowa, NJ: Humana Press; 2000:403–437.
10. Harvey DJ, Scarff CA, Crispin M, Scanlan CN, Bonomelli C, Scrivens JH. MALDI-MS/MS with traveling wave ion mobility for the structural analysis of N-linked glycans. *J Am Soc Mass Spectrom*. 2012;23:1955–1966.
11. Zaia J. Mass spectrometry and glycomics. *Omics*. 2010;14:401–418.

12. Gaunitz S, Nagy G, Pohl NLB, Novotny MV. Recent advances in the analysis of complex glycoproteins. *Anal Chem*. 2017;89:389–413.
13. Drake RR, Powers TW, Norris-Caneda K, Mehta AS, Angel PM. In situ imaging of N-glycans by MALDI imaging mass spectrometry of fresh or formalin-fixed paraffin-embedded tissue. *Curr Protoc Protein Sci*. 2018;94:e68.
14. Powers TW, Jones EE, Betesh LR, et al. Matrix assisted laser desorption ionization imaging mass spectrometry workflow for spatial profiling analysis of N-linked glycan expression in tissues. *Anal Chem*. 2013;85:9799–9806.
15. Powers TW, Neely BA, Shao Y, et al. MALDI imaging mass spectrometry profiling of N-glycans in formalin-fixed paraffin embedded clinical tissue blocks and tissue microarrays. *PloS One*. 2014;9:e106255.
16. Drake RR, West CA, Mehta AS, Angel PM. MALDI mass spectrometry imaging of N-linked glycans in tissues. In: Yamaguchi Y, Kato K, eds. *Glycobiophysics*: Springer Singapore: Singapore; 2018:59–76.
17. Briggs MT, Ho YY, Kaur G, et al. N-Glycan matrix-assisted laser desorption/ionization mass spectrometry imaging protocol for formalin-fixed paraffin-embedded tissues. *Rapid Commun Mass Spectrom*. 2017;31:825–841.
18. Stanback AE, Conroy LR, Young LEA, et al. Regional N-glycan and lipid analysis from tissues using MALDI-mass spectrometry imaging. *STAR Protocols*. 19 March 2021;2:100304.
19. Angel PM, Mehta A, Norris-Caneda K, Drake RR. MALDI imaging mass spectrometry of n-glycans and tryptic peptides from the same formalin-fixed, paraffin-embedded tissue section. In: Sarwal MM, Sigdel TK, eds. *Tissue Proteomics: Methods and Protocols*. New York, NY: Springer New York; 2018:225–241.
20. Gustafsson OJR, Briggs MT, Condina MR, et al. MALDI imaging mass spectrometry of N-linked glycans on formalin-fixed paraffin-embedded murine kidney. *Anal Bioanal Chem*. 2015;407:2127–2139.
21. Holst S, Wuhrer M, Rombouts Y. Chapter six - glycosylation characteristics of colorectal cancer. In: Drake RR, Ball LE, eds. *Advances in Cancer Research*: Academic Press; 2015:203–256 Vol. 126.
22. Drake RR, Jones EE, Powers TW, Nyalwidhe JO. Chapter ten - altered glycosylation in prostate cancer. In: Drake RR, Ball LE, eds. *Advances in Cancer Research*: Academic Press; 2015:345–382 Vol. 126.
23. Soltwisch J, Kettling H, Vens-Cappell S, Wiegelmann M, Müthing J, Dreisewerd K. Mass spectrometry imaging with laser-induced postionization. *Science*. 2015;348:211–215.
24. Heijs B, Potthoff A, Soltwisch J, Dreisewerd K. MALDI-2 for the enhanced analysis of N-linked glycans by mass spectrometry imaging. *Anal Chem*. 2020;92(20):13904–13911.
25. Zhang H, Shi X, Vu NQ, et al. On-tissue derivatization with Girard's reagent P enhances N-glycan signals for formalin-fixed paraffin-embedded tissue sections in MALDI mass spectrometry imaging. *Anal Chem*. 2020.
26. Heijs B, Holst S, Briaire-de Bruijn IH, et al. Multimodal mass spectrometry imaging of N-glycans and proteins from the same tissue section. *Anal Chem*. 2016;88:7745–7753.
27. Choi H-J, Chung T-W, Choi H-J, et al. Increased α2-6 sialylation of endometrial cells contributes to the development of endometriosis. *Exp Mol Med*. 2018;50:1–12.
28. Holst S, Heijs B, de Haan N, et al. Linkage-specific in situ sialic acid derivatization for n-glycan mass spectrometry imaging of formalin-fixed paraffin-embedded tissues. *Anal Chem*. 2016;88:5904–5913.
29. West CA, Liang H, Drake RR, Mehta AS. New enzymatic approach to distinguish fucosylation isomers of N-linked glycans in tissues using MALDI imaging mass spectrometry. *J Proteome Res*. 2020;19:2989–2996.

Chapter 16

Imaging mass spectrometry: steroids mapping using on-tissue chemical derivatization

Chapter Outline

Steroids are clinically significant hormones derived from cholesterol in the adrenal cortex, the gonads, and the placenta.[1] Steroids are routinely analyzed in biofluids, such as plasma and urine, to diagnose endocrine disorders.[2] Steroids can be measured by immunoassay, or mass spectrometry (MS) coupled with gas chromatography or liquid chromatography (LC).[3-4] Tandem MS is an ideal tool for detecting steroids from biological samples because it has high specificity. In contrast, immunoassays lack the specificity of antibodies to measure a specific steroid.[5] Ionization efficiency or sensitivity of steroid detection can depend on the type of ion source used. For example, electrospray ionization was sensitive for steroids containing a 3-one, 4-ene functional group, and ones containing at least one ketone group without conjugation, but not for ones containing hydroxy groups only. The atmospheric pressure chemical ionization had the highest sensitivity for all structural groups.[6]

Nanostructure-initiator mass spectrometry, a laser desorption ionization platform using porous silicon surface, has been used to analyze steroids. Nanostructure-initiator mass spectrometry was spray deposited with $AgNO_3$ to provide cationization agents before desorption of the fluorinated polymer initiator for the Ag^+ cationization of steroids.[7] Desorption atmospheric pressure photoionization is an ambient photoionization technique, where ionization occurs in gas-phase ion–molecule reactions using a heated jet of gas and vaporized spray solvent.[8] Desorption atmospheric pressure photoionization has been found to enhance the efficiency of ionization of steroids, such as testosterone, by proton transfer.[9] Atmospheric solids analysis probe is another ion source that can be used to rapidly detect anabolic steroid esters.[10]

Introduction to Spatial Mapping of Biomolecules by Imaging Mass Spectrometry. DOI: 10.1016/C2018-0-03962-6

Imaging MS can plot any detected molecules' spatial distribution in a tissue section. Steroid imaging by MS suffers from suboptimal detection of molecules from the tissue sections. Chemical derivatization has been utilized to improve the detection of molecules-of-interest, including steroids, challenging to ionize, labile to fragmentation during analysis, or masked by interfering ions.[11] Chemical derivatization performs chemical modification of a specific group of molecules to their derivatized counterparts within tissue using organic chemical reactions. Molecules in tissue may also undergo other chemical changes, such as enzymatic digestion, immunohistochemistry tagging, the breaking of bonds, etc. For example, antibody labeling has been used for imaging of protein by detecting the elemental mass tags in the antibodies labels by imaging mass cytometry and multiplexed ion-beam imaging.[12-13]

On-tissue chemical derivatization workflow

Chemical derivatization strategies are often utilized in LC–MS to improve ionization efficiency and selectivity, simplify structure elucidation by unique fragmentation, and enhance chromatographic separation.[14] Chemical derivatization is a precise chemical reaction between a reactive functional group of the analyte and the derivatization reagent. For example, *N*-methyl-*N*-trimethylsilyltrifluoroacetamide was found to be a steroid derivatizing reagent for gas chromatography–MS analysis.[15] The resulting derivatized analyte has different chemical properties than the underivatized analyte, such as solubility, liable nature, ionization efficiency, and mass. Chemical derivatization of steroids can enhance their sensitivity for detection by LC–MS.[16]

On-tissue chemical derivatization is implemented by adding a reactive reagent to the tissue to selectively modify a group of molecules within the tissue into a surrogate molecular species that readily undergo ionization and detection by a mass spectrometer. On-tissue chemical derivatization improves the detection of a targeted group of molecules and enables their imaging using an MS technique. For example, on-tissue chemical derivatization of N-terminal peptide by sulfonation agents or *N*-succinimidyloxycarbonylmethyl) tris(2,4,6-trimethoxyphenyl) phosphonium bromide has found to improve protein identification by producing characteristic MS/MS spectra in MALDI imaging MS.[17] On-tissue chemical derivatization reactions for three common derivatization methods used to increase metabolite coverage of imaging MS—coniferyl aldehyde for primary amines, Girard's reagent T for carbonyl groups, and 2-picolylamine for carboxylic acids—are shown in Fig. 16.1.[18]

On-tissue chemical derivatization can be performed by depositing the reactant on the tissue sections, pretreating slides before mounting of the tissue section, or by submerging the tissue section in the reactant dissolved in the solvent. For MALDI, the reactive derivatizing reagent is added before the matrix application unless the matrix itself acts as the reagent. The derivatization reagent is deposited on tissue, generally using a pneumatic sprayer like the

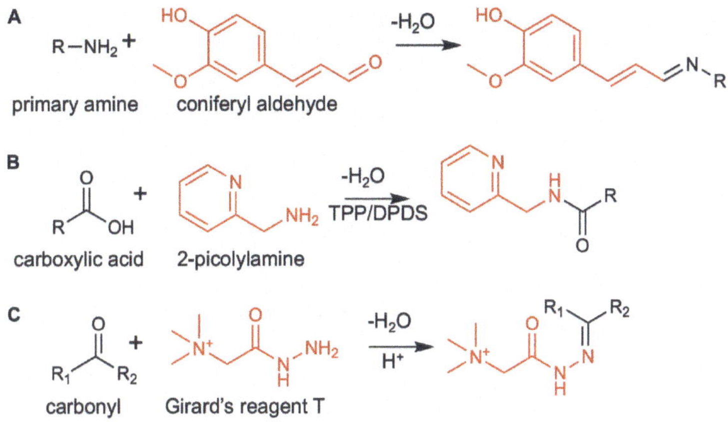

FIG. 16.1 Three commonly used on-tissue chemical derivatization reactions with the reagents in red is shown, (A) coniferyl aldehyde (CA) and a primary amine, (B) 2-picolylamine (2-PA) and a carboxylic acid, and (C) Girard's reagent T (GT) and a carbonyl group. Reprinted with permission from Dueñas ME, Larson EA, Lee YJ. Toward mass spectrometry imaging in the metabolomics scale: increasing metabolic coverage through multiple on-tissue chemical modifications. Front Plant Sci. 2019;10:860–860.[18] Creative commons.

MALDI matrix. In a study, electrospray deposition of reagents compared with airbrush showed improved sensitivity and decreased analyte delocalization for imaging.[19] Other mechanism of deposition includes placing tissue in the vapor chamber, sublimation, or by using electrospray.

A transfer of a thin layer of tissue using laser has been found to improve on-tissue chemical derivatization. The significant increase in signal intensity for derivatized analytes for the laser-assisted tissue transfer section in rat brain tissues can be attributed to the minimization of ion suppression effects due to the reduced thickness of tissue.[20] Fig. 16.2 shows a conceptual workflow for several on-tissue derivatization workflows. Detailed protocols on the specific on-tissue chemical derivatization step refer to the respective scientific literature on that procedure.

On-tissue chemical derivatization and imaging of steroids

Steroids are challenging to detect directly off tissue due to low ionization. Steroids with a ketone group can be derivatized with the Girard P reagent into hydrazones contain a quaternary nitrogen atom that can be readily ionized for analysis by MALDI—with two orders of magnitude increase in sensitivity. Steroids without a ketone group, but with a 3β-hydroxy-Δ^5 function, can be modified to 3-oxo-Δ^4 steroids, which can be then derivatized Girard P reagent.[21]

Girard T reagent can be used to perform on-tissue chemical derivatization of steroids. The Girard T reagent derivatives of steroids, such as 11-dehydrocorticosterone and corticosterone, were mapped in rat adrenal gland and mouse

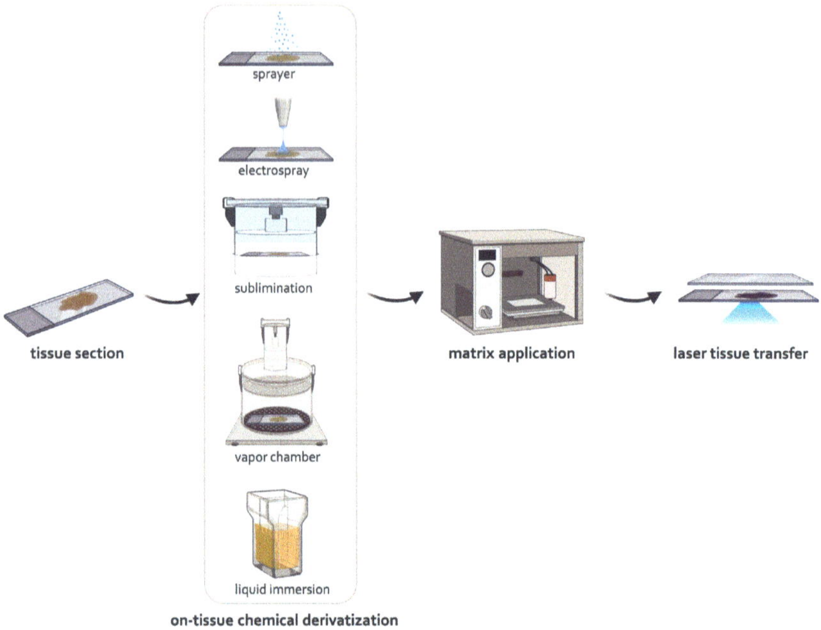

FIG. 16.2 On-tissue chemical derivatization is performed on a tissue section before matrix application for MALDI unless the derivatization reagent is also a matrix. The chemical reagent can be performed by spraying tissue with a pneumatic sprayer or electrospray, depositing the reagent in a vapor chamber, or sublimating or submerging tissue in an aqueous solution of reagent.

brain sections by high-resolution MALDI MS as shown in Fig. 16.3.[22] The identification of steroids was confirmed with collision-induced dissociation and liquid extraction surface analysis. On-tissue chemical derivatization by Girard's reagent T was used to improve the ionization efficiency of steroids, such as testosterone and 5α-dihydrotestosterone in testis, by adding a polar group. The derivatized steroids, such as testosterone, was easily detected off tissue and imaged in testis tissues of mice treated with human chorionic gonadotropin.[23-24] On-tissue chemical derivatization enhances the ionization efficiency of steroids. However, in order to discriminate between clinically relevant structural isomers will require the utilization of tandem MS. An ion trap MALDI was used to distinguish between derivatized mineralocorticoid aldosterone and glucocorticoid cortisol from structural isomers based on unique fragmentation as shown in Fig. 16.4.[25]

Vitamin D is a group of steroids that have an open ring chemical structure. Vitamin D is an essential metabolite clinically associated with infection, reproduction, cardiovascular condition, cancer. Various forms of vitamin D can be measured in biological samples by immunoassays, such as enzyme-linked immunosorbent assay or chemiluminescent immunoassay, or using mass

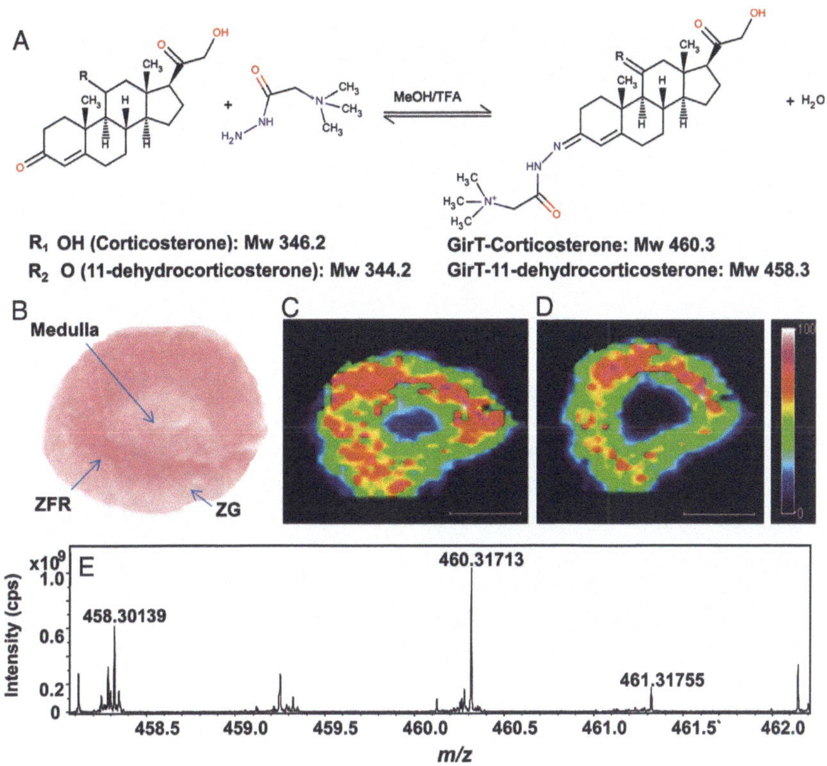

FIG. 16.3 (A) Chemical derivatization of steroids (corticosterone and 11-dehydrocorticosterone) with Girard T reagent, (B) hematoxylin and eosin image of rat adrenal gland stained (ZG, zona glomerulosa; ZFR, zona fasciculata reticularis), MALDI imaging MS of chemical derivatives of (C) corticosterone at *m/z* 460.3 and (D) 11-dehydrocorticosterone at *m/z* 458.3, (E) corresponding mass spectrum. The scale bar is 2 mm wide. Adapted with permission from Cobice DF, Mackay CL, Goodwin RJA, et al. Mass spectrometry imaging for dissecting steroid intracrinology within target tissues. Anal Chem. 2013;85:11576–11584. Copyright 2013 American Chemical Society.[22]

spectrometer using electrospray ionization.[26-27] Vitamin D metabolites are often measured LC–MS after chemical derivatization using commercial reagents such as Amplifex, which enhances the detection of any molecule with a *cis*-diene system, such as vitamin D_3, vitamin D_2, and their analogs, by up to three orders of magnitude.[28] On-tissue chemical derivatization with Amplifex was utilized to spatial map distribution of vitamin D metabolites in murine kidney tissue sections as shown in Fig. 16.5.[29]

An alternate way of detecting steroids includes novel MALDI matrix composition. Steroids, such as estrogens, androgens, and progestogens, can be detected by MALDI using cyclodextrin-supported 2,5-dihydroxybenzoic acid as the matrix, where cyclodextrin and the matrix is heated with sonication resulting in a complex formation that acts as a matrix.[30]

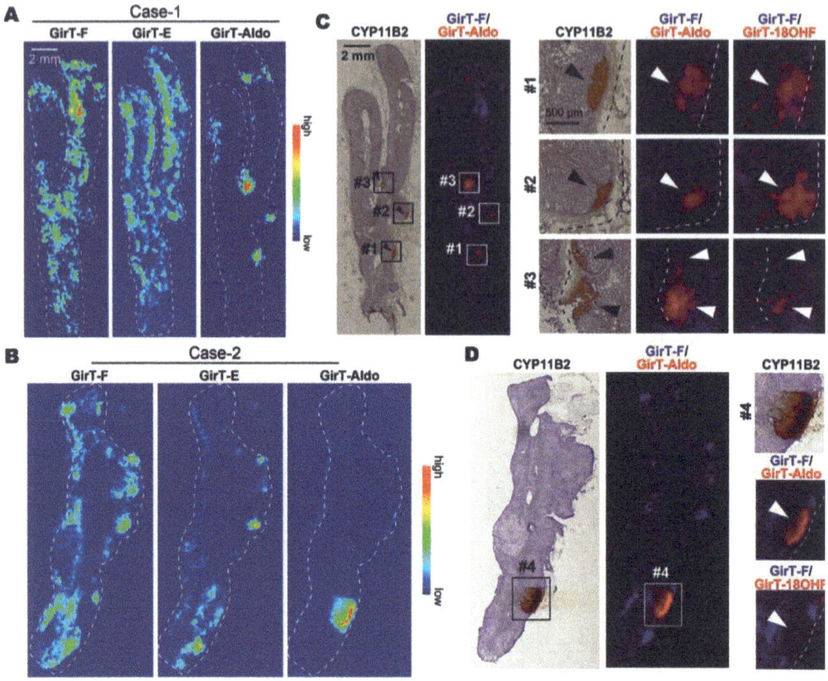

FIG. 16.4 Tandem MALDI imaging can show differential distribution structural isomers mineralocorticoid aldosterone (Aldo) and glucocorticoid cortisol (F) using MS3 fragments of Girard's T (GirT) derivatized steroids in the healthy human adrenal gland. Imaging results in whole sections with CYP11B2 immunoreactivity are shown along with MALDI images of GirT-F (*m/z* 387.2), GirT-E (*m/z* 385.2), GirT-Aldo (*m/z* 397.2), and GirT-18-OHF (*m/z* 413.2). The arrowheads represent aldo-producing cell clusters regions. Reprinted with permission from Takeo E, Sugiura Y, Uemura T, et al. Tandem mass spectrometry imaging reveals distinct accumulation patterns of steroid structural isomers in human adrenal glands. Anal Chem. 2019;91:8918–8925. Copyright 2013 American Chemical Society.[25]

Outlook

Imaging MS of steroids is mainly hindered by poor detection sensitivity due to low ionization efficiency. Chemical derivatization for mass spectrometric analysis is a well-established technique to analyze hard-to-detect molecules as derivatized molecular entities. The chemistry can be easily adapted for on-tissue derivatization workflow to selectively target steroids that are hard to ionize or masked by interference. On-tissue chemical derivatization has been shown to help detect and image steroid molecules on the tissue by imaging MS without any loss of sensitivity or delocalization. These on-tissue derivatization workflows and the novel matrix will expand imaging MS's utility beyond imaging of the limited number of abundant and easily ionizable compounds and more clinically relevant molecules such as steroids at physiological concentrations.

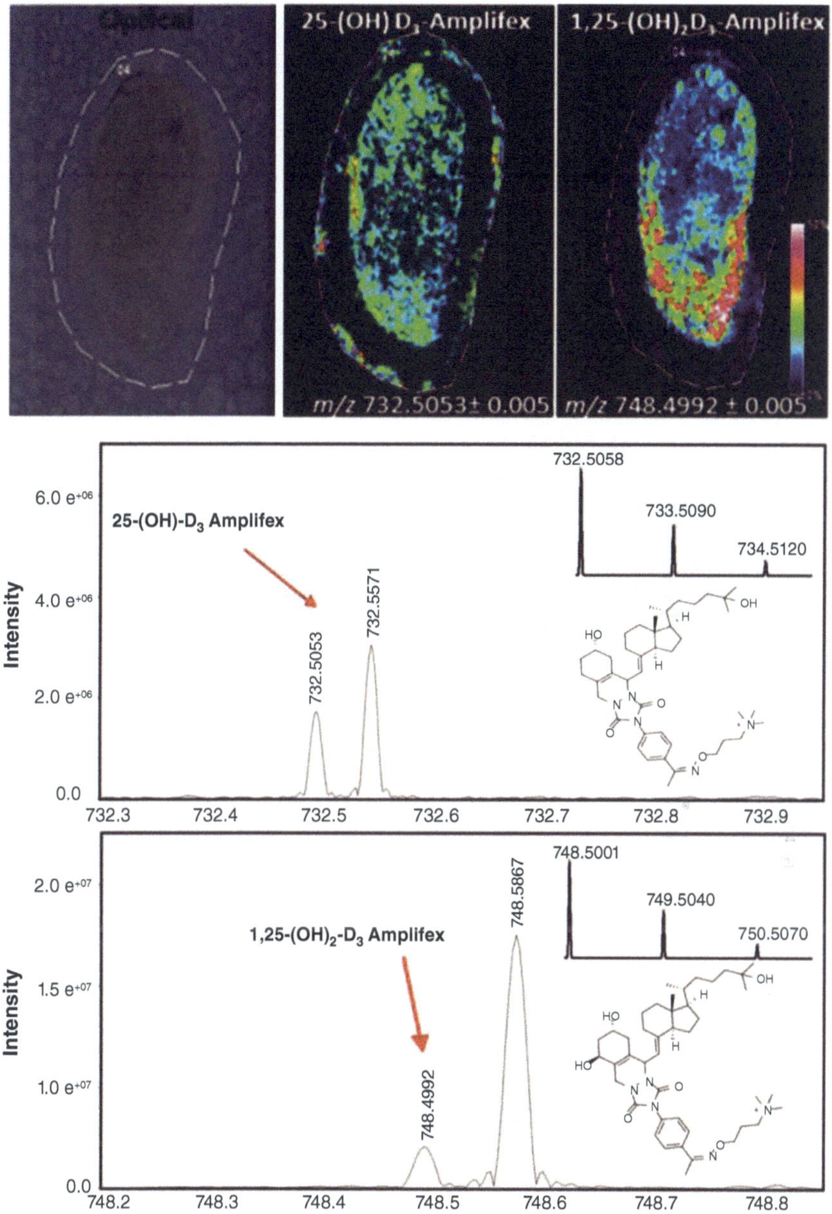

FIG. 16.5 Spatial distribution of endogenous vitamin D metabolites, 25-OH-D3 and 1,25-(OH)2-D3, in a mouse kidney section after on-tissue chemical derivatization by Amplifex is shown on the top. The mass spectra for the derivatives are shown at the bottom. MALDI MS images were acquired at 100 μm pixel size and visualized with TIC normalization. Adapted with permission from Smith KW, Flinders B, Thompson PD, et al. Spatial localization of vitamin D metabolites in mouse kidney by mass spectrometry imaging. ACS Omega. 2020;5:13430–13437.[29] Creative commons.

References

1. Holst JP, Soldin OP, Guo T, Soldin SJ. Steroid hormones: relevance and measurement in the clinical laboratory. *Clin Lab Med*. 2004;24:105–118.
2. Andrew R. Clinical measurement of steroid metabolism. *Best Pract Res Clin Endocrinol Metab*. 2001;15:1–16.
3. Hsing AW, Stanczyk FZ, Bélanger A, et al. Reproducibility of serum sex steroid assays in men by RIA and mass spectrometry. *Cancer Epidemiol Biomarkers Prev*. 2007;16:1004–1008.
4. Dorgan JF, Fears TR, McMahon RP, Aronson Friedman L, Patterson BH, Greenhut SF. Measurement of steroid sex hormones in serum: a comparison of radioimmunoassay and mass spectrometry. *Steroids*. 2002;67:151–158.
5. Soldin SJ, Soldin OP. Steroid hormone analysis by tandem mass spectrometry. *Clin Chem*. 2009;55:1061–1066.
6. Ma Y-C, Kim H-Y. Determination of steroids by liquid chromatography/mass spectrometry. *J Am Soc Mass Spectrom*. 1997;8:1010–1020.
7. Patti GJ, Woo H-K, Yanes O, et al. Detection of carbohydrates and steroids by cation-enhanced nanostructure-initiator mass spectrometry (NIMS) for biofluid analysis and tissue imaging. *Anal Chem*. 2010;82:121–128.
8. Kauppila TJ, Kuuranne T, Meurer EC, Eberlin MN, Kotiaho T, Kostiainen R. Atmospheric pressure photoionization mass spectrometry. ionization mechanism and the effect of solvent on the ionization of naphthalenes. *Anal Chem*. 2002;74:5470–5479.
9. Suni NM, Aalto H, Kauppila TJ, Kotiaho T, Kostiainen R. Analysis of lipids with desorption atmospheric pressure photoionization-mass spectrometry (DAPPI-MS) and desorption electrospray ionization-mass spectrometry (DESI-MS). *J Mass Spectrom*. 2012;47:611–619.
10. Doué M, Dervilly-Pinel G, Gicquiau A, Pouponneau K, Monteau F, Le Bizec B. High throughput identification and quantification of anabolic steroid esters by atmospheric solids analysis probe mass spectrometry for efficient screening of drug preparations. *Anal Chem*. 2014;86:5649–5655.
11. Higashi T, Ogawa S. Chemical derivatization for enhancing sensitivity during LC/ESI–MS/MS quantification of steroids in biological samples: a review. *J Steroid Biochem Mol Biol*. 2016;162:57–69.
12. Giesen C, Wang HA, Schapiro D, et al. Highly multiplexed imaging of tumor tissues with subcellular resolution by mass cytometry. *Nat Methods*. 2014;11:417–422.
13. Angelo M, Bendall SC, Finck R, et al. Multiplexed ion beam imaging of human breast tumors. *Nat Med*. 2014;20:436–442.
14. Qi B-L, Liu P, Wang Q-Y, Cai W-J, Yuan B-F, Feng Y-Q. Derivatization for liquid chromatography-mass spectrometry. *Trends Anal Chem*. 2014;59:121–132.
15. Bowden JA, Colosi DM, Mora-Montero DC, Garrett TJ, Yost RA. Enhancement of chemical derivatization of steroids by gas chromatography/mass spectrometry (GC/MS). *J Chromatogr B*. 2009;877:3237–3242.
16. Higashi T, Shimada K. Derivatization of neutral steroids to enhance their detection characteristics in liquid chromatography–mass spectrometry. *Anal Bioanal Chem*. 2004;378:875–882.
17. Franck J, El Ayed M, Wisztorski M, Salzet M, Fournier I. On-tissue N-terminal peptide derivatizations for enhancing protein identification in MALDI mass spectrometric imaging strategies. *Anal Chem*. 2009;81:8305–8317.
18. Dueñas ME, Larson EA, Lee YJ. Toward mass spectrometry imaging in the metabolomics scale: increasing metabolic coverage through multiple on-tissue chemical modifications. *Front Plant Sci*. 2019;10:860 860.

19. Wu Q, Comi TJ, Li B, Rubakhin SS, Sweedler JV. On-tissue derivatization via electrospray deposition for matrix-assisted laser desorption/ionization mass spectrometry imaging of endogenous fatty acids in rat brain tissues. *Anal Chem.* 2016;88:5988–5995.
20. Guo S, Tang W, Hu Y, et al. Enhancement of on-tissue chemical derivatization by laser-assisted tissue transfer for MALDI MS imaging. *Anal Chem.* 2020;92:1431–1438.
21. Khan MA, Wang Y, Heidelberger S, et al. Analysis of derivatised steroids by matrix-assisted laser desorption/ionisation and post-source decay mass spectrometry. *Steroids.* 2006;71:42–53.
22. Cobice DF, Mackay CL, Goodwin RJA, et al. Mass spectrometry imaging for dissecting steroid intracrinology within target tissues. *Anal Chem.* 2013;85:11576–11584.
23. Cobice DF, Livingstone DEW, Mackay CL, et al. Spatial localization and quantitation of androgens in mouse testis by mass spectrometry imaging. *Anal Chem.* 2016;88:10362–10367.
24. Shimma S, Kumada H-O, Taniguchi H, et al. Microscopic visualization of testosterone in mouse testis by use of imaging mass spectrometry. *Anal Bioanal Chem.* 2016;408:7607–7615.
25. Takeo E, Sugiura Y, Uemura T, et al. Tandem mass spectrometry imaging reveals distinct accumulation patterns of steroid structural isomers in human adrenal glands. *Anal Chem.* 2019;91:8918–8925.
26. Jukic AMZ, Hoofnagle AN, Lutsey PL. Measurement of vitamin D for epidemiologic and clinical research: shining light on a complex decision. *Am J Epidemiol.* 2017;187:879–890.
27. Farrell C-JL, Martin S, McWhinney B, Straub I, Williams P, Herrmann M. State-of-the-art vitamin A assays: a comparison of automated immunoassays with liquid chromatography–tandem mass spectrometry methods. *Clin Chem.* 2012;58:531–542.
28. Hedman CJ, Wiebe DA, Dey S, Plath J, Kemnitz JW, Ziegler TE. Development of a sensitive LC/MS/MS method for vitamin D metabolites: 1, 25 Dihydroxyvitamin D2&3 measurement using a novel derivatization agent. *J Chromatogr B.* 2014;953:62–67.
29. Smith KW, Flinders B, Thompson PD, et al. Spatial localization of vitamin D metabolites in mouse kidney by mass spectrometry imaging. *ACS Omega.* 2020;5:13430–13437.
30. Son J, Cha S. MALDI mass spectrometric analysis of nonderivatized steroids using cyclodextrin-supported 2, 5-dihydroxybenzoic acid as matrix. *Bull Korean Chem Soc.* 2014;35:1409–1412.

Chapter 17

Imaging mass spectrometry: neurotransmitter distribution using reactive matrix and chemical derivatization

Chapter Outline

Many types of neurotransmitters can be classified chemically as amino acids (e.g., γ-aminobutyric acid [GABA], glutamate, aspartate, D-serine), monoamines (e.g., dopamine, adrenaline, norepinephrine, histamine), small peptides (e.g., substance P, oxytocin, somatostatin), or others. Neurotransmitters are an integral part of many biological processes. For instance, catecholamines are associated with the regulation of memory, learning, sleep, emotion, stress response, and psychomotor coordination.[1]

Neurotransmitters are routinely detected by high-performance liquid chromatography with electrochemical and fluorescence detectors, capillary electrophoresis with ultraviolet detectors, or mass spectrometry (MS).[2] Liquid chromatography (LC) electrospray ionization tandem MS has been routinely utilized to detect and quantitate many neurotransmitters such as dopamine, glutamate, and GABA.[3] Several imaging MS techniques, such as matrix-assisted laser desorption/ionization (MALDI), desorption electrospray ionization (DESI), and laser-ablation electrospray ionization, can directly ionize, detect, and image some neurotransmitters from tissue sections.[4–6]

Molecular images from mass spectrometers are generated by directly detecting ionized molecules in each pixel of tissue. The detection of neurotransmitter molecules by MS is susceptible to isobaric interferences and ion suppression due to a lack of extensive extraction and chromatography separation. Ion suppression increases due to competing endogenous lipids or high salt levels in the tissue can drastically decrease the MS signal for the metabolites.[7] Some of

Introduction to Spatial Mapping of Biomolecules by Imaging Mass Spectrometry. DOI: 10.1016/C2018-0-03962-6

the decreases in ion signal can be mitigated by using ion mobility separation or using an ion source more amenable to that molecule. However, if a molecule cannot undergo ionization efficiently or fragments during MS analysis, addressing ion suppression or isobaric interference will not improve its detection.

On-tissue chemical derivatization can be employed to improve the ionization efficiency of neurotransmitters that are difficult to ionize, labile to fragmentation during analysis, or masked by interfering ions. During on-tissue chemical derivatization, a targeted group of chemical species is modified by a reagent to their derivatized counterparts within the tissue. Typically, on-tissue chemical derivatization can be performed by either immersing the tissue section in the chemical reagent or by pneumatically spraying the reagent similar to matrix application. Electrospray deposition of liquid-phase reagents compared with airbrush spraying tissue for 2-picolylamine reagent application for endogenous fatty acids showed a threefold improvement in sensitivity, as well as, decrease in analyte delocalization for imaging MS by MALDI.[8]

On-tissue chemical derivatization has been used to detect small molecules like metabolites and drugs. Many on-tissue chemical derivatization protocols take advantage of known organic reagents, especially ones used for gas chromatography or LC mass spectrometric analysis. For instance, Girard's reagents have been used to detect and image many small molecules using in situ on-tissue chemical derivatization.[9] On-tissue chemical derivatization using Girard's reagent P has also been found to enhance detection of N-glycan detection for MALDI imaging.[10] For imaging MS, an optimal chemical derivatization process should be rapid, efficient, specific to the target analyte, not cause delocalization, and form stable, easily ionizable compound by the imaging ion source. On-tissue chemical derivatization is carried out before the matrix application unless a reactive matrix can also act as a derivatization agent.

Reactive matrix for imaging neurotransmitters

Ideally, on-tissue chemical derivatization reagent should rapidly react with a selective group of targeted compounds to form stable derivative in ambient conditions. Reactive matrices are molecules that can serve as on-tissue chemical derivatizing reagents while also dually functioning as a matrix for those derivatized molecules. Using a reactive matrix simplifies the imaging MS workflow and reduces any delocalization or loss of molecules in tissue due to repetitive spraying. For example, aromatic carbonyl reagents (2,4-dihydroxybenzaldehyde and 2,5-dihydroxyacetophenone) react with primary amines to form Schiff bases and also serve as a matrix to analyze that derivatized mono-, di-, and polyamines by MALDI.[11]

Reactive matrix has been used to detect corticosteroids, such as triamcinolone, prednisone, cortisone, fludrocortisone, dexamethasone, deoxycorticosterone, and budesonide, by MALDI and LC–MS.[12] On-tissue chemical derivatization of primary amines, such as naturally occurring dopamine or exogenous amphetamine, with pyrylium salts, can be used for imaging.[13] Reactive matrix was used to imaging biogenic amines and amino acids, such as tyrosine, tryptamine, tyramine, phenethylamine, 3-methoxytyramine, serotonin (5-HT), and GABA, in brain tissue sections.[14] A benzophenone was used as both derivatization reagent and matrix for structural characterization of double bonds in unsaturated phospholipids by the photochemical reaction by MALDI.[15–16]

Organic compounds, hydrazides (e.g., 2-hydroxybenzohydrazide, 3-hydroxy-2-naphthoic acid hydrazide), and hydrazine reagents (e.g., 2-hydrazinoquinoline, 2,4-dinitrophenylhydrazine) were found to react with carbonyl groups of gaseous aldehydes such as formaldehyde, acetaldehyde, and propionaldehyde and also serve as a reactive matrix for MALDI.[17] 2,4-Dinitrophenylhydrazine can be utilized as a reactive matrix to analyze oxidized phospholipids by converting small aldehydes into hydrazones that can be detected by MALDI.[18]

Neuroactive molecules, such as endogenous neurotransmitters or catecholamine, exogenous neurotoxins, and drugs, contain primary amine functional groups.[13] Mass spectrometric analysis of many of these compounds, such as neurotransmitters (e.g., dopamine), recreational phenethylamines (e.g., amphetamine), amino acids, neuropeptides, is challenging at physiological trace concentration due to low ionization. Pyrylium salts can be used for derivatizing these biologically relevant primary amines into corresponding pyridinium salts at room temperature without heat or extensive incubation. MALDI can then detect these pyridinium salts without using any a separate matrix. On-tissue chemical derivatization using pyrylium salts enabled the MALDI imaging of primary amines, such as dopamine, amphetamine, and neurotoxin β-*N*-methylamino-L-alanine, in brain tissue sections.[13] Chemical reaction of dopamine with a few pyrylium salts and MALDI images on on-tissue chemically derivatized dopamine are shown in Fig. 17.1.

Several neurotransmitters were imaged using a reactive matrix based on a 2-fluoro-1-methyl pyridinium (FMP) compound. On-tissue chemical derivatization occurs by a nucleophilic aromatic substitution reaction of FMP cation with phenolic hydroxyl and primary/secondary amine groups of neurotransmitters and their related metabolites.[19] The FMP reactive matrix increased the detection sensitivity of low-abundance neurotransmitters directly from tissue enabling their mapping within the brain's fine structures at 10-μm pixel size. Molecular images of several FMP derivatized neurotransmitters imaged by MALDI in a Parkinson rat model brain are shown in Fig. 17.2.

On-tissue chemical derivatization by charge-tagging of primary amine using pyrylium salts improves their sensitivity and detection. However, the on-tissue chemical derivatization adds carbon and hydrogen to the compounds resulting in similar isotopic patterns for both derivatized and nonderivatized species. The reactive pyrylium ion can be chemically modified to introduce the distinct isotopic signature of bromine in mass spectra that can be passed on to the derivatized molecule for improved identification.[20]

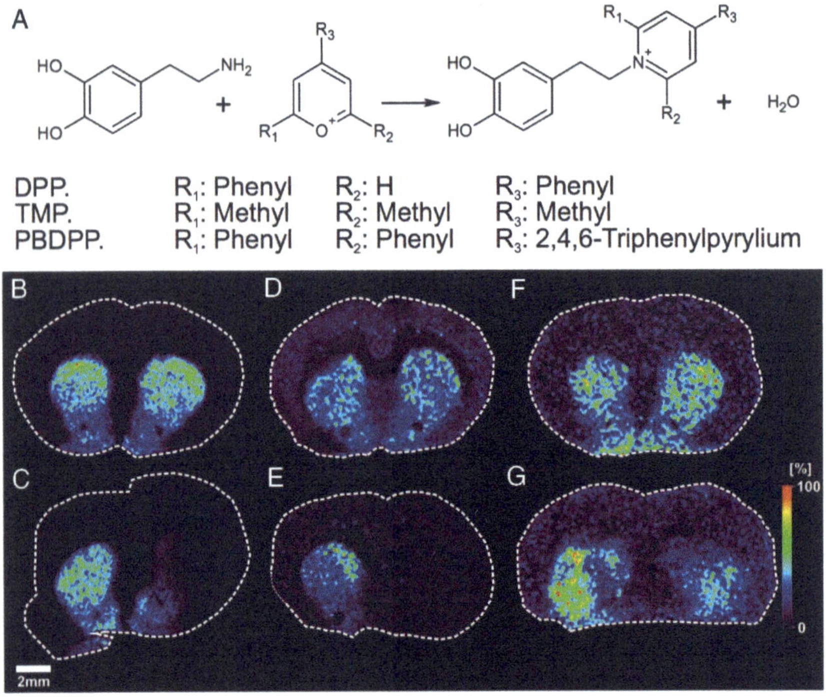

FIG. 17.1 (A) Reaction of dopamine with pyrylium salts, (B, D, F) MALDI MS images of derivatized dopamine in coronal rat tissue sections of control, and (C, E, G) unilaterally lesioned 6-hydroxydopamine after derivatization with 2,4-diphenyl-pyranylium (DPP), (B, C) 1,4-phenylene-4,4′-bis (2,6-diphenyl-4-pyrylium) (PBDPP), (D, E) or (F, G) 2,4,6-trimethylpyrylium (TMP). Reprinted with permission from Shariatgorji M, Nilsson A, Källback P, et al. Pyrylium salts as reactive matrices for MALDI-MS imaging of biologically active primary amines. J Am Soc Mass Spectrom. 2015;26:934–939. Copyright 2015 American Chemical Society.[13]

Inorganic nanomaterials have often used in laser desorption ionization in lieu of traditional organic matrices, often enabling the detection of complementary sets of molecules from tissues.[21–24] Laser desorption ionization of pneumatically sprayed citrate-capped gold nanoparticles enabled the detection of neurotransmitters, such as dopamine, epinephrine, glutamine, GABA, norepinephrine, octopamine, and serotonin in physiologically relevant concentrations in serum and homogenized tissue.[25] The gold nanoparticles also facilitated the imaging of several neurotransmitters without any chemical derivatization in rabbit brain tissue sections, zebrafish embryos, and neuroblastoma.

On-tissue chemical derivatization for imaging neurotransmitters

On-tissue chemical derivatization is generally performed before the application of the matrix. Other sample preparation steps can affect the sensitivity of the detected

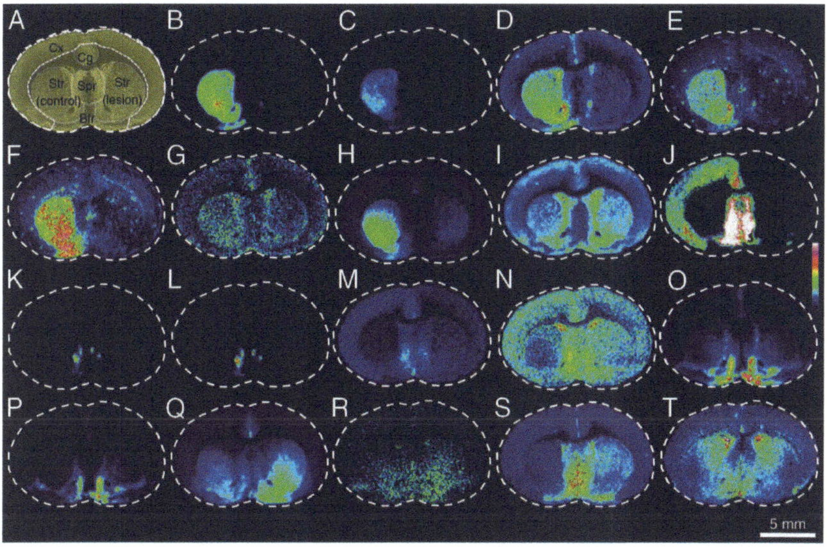

FIG. 17.2 MALDI MS of neurotransmitters and metabolites in coronal brain sections of Parkinson's model rat treated with L-3,4-dihydroxyphenylalanine or L-DOPA. (A) Microscope image of the tissue section. MS images acquired after using 2-fluoro-1-methyl pyridinium (FMP)-10 reactive matrix, and show the distributions of (B) endogenous dopamine, (C) 3-methoxytyramine, (D) 3,4-dihydroxyphenylacetaldehyde or DOPAL, (E) DOPAL (double derivatized), (F) 3,4-dihydroxyphenylacetic acid or DOPAC (double derivatized), (G) 3-methoxy-4-hydroxyphenylacetaldehyde or MOPAL, (H) homovanillic acid (HVA), (I) 3-*O*-methyldopa (3-OMD), (J) norepinephrine, (K) epinephrine, (L) epinephrine (double derivatized), (M) 3,4-dihydroxyphenylacetic acid (DOPAC), (N) 3-methoxy-4-hydroxyphenylethyleneglycol or MOPEG, (O) serotonin, (P) 5-hydroxyindolealdehyde, (Q) 5-hydroxyindoleacetic acid or 5-HIAA, (R) 5-hydroxytryptophol or 5-HTOL, (S) GABA, and (T) lysine (t). Bfr, basal forebrain region; Cx, cortex; Cg, cingulate cortex; Str, striatum; Spr, septal region. Pixel size is 80 μm. Reprinted with permission from Shariatgorji M, Nilsson A, Fridjonsdottir E, et al. Comprehensive mapping of neurotransmitter networks by MALDI-MS imaging. Nat Methods 2019;16:1021–1028.[19]

molecule. Tissue transfer using a laser was utilized to improve on-tissue chemical derivatization of small molecules for MALDI imaging. About 1 micron thick tissue film was transferred to a slide from a dry tissue section already coated derivatization reagent and matrix using a focused blue laser beam. For example, the laser-assisted tissue transfer section showed 20- to more than 200-folds increase in signal intensity for derivatized metabolites (e.g., amino acids, neurotransmitters) in rat brain tissues compared with conventional incubation-based derivatization with amino-specific derivatization reagent 4-hydroxy-3-methoxycinnamaldehyde.[26] This laser transfer technique was also used to derivatize steroids with Girard reagent T with shorter derivatization time with no noticeable metabolite delocalization. The improved detection can be attributed to the minimization of ion suppression effects due to the tissue's reduced thickness.[26]

On-tissue chemical derivatization and underivatized tissue were used to obtain a complimentary neurotransmitter molecular coverage for MALDI imaging MS. A pyrylium salt, 2,4-diphenylpyranylium tetrafluoroborate, was used as a primary amine derivatization reagent leading detection of neurotransmitters (e.g., dopamine, serotonin, GABA, and histamine) with reduced signal complexity on crab brain sections.[27] While neurotransmitters with other functional groups, such as acetylcholine and glutamine, were directly imaged after matrix application without on-tissue chemical derivatization. Molecular visualization of some of the derivatized neurotransmitters in the crustacean brain section is shown in Fig. 17.3.

A commercial LC/MS reagent kit, mTRAQ, consisting of a group of primary amine-specific derivatization agents with stable-isotope-labeled analogs can be sprayed in succession to perform on-tissue derivatization and quantitatively image small neurotransmitters like GABA.[28] Any diol-containing molecules such as aminols, vicdiols, saccharides, glycans, and catecholamines can be derivatized by boronate ester formation (boronic acid–diol reaction) with a pyridinium cation-containing boronic acid molecule called 4-(*N*-methyl) pyridinium boronic acid. The derivatized molecule can be sensitively detected and imaged by secondary ion MS and MALDI, such as imaging of dopamine, epinephrine, and norepinephrine in porcine adrenal gland tissue sections down to 3-μm pixel size by secondary ion MS.[29]

p-*N*,*N*,*N*-Trimethylammonioanilyl *N′*-hydroxysuccinimidyl carbamate iodide (TAHS), an activated carbamate, was developed for analyzing of amino acids using LC/MS method by reacting with the amino group to form a ureide bond. The amino acid TAHS derivatives can be cleaved at the binding site by tandem MS producing a characteristic fragment.[30] Catecholamine species, such as adrenaline, was mapped on pheochromocytoma tumor in adrenal glands by MALDI tandem MS after derivatization with TAHS reagent.[31]

On-tissue derivatization using a novel reagent, *N*,*N*,*N*-trimethyl-2-(piperazin-1-yl)ethan-1-aminium iodide, involves adding a quaternary ammonium group into metabolites containing carboxyl group (e.g., tricarboxylic acid cycle intermediates, fatty acids, and bile acids) leading to their sensitive detection by MALDI.[32] Small metabolites, including amino acids (e.g., glutamate, glutamine, glycine, leucine/isoleucine, phenylalanine, and alanine), were derivatized using a reagent, TAHS, to increase their ionization efficiency and detection sensitivity for MALDI imaging.[33] On-tissue chemical derivatization with 4-hydroxy-3-methoxycinnamaldehyde has been used to detect endogenous amine metabolites, such as amino acids, neurotransmitters by MALDI.[34]

On-tissue chemical derivatization of three derivatization reagents, TAHS, 4-hydroxy-3-methoxycinnamaldehyde, and 2,3-diphenylpyranylium tetrafluoroborate, was compared for imaging neurotransmitter in brain tissue.[35] Total of 18 neurotransmitters and related compounds were detected and imaged. A few of them were detected only using one on-tissue derivatization reagent, while others were detected with more than one on-tissue derivatization reagents, as shown in Fig. 17.4.

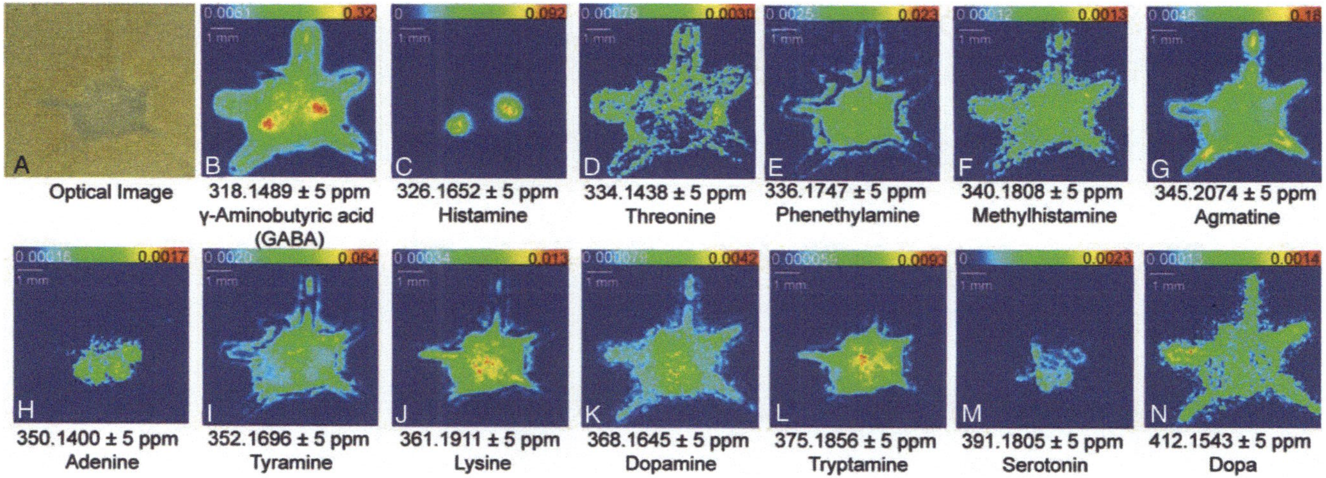

FIG. 17.3 Ion images showing the spatial distribution of biogenic amines neurotransmitters after on-tissue chemical derivatization with 2,4-diphenyl-pyranylium tetrafluoroborate (DPP-TFB). Reprinted with permission from Cao Q, Wang Y, Chen B, et al. Visualization and identification of neurotransmitters in crustacean brain via multifaceted mass spectrometric approaches. ACS Chem Neurosci. 2019;10:1222–1229. Copyright 2019 American Chemical Society.[29]

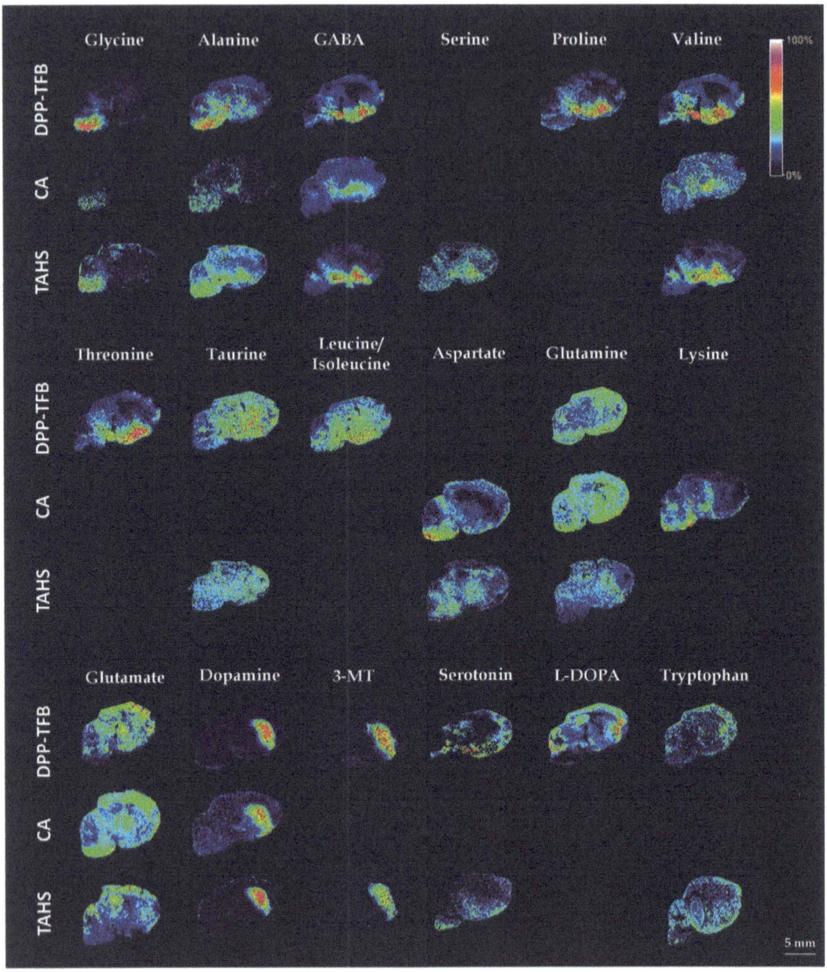

FIG. 17.4 MS images of derivatives of neurometabolites in wild type mouse sagittal brain sections after on-tissue chemical derivatization by TAHS, CA, and DPP-TFB are shown. The ions were images in positive ion mode using 2,5-dihydroxybenozic acid (DHB) as matrix. Reprinted with permission from Esteve C, Tolner EA, Shyti R, van den Maagdenberg AM, McDonnell LA. Mass spectrometry imaging of amino neurotransmitters: a comparison of derivatization methods and application in mouse brain tissue. Metabolomics. 2016;12:30.[36] Creative commons.

Imaging MS using DESI can map localization of neurotransmitters/metabolites (e.g., dopamine, serotonin, glutamate, glutamine, dihydroxyphenylacetic acid, 3-methoxytyramine, aspartate, GABA, and adenosine) and neuroactive drugs (amphetamine, sibutramine, and fluvoxamine) in brain tissue sections as shown in Fig. 17.5.[36] Although DESI can image neurotransmitters in native tissue, on-tissue chemical derivatization of the primary amines increased in the sensitivity of low abundant neurotransmitters. Pyrylium salts react with primary

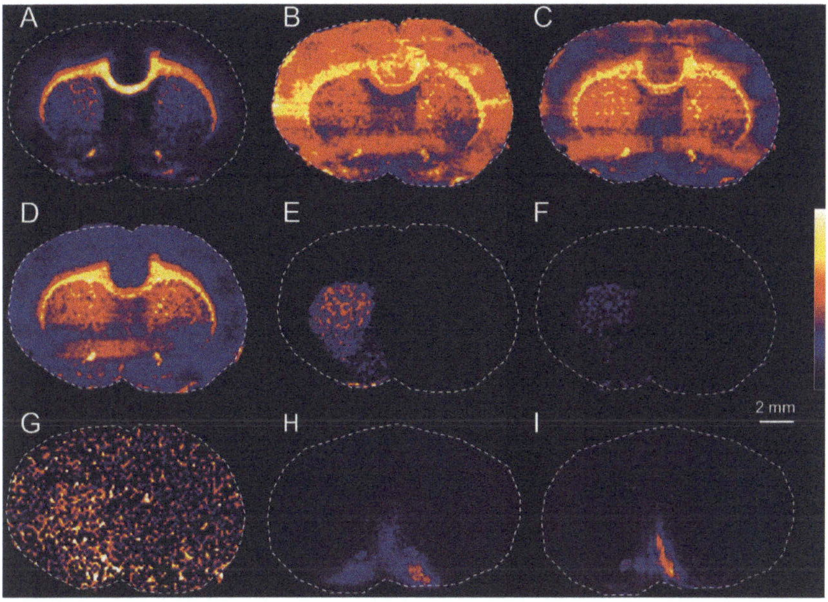

FIG. 17.5 DESI-MS images of neurotransmitters in coronal rat brain tissue sections of (A) adenosine, (B) aspartate, (C) glutamate, (D) glutamine, (E) dopamine, (F) D3-dopamine, and (G) dihydroxyphenylacetic acid or DOPAC in negative ion mode, (H) GABA, (I) MS/MS fragment of GABA in positive ion mode. A brain tissue section from 6-hydroxydopamine unilaterally lesioned rat treated with D3-L-DOPA in the negative ion mode and untreated in the positive ion mode. Adapted from Shariatgorji M, Strittmatter N, Nilsson A, et al. Simultaneous imaging of multiple neurotransmitters and neuroactive substances in the brain by desorption electrospray ionization mass spectrometry. NeuroImage. 2016;136:129–138.[37]

amines to form pyridinium cations stabilizing charge on it and reducing the charge competition in the electrospray mechanism.[36]

Outlook

Imaging MS of many neurotransmitters is hindered by poor detection sensitivity, low ionization efficiency, ion suppression from other endogenous molecules, in-source fragmentation of labile molecules, and isobaric spectral interferences from background chemicals or endogenous molecules. Using on-tissue chemical derivatization, the number of molecules detected by imaging MS technique is not limited to a handful, just abundant and easily ionizable compounds. Reactive matrix is a compound that can be used as both derivatization reagent and a matrix for the same derivatized analyte—thus eliminating additional time-consuming derivatization steps and any potential analyte delocalization. On-tissue chemical derivatization using the reactive matrix is expected to become more routine with more development of new reactive matrices that selectively target a group of clinically and biologically relevant molecules, such as neurotransmitters, for imaging MS.

References

1. Bergquist J, Ściubisz A, Kaczor A, Silberring J. Catecholamines and methods for their identification and quantitation in biological tissues and fluids. *J Neurosci Methods*. 2002;113:1–13.
2. Bourcier S, Benoist J - F, Clerc F, Rigal O, Taghi M, Hoppilliard Y. Detection of 28 neurotransmitters and related compounds in biological fluids by liquid chromatography/tandem mass spectrometry. *Rapid Commun Mass Spectrom*. 2006;20:1405–1421.
3. Cai H - L, Zhu R - H, Li H - D. Determination of dansylated monoamine and amino acid neurotransmitters and their metabolites in human plasma by liquid chromatography–electrospray ionization tandem mass spectrometry. *Anal Biochem*. 2010;396:103–111.
4. Romero-Perez G A, Takei S, Yao I, Imaging Mass Spectrometric Analysis of Neurotransmitters. A Review. Mass Spectrom (Tokyo). 2014;3(4):S0049. doi:10.5702/massspectrometry.S0049.
5. Fernandes A M A, Vendramini P H, Galaverna R, et al. Direct visualization of neurotransmitters in rat brain slices by desorption electrospray ionization mass spectrometry imaging (DESI-MS). *J Am Soc Mass Spectrom*. 2016;27:1944–1951.
6. Shrestha B, Nemes P, Nazarian J, Hathout Y, Hoffman E P, Vertes A. Direct analysis of lipids and small metabolites in mouse brain tissue by AP IR-MALDI and reactive LAESI mass spectrometry. *Analyst*. 2010;135:751–758.
7. Midey A, Olivos H, Shrestha B. Spatial mapping of cellular metabolites using DESI ion mobility mass spectrometry. In *Single Cell Metabolism*: Springer; 2020:181–190.
8. Tomlinson L, Fuchser J, Fütterer A, et al. Using a single, high mass resolution mass spectrometry platform to investigate ion suppression effects observed during tissue imaging. *Rapid Commun Mass Spectrom*. 2014;28:995–1003.
9. Wu Q, Comi T J, Li B, Rubakhin S S, Sweedler J V. On-tissue derivatization via electrospray deposition for matrix-assisted laser desorption/ionization mass spectrometry imaging of endogenous fatty acids in rat brain tissues. *Anal Chem*. 2016;88:5988–5995.
10. Barré F P Y, Flinders B, Garcia J P, et al. Derivatization strategies for the detection of triamcinolone acetonide in cartilage by using matrix-assisted laser desorption/ionization mass spectrometry imaging. *Anal Chem*. 2016;88:12051–12059.
11. Zhang H, Shi X, Vu N Q, et al. On-tissue derivatization with Girard's reagent P enhances N-glycan signals for formalin-fixed paraffin-embedded tissue sections in MALDI mass spectrometry imaging. *Anal Chem*. 2020;92:13361–13368.
12. Zaikin V G, Borisov R S, Polovkov N Y, Slyundina M S. Reactive matrices for matrix-assisted laser desorption/ionization mass spectrometry of primary amines. *Eur J Mass Spectrom*. 2015;21:403–411.
13. Brombacher S, Owen S J, Volmer D A. Automated coupling of capillary-HPLC to matrix-assisted laser desorption/ionization mass spectrometry for the analysis of small molecules utilizing a reactive matrix. *Anal Bioanal Chem*. 2003;376:773–779.
14. Shariatgorji M, Nilsson A, Källback P, et al. Pyrylium salts as reactive matrices for MALDI-MS imaging of biologically active primary amines. *J Am Soc Mass Spectrom*. 2015;26:934–939.
15. Shariatgorji M, Nilsson A, Goodwin R J A, et al. Direct targeted quantitative molecular imaging of neurotransmitters in brain tissue sections. *Neuron*. 2014;84:697–707.
16. Wäldchen F, Spengler B, Heiles S. Reactive matrix-assisted laser desorption/ionization mass spectrometry imaging using an intrinsically photoreactive Paternò–Büchi matrix for double-bond localization in isomeric phospholipids. *J Am Chem Soc*. 2019;141:11816–11820.
17. Wäldchen F, Mohr F, Wagner A H, Heiles S. Multifunctional reactive MALDI matrix enabling high-lateral resolution dual polarity MS imaging and lipid C=C position-resolved MS2 imaging. *Anal Chem*. 2020;92(20):143130–14138.

18. Shigeri Y, Ikeda S, Yasuda A, Ando M, Sato H, Kinumi T. Hydrazide and hydrazine reagents as reactive matrices for MALDI-MS to detect gaseous aldehydes. *J Mass Spectrom.* 2014;49:742–749.
19. Teuber K, Fedorova M, Hoffmann R, Schiller J. 2,4-dinitrophenylhydrazine as a new reactive matrix to analyze oxidized phospholipids by MALDI-TOF mass spectrometry. *Anal Lett.* 2012;45:968–976.
20. Shariatgorji M, Nilsson A, Fridjonsdottir E, et al. Comprehensive mapping of neurotransmitter networks by MALDI–MS imaging. *Nat Methods.* 2019;16:1021–1028.
21. Shariatgorji R, Nilsson A, Strittmatter N, et al. Bromopyrylium derivatization facilitates identification by mass spectrometry imaging of monoamine neurotransmitters and small molecule neuroactive compounds. *J Am Soc Mass Spectrom.* 2020;31(12):2553–2557.
22. McLean J A, Stumpo K A, Russell D H. Size-selected (2–10 nm) gold nanoparticles for matrix assisted laser desorption ionization of peptides. *J Am Chem Soc.* 2005;127:5304–5305.
23. Jackson S N, Baldwin K, Muller L, et al. Imaging of lipids in rat heart by MALDI-MS with silver nanoparticles. *Anal Bioanal Chem.* 2014;406:1377–1386.
24. Sacks C D, Stumpo K A. Gold nanoparticles for enhanced ionization and fragmentation of biomolecules using LDI-MS. *J Mass Spectrom.* 2018;53:1070–1077.
25. Mohammadi A S, Phan N T, Fletcher J S, Ewing A G. Intact lipid imaging of mouse brain samples: MALDI, nanoparticle-laser desorption ionization, and 40 keV argon cluster secondary ion mass spectrometry. *Anal bioanal Chem.* 2016;408:6857–6868.
26. McLaughlin N, Bielinski T M, Tressler C M, Barton E, Glunde K, Stumpo K A. Pneumatically sprayed gold nanoparticles for mass spectrometry imaging of neurotransmitters. *J Am Soc Mass Spectrom.* 2020;31(12):2452–2461.
27. Guo S, Tang W, Hu Y, et al. Enhancement of on-tissue chemical derivatization by laser-assisted tissue transfer for MALDI MS imaging. *Anal Chem.* 2020;92:1431–1438.
28. Cao Q, Wang Y, Chen B, et al. Visualization and identification of neurotransmitters in crustacean brain via multifaceted mass spectrometric approaches. *ACS Chem Neurosci.* 2019;10:1222–1229.
29. Ito T, Hiramoto M. Use of mTRAQ derivatization reagents on tissues for imaging neurotransmitters by MALDI imaging mass spectrometry: the triple spray method. *Anal Bioanal Chem.* 2019;411:6847–6856.
30. Kaya I, Brülls S M, Dunevall J, et al. On-tissue chemical derivatization of catecholamines using 4-(n-methyl)pyridinium boronic acid for ToF-SIMS and LDI-ToF mass spectrometry imaging. *Anal Chem.* 2018;90:13580–13590.
31. Shimbo K, Yahashi A, Hirayama K, Nakazawa M, Miyano H. Multifunctional and highly sensitive precolumn reagents for amino acids in liquid chromatography/tandem mass spectrometry. *Anal Chem.* 2009;81:5172–5179.
32. Takeo E, Sugiura Y, Uemura T, et al. Tandem mass spectrometry imaging reveals distinct accumulation patterns of steroid structural isomers in human adrenal glands. *Anal Chem.* 2019;91:8918–8925.
33. Sun C, Liu W, Geng Y, Wang X. On-tissue derivatization strategy for mass spectrometry imaging of carboxyl-containing metabolites in biological tissues. *Anal Chem.* 2020;92:12126–12131.
34. Toue S, Sugiura Y, Kubo A, et al. Microscopic imaging mass spectrometry assisted by on-tissue chemical derivatization for visualizing multiple amino acids in human colon cancer xenografts. *Proteomics.* 2014;14:810–819.
35. Manier M L, Spraggins J M, Reyzer M L, Norris J L, Caprioli R M. A derivatization and validation strategy for determining the spatial localization of endogenous amine metabolites in tissues using MALDI imaging mass spectrometry. *J Mass Spectrom.* 2014;49:665–673.

36. Esteve C, Tolner E A, Shyti R, van den Maagdenberg A M, McDonnell L A. Mass spectrometry imaging of amino neurotransmitters: a comparison of derivatization methods and application in mouse brain tissue. *Metabolomics*. 2016;12:30.
37. Shariatgorji M, Strittmatter N, Nilsson A, et al. Simultaneous imaging of multiple neurotransmitters and neuroactive substances in the brain by desorption electrospray ionization mass spectrometry. *NeuroImage*. 2016;136:129–138.

Chapter 18

Imaging mass spectrometry: small drugs and metabolites in tissue

Chapter Outline

Mapping the spatial distribution of pharmaceutical compounds and their metabolites is insightful during the research and development of new drugs. Molecular imaging of drugs and their associated molecules impacts several aspects of drug development—pharmacology, pharmacokinetics, drug target localization, drug–drug interactions, therapeutic efficacy, safety, transport, and metabolism.[1-2]

The measurement of drug concentrations in plasma by liquid chromatography (LC) mass spectrometry (MS) is routinely used in pharmacokinetic studies.[3] The molecular distribution of the drugs and its metabolites at the body's specific organ informs if the drug reached its target at the intended site in a sufficient concentration—with valuable insight on the drug's safety and efficacy. Quantitative whole-body autoradiography map radiolabeled drugs can be expensive and suffer from ambiguous spatial distribution due to the inability to distinguish between drugs and their metabolites.[4-5]

Imaging MS can map any detected molecules' spatial distribution in the tissue section. Imaging MS of drug and associated molecules for pharmaceutical research has been extensively reviewed and discussed in the referenced publications.[6-28] Although it is impossible to completely catalog all the drugs imaged by MS, a few reviews have published a comprehensive summary of drugs studied by imaging MS.[11-12, 25] The summary of drug imaging MS tabulates drug compounds, if metabolites were analyzed, tissue types, ionization methods, detection modes, type of mass analyzers, and references of the related publications. Most of the imaging MS studies have been focused on small drug molecules. Fig. 18.1 summarizes the status of drug imaging by imaging MS based

Introduction to Spatial Mapping of Biomolecules by Imaging Mass Spectrometry. DOI: 10.1016/C2018-0-03962-6

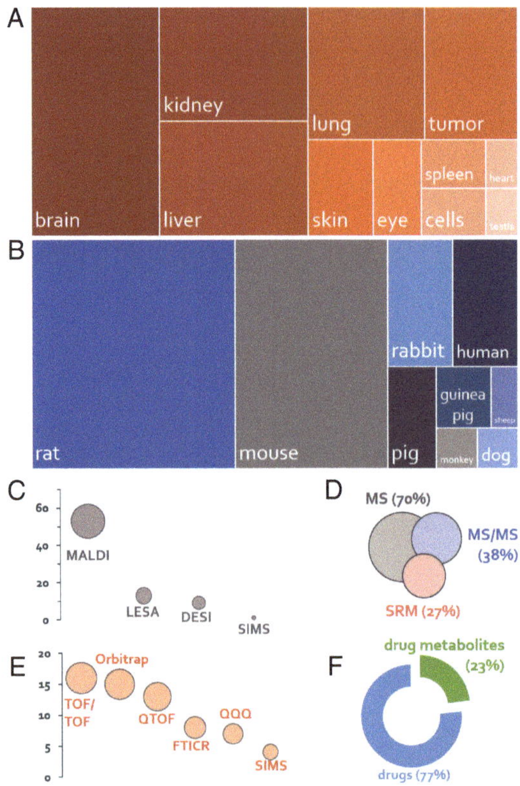

FIG. 18.1 Data visualization of drug molecules imaged by imaging MS published in the scientific literature is shown. The graph highlights types of (A) tissue imaged, (B) animal studied, (C) ion source, (D) MS detection mode used, (E) mass analyzer used for imaging, and (F) if metabolites were also imaged. Based on data from the reference Swales JG, Hamm G, Clench MR, Goodwin RJ. Mass spectrometry imaging and its application in pharmaceutical research and development: a concise review. Int J Mass Spectrom. 2019;437:99–112.[25]

on that table.[4] Brain, kidney, liver, and lungs are the four most popular organ targets. Rodents, i.e., rats and mice tissues, were imaged predominantly followed by human and rabbit. About one-fourth of the imaging MS also imaged drug metabolites along with the parent drugs. Matrix-assisted laser desorption/ionization (MALDI) is the most popular ionization method, followed by liquid extraction surface analysis (LESA) and desorption electrospray ionization (DESI). About 70% of analysis had MS mode, while 38% had MS/MS mode, and 27% had selective reaction monitoring.

In this chapter, only the detection approach for imaging drugs is discussed. A general overview of quantitative imaging MS of drugs is covered in another chapter titled, "Strategies for Quantitative Imaging Mass Spectrometry."[29] The topics covered will elaborate on (a) "relative quantitation" done without internal standard showing fold change or general trends, (b) "specific quantitation" performed using general internal standards that can inform on molar concentration range or percentage, and (c) "absolute quantitation" done ideally with isotopically labels or other suitable internal standards providing the most accurate molar concentration. Additionally, there are many reviews on the quantitative imaging MS in the literature as referenced here.[16, 21, 30-33]

A generic method development workflow for imaging MS the drug is outlined in Fig. 18.2. The first step is to establish if the chosen imaging MS method can detect the drug standard. If drug standard cannot be detected as a pure aliquot, it can be detected in pharmacological concentration in tissue. If the drug is not detected, the matrix for MALDI can be changed, or if possible, the ionization source can be changed. If the drug is still not detected, a chemical derivatization step may be required to ionize and detect the drug molecule. After detecting drugs, a limit of detection of the drug in pure aliquot needs to be established to evaluate if drugs can be detected in pharmacological concentration established by plasma studies done by LC/MS. A limit of detection of the drug mixed or spotted on the undosed biological tissue will also be needed to weigh in any decrease in sensitivity due to ion suppression from the tissue. The dosed tissue can be analyzed if all studies so far indicate drugs can be analyzed.

Increasing sensitivity for drug imaging by MS

The molecular imaging MS of the drug can be difficult if the drug is not detected of the tissue due to low ionization of the molecule, ion suppression from abundant endogenous molecules such as lipids, interferences of other molecules in tissue, low concentration of the drug, covalent binding of the drug to other molecules, etc. Imaging MS of drug molecule on the tissue can be enhanced by similar to metabolite imaging by selective removing interfering lipids by washing, on-tissue chemical derivatization to boost ionization of drugs, using more optimal matrix molecules, employing compatible ionization source, separating drug molecule using a gas-phase ion mobility separation or with higher mass resolution.

When using MALDI, matrix screening will be the first step. Since the introduction of MALDI, considerable effort has been devoted to finding the right matrix for the drug molecule. An excellent overview of MALDI matrices for small molecules, such as drug, with a list of compatible matrices for the molecular class, is given in the referenced publication.[34] Screening a new drug with several MALDI matrices is important because many drugs have been found to ionize well using only a few or single matrix. One thing to note is a matrix may be excellent at detecting the analyte in a dried standard aliquot of the drug but may fail to detect the same drug at the higher concentration in the tissue section due to ion suppression brought in by the biological matrix effect. Using electrospray-based ambient ionization modes, such as DESI and laser-ablation electrospray ionization (LAESI), changing solvent composition seemed to change the coverage of extracted ions from the tissue.

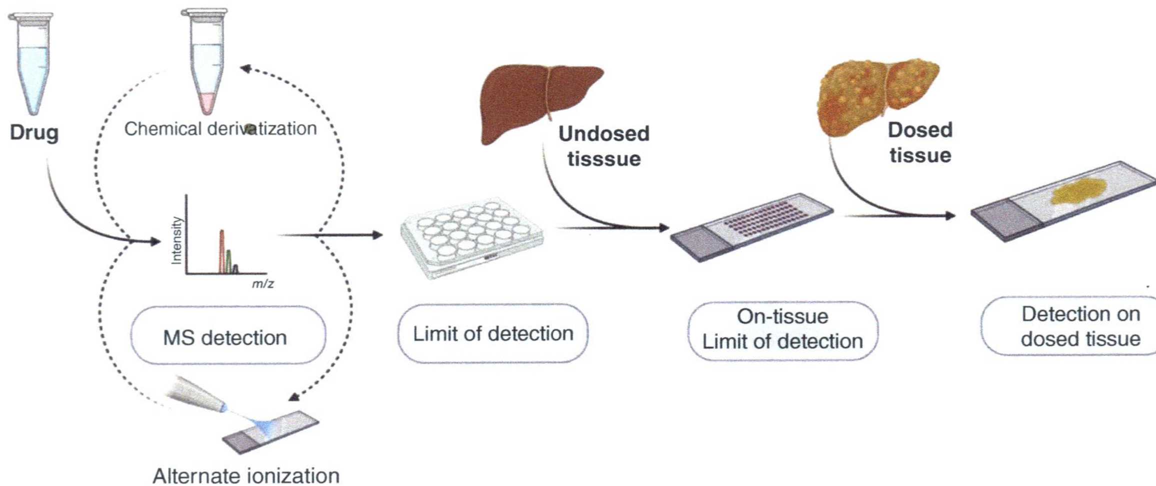

FIG. 18.2 A generic method development workflow for imaging any new drug molecule by mass spectrometry is illustrated. The first step is to detect the drug by chosen imaging MS method. If a drug cannot be detected, modify the MALDI matrix, use different ionization sources, or perform chemical derivatization. After the drug is detected, establish a limit of drug detection as pure standard and on tissue in pharmacological concentrations.

On-tissue chemical derivatization enables the detection of drugs by structurally modifying the drugs into a compound that can be easily ionized and detected by a mass spectrometer. Isoniazid, a tuberculosis drug, can be detected in serum but cannot be detected in tissue, thus hindering studying drug penetration in the tuberculosis region. On-tissue chemical derivatization with *trans*-cinnamaldehyde reagent was utilized to directly map the isoniazid cinnamaldehyde derivative in tissue by MALDI MS/MS. The reagent was applied to MALDI targets before mounting the tissue sections.[35] Organic molecule, 3-methoxysalicylamine (3-MoSA), is a scavenger of reactive γ-ketoaldehydes formed during oxidative stress and inflammation. On-tissue chemical derivatization of 3-MoSA with 1,1′-thiocarbonyldiimidazole reagent results in an oxothiazolidine derivative that can be detected with higher sensitivity in different organ tissues by MALDI.[36] In a similar manner, 3-MoSA levels were determined in mouse plasma and tissue by using the same derivatization reagent using LC–MS/MS for pharmacokinetics studies.[37] Intra-articular administration of triamcinolone acetonide (TAA) drug intends to relieve the inflammation and pain from osteoarthritis due to the degeneration of the cartilaginous tissue in articular joints leading to severely impaired mobility. TAA is not easily ionized by electrospray ionization or MALDI, hindering study on the drug penetration and distribution of the avascular cartilage. On-tissue chemical derivatization of TAA with Girard's reagent T incubation enabled detection of TAA in human cartilage in low ng/μL concentration.[38]

Increasing selectivity for drug imaging MS

Imaging of drug molecules in tissue by MS is often complicated due to spectral interference from other ions, background ions, or endogenous ions. A higher mass resolution helps in separating matrix ions from small molecules, such as drugs. An ultrahigh mass resolving power of Fourier transform ion cyclotron resonance mass spectrometer has been extensively used to detect drugs using MALDI.[39] Imaging ion source has also been used in other high-resolution mass spectrometers such as linear ion trap–Orbitrap, quadrupole time-of-flight (q-ToF), or quadrupole ion trap time-of-flight mass spectrometers.[40-42]

An anesthetic used as a designer drug, diphenidine, and its metabolites was detected in blood and urine using a MALDI q-ToF.[43] Similarly, several tricyclic antidepressants and related drugs, such as amitriptyline, carbamazepine, clomipramine, imipramine, nortriptyline, quetiapine, mianserin, and mirtazapine, were simultaneously detected in human whole blood and urine by MALDI q-ToF.[44] The distribution of chloroquine in the ocular tissue of pigmented rats was mapped by using MALDI q-ToF tandem MS (MS/MS).[45] MALDI analysis of liver tissue sections of an erlotinib-dosed rat using MALDI coupled with an orthogonal Q-TOF mass spectrometer was able to detect parent compound and metabolites of the drug after O-dealkylation.[46]

Imaging MS techniques directly extract and ionize analytes from the tissue without separation due to the technical inability of coupling with the LC or gas chromatography separation. However, imaging MS techniques can employ ion mobility separation after the ionization and before detecting ions by a mass spectrometer. Ion mobility separation is a gas-phase separation of the ions according to their drift in the buffer gas. Also, drift times of molecules during ion mobility can be used to measure its collision cross sections, which can be used as an additional identifier of a molecule. MALDI imaging of drug vinblastine in whole-body tissue sections in ion mobility q-ToF showed the ion mobility separation provided higher confidence in both MS and MS/MS imaging experiments by separating vinblastine from an endogenous isobaric lipid.[47]

Direct ion sources such as MALDI have been coupled with a triple quadrupole mass spectrometer for the rapid quantification of small pharmaceutical drugs using multiple reaction monitoring (MRM) or selected reaction monitoring.[32, 48] Quantitative analysis of drug, talinolol, in human plasma was demonstrated using MALDI MRM after liquid–liquid extraction and protein precipitation.[49] MALDI MRM method in triple quadrupole mass spectrometer was used for the determination of 3,4-methylenedioxymethamphetamine in oral fluid.[50] Imaging MS using MRM mode in a triple quadrupole and ion trap mass spectrometers can measures unique precursor ion–product ion transitions—increasing both selectivity and sensitivity of the analysis. MALDI-MRM imaging of moxifloxacin in tuberculosis-infected rabbit lungs shows MRM can be a sensitive method for mapping the distribution of the drug.[51] MALDI MS/MS imaging using MRM was used to rapidly analyze recreational drugs, such as cocaine and a range of its metabolites, in hair samples.[52] DESI coupled with triple quadrupole mass spectrometer was able to image a drug candidate and metabolite in canine liver tissue using MRM.[53]

MALDI and electrospray ionization ion sources utilized in DESI and LAESI are believed to detect a complementary set of molecules due to differences in ionization mechanism and ion suppression.[54-55] For example, images of several dosed drugs shown in Fig. 18.3 using LESA, MALDI, and DESI display the advantage of using multiple imaging techniques to expand coverage of detected drugs.[56] Other imaging MS ion source, such as laser-ablation atmospheric pressure photoionization, utilizes orthogonal hot solvent plume (e.g., toluene or anisole) and photoionization to detect complementary sets of molecules such as ionizes neutral and nonpolar compound.[57] Laser desorption/ionization based on nano- and micromaterials, such as nanophotonic laser desorption ionization on a highly uniform silicon nanopost array and nanostructure-initiator mass spectrometry, has shown enhanced capability to detect and image of small molecules.[58-59]

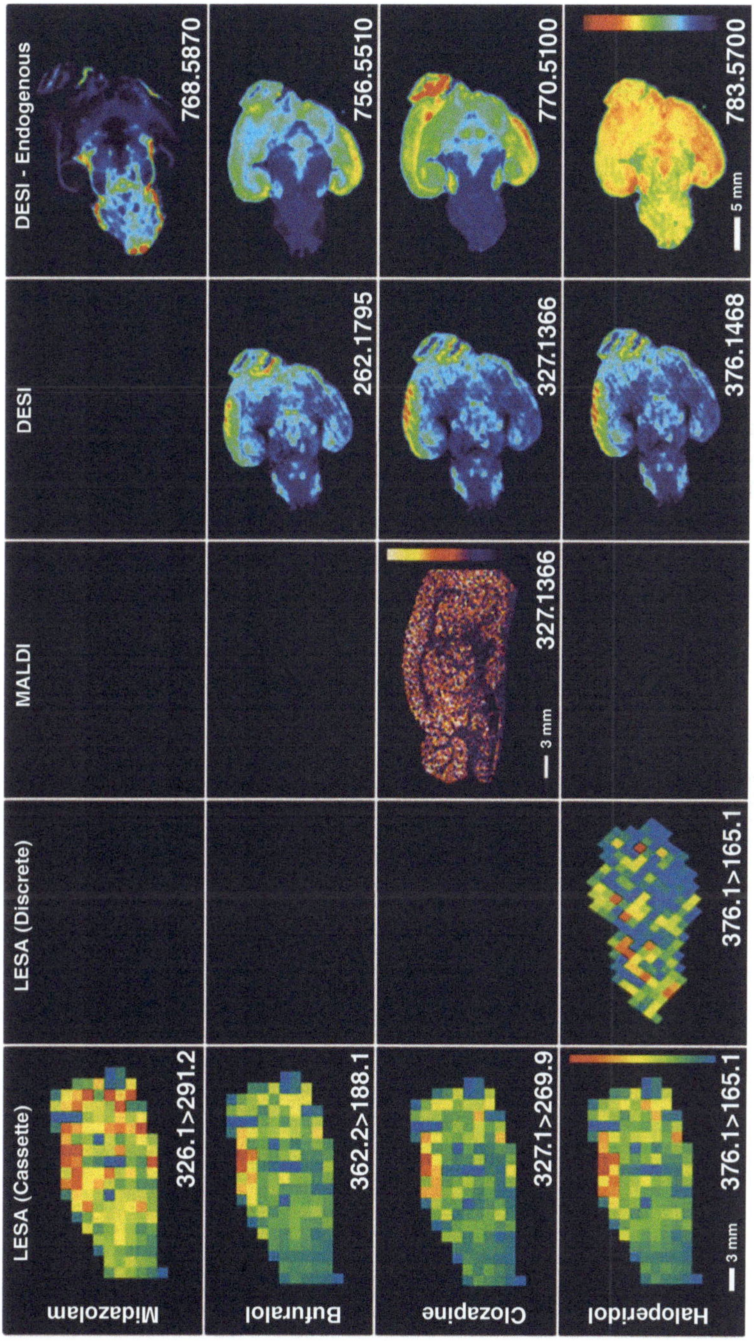

FIG. 18.3 MS images of intravenously dosed drugs imaged using LESA-MS/MS, MALDI-MS, and DESI-MS after cassette dosing of multiple drugs in a single time or discrete dosing of a single drug (haloperidol). The changing coverage shows the benefit of using more than one imaging technique to expand coverage and resolution of the drug. Reprinted with permission from Swales JG, Tucker JW, Strittmatter N, et al. Mass spectrometry imaging of cassette-dosed drugs for higher throughput pharmacokinetic and biodistribution analysis. Anal Chem. 2014;86:8473–8480. Copyright 2014 American Chemical Society.[56]

Outlook

Imaging MS is a powerful analytical tool to map spatial distributions of many drugs within the variety of the tissue. Imaging MS enables users to map the distribution of drugs without any labeling and can map the distribution of both parent drug and their metabolites. MALDI remains a popular choice for drug imaging, but other ambient techniques (e.g., DESI, LESA, LAESI) show utility for drug imaging by MS. Since its initial demonstration, many new tools have been developed to detect and image drugs by a mass spectrometer. Notable advancements include: the development of several ambient ionization sources, on-tissue chemical derivatization strategies, the coupling of ion mobility separation, and general improvements in mass analyzers performance. With more improvements in analytical techniques and workflows, we can anticipate increased utilization of imaging MS pharmacokinetics and the pharmacodynamics studies of the drug, not possible by plasma or bulk tissue analysis by LC/MS.

References

1. Rudin M, Weissleder R. Molecular imaging in drug discovery and development. *Nat Rev Drug Discov*. 2003;2:123–131.
2. Willmann J K, Van Bruggen N, Dinkelborg L M, Gambhir S S. Molecular imaging in drug development. *Nat Rev Drug Discov*. 2008;7:591–607.
3. Korfmacher W A. *Using Mass Spectrometry for Drug Metabolism studies*: CRC Press; 2009.
4. Kertesz V, Van Berkel G J, Vavrek M, Koeplinger K A, Schneider B B, Covey T R. Comparison of drug distribution images from whole-body thin tissue sections obtained using desorption electrospray ionization tandem mass spectrometry and autoradiography. *Anal Chem*. 2008;80:5168–5177.
5. Goodwin R J, Nilsson A, Mackay C L, et al. Exemplifying the screening power of mass spectrometry imaging over label-based technologies for simultaneous monitoring of drug and metabolite distributions in tissue sections. *J Biomol Screen*. 2016;21:187–193.
6. Rubakhin S S, Jurchen J C, Monroe E B, Sweedler J V. Imaging mass spectrometry: fundamentals and applications to drug discovery. *Drug Discov Today*. 2005;10:823–837.
7. Trim P J, Francese S, Clench M R. Imaging mass spectrometry for the assessment of drugs and metabolites in tissue. *Bioanalysis*. 2009;1:309–319.
8. Goodwin R J, Pitt A R. Mass spectrometry imaging of pharmacological compounds in tissue sections. *Bioanalysis*. 2010;2:279–293.
9. Sugiura Y, Setou M. Imaging mass spectrometry for visualization of drug and endogenous metabolite distribution: toward in situ pharmacometabolomes. *J Neuroimmune Pharmacol*. 2010;5:31–43.
10. Castellino S, Groseclose M R, Wagner D. MALDI imaging mass spectrometry: bridging biology and chemistry in drug development. *Bioanalysis*. 2011;3:2427–2441.
11. Greer T, Sturm R, Li L. Mass spectrometry imaging for drugs and metabolites. *J Proteomics*. 2011;74:2617–2631.
12. Prideaux B, Stoeckli M. Mass spectrometry imaging for drug distribution studies. *J Proteomics*. 2012;75:4999–5013.
13. Takai N, Tanaka Y, Inazawa K, Saji H. Quantitative analysis of pharmaceutical drug distribution in multiple organs by imaging mass spectrometry. *Rapid Commun Mass Spectrom*. 2012;26:1549–1556.

14. Lietz C B, Gemperline E, Li L. Qualitative and quantitative mass spectrometry imaging of drugs and metabolites. *Adv Drug Deliv Rev*. 2013;65:1074–1085.
15. Morosi L, Zucchetti M, D'Incalci M, Davoli E. Imaging mass spectrometry: challenges in visualization of drug distribution in solid tumors. *Curr Opin Pharmacol*. 2013;13:807–812.
16. Sun N, Walch A. Qualitative and quantitative mass spectrometry imaging of drugs and metabolites in tissue at therapeutic levels. *Histochem Cell Biol*. 2013;140:93–104.
17. Nilsson A, Goodwin R J A, Shariatgorji M, Vallianatou T, Webborn P J H, Andrén P E. Mass spectrometry imaging in drug development. *Anal Chem*. 2015;87:1437–1455.
18. Végvári Á. Drug localizations in tissue by mass spectrometry imaging. *Biomark Med*. 2015;9:869–876.
19. Oppenheimer S R, Wehr A Y. Imaging mass spectrometry in drug discovery and development. *Bioanalysis*. 2015;7:2609–2610.
20. Karlsson O, Hanrieder J. Imaging mass spectrometry in drug development and toxicology. *Arch Toxicol*. 2017;91:2283–2294.
21. Prideaux B, Lenaerts A, Dartois V. Imaging and spatially resolved quantification of drug distribution in tissues by mass spectrometry. *Curr Opin Chem Biol*. 2018;44:93–100.
22. Jove M, Spencer J, Clench M, Loadman P M, Twelves C. Precision pharmacology: mass spectrometry imaging and pharmacokinetic drug resistance. *Crit Rev Oncol Hematol*. 2019;141:153–162.
23. Nishidate M, Hayashi M, Aikawa H, et al. Applications of MALDI mass spectrometry imaging for pharmacokinetic studies during drug development. *Drug Metab Pharmacokinet*. 2019;34:209–216.
24. Schulz S, Becker M, Groseclose M R, Schadt S, Hopf C. Advanced MALDI mass spectrometry imaging in pharmaceutical research and drug development. *Curr Opin Biotechnol*. 2019;55:51–59.
25. Swales J G, Hamm G, Clench M R, Goodwin R J. Mass spectrometry imaging and its application in pharmaceutical research and development: a concise review. *Int J Mass Spectrom*. 2019;437:99–112.
26. Grégoire S, Luengo G S, Hallegot P, et al. Imaging and quantifying drug delivery in skin – Part 1: Autoradiography and mass spectrometry imaging. *Adv Drug Deliv Rev*. 2020;153:137–146.
27. Goodwin R J, Takats Z, Bunch J. A critical and concise review of mass spectrometry applied to imaging in drug discovery. *SLAS Discov*. 2020;25:963–976.
28. Bonnel D, Legouffe R, Willand N, et al. MALDI imaging techniques dedicated to drug-distribution studies. *Bioanalysis*. 2011;3:1399–1406.
29. Shrestha B. Introduction to spatial mapping of biomolecules by imaging mass spectrometry-*Introduction to Spatial Mapping of Biomolecules by Imaging Mass Spectrometry*: Elsevier: Strategies for Quantitative Imaging Mass Spectrometry; 2021.
30. Rzagalinski I, Volmer D A. Quantification of low molecular weight compounds by MALDI imaging mass spectrometry–a tutorial review. *Biochim Biophys Acta Proteins Proteomics*. 2017;1865:726–739.
31. Ellis S R, Bruinen A L, Heeren R M. A critical evaluation of the current state-of-the-art in quantitative imaging mass spectrometry. *Anal Bioanal Chem*. 2014;406:1275–1289.
32. Porta T, Lesur A, Varesio E, Hopfgartner G. Quantification in MALDI-MS imaging: what can we learn from MALDI-selected reaction monitoring and what can we expect for imaging?. *Anal Bioanal Chem*. 2015;407:2177–2187.
33. Tobias F, Hummon A B. Considerations for MALDI-based quantitative mass spectrometry imaging studies. *J Proteome Res*. 2020;19:3620–3630.

34. Calvano C D, Monopoli A, Cataldi T R I, Palmisano F. MALDI matrices for low molecular weight compounds: an endless story?. *Anal Bioanal Chem.* 2018;410:4015–4038.
35. Manier M L, Reyzer M L, Goh A, et al. Reagent precoated targets for rapid in-tissue derivatization of the anti-tuberculosis drug isoniazid followed by MALDI imaging mass spectrometry. *J Am Soc Mass Spectrom.* 2011;22:1409–1419.
36. Chacon A, Zagol-Ikapitte I, Amarnath V, et al. On-tissue chemical derivatization of 3-methoxysalicylamine for MALDI-imaging mass spectrometry. *J Mass Spectrom.* 2011;46:840–846.
37. Zagol-Ikapitte I, Amarnath V, Jadhav S, Oates J A, Boutaud O. Determination of 3-methoxysalicylamine levels in mouse plasma and tissue by liquid chromatography-tandem mass spectrometry: application to in vivo pharmacokinetics studies. *J Chromatogr B Analyt Technol Biomed Life Sci.* 2011;879:1098–1104.
38. Barré F P Y, Flinders B, Garcia J P, et al. Derivatization strategies for the detection of triamcinolone acetonide in cartilage by using matrix-assisted laser desorption/ionization mass spectrometry imaging. *Anal Chem.* 2016;88:12051–12059.
39. Cornett D S, Frappier S L, Caprioli R M. MALDI-FTICR imaging mass spectrometry of drugs and metabolites in tissue. *Anal Chem.* 2008;80:5648–5653.
40. Goodwin R J A, Mackay C L, Nilsson A, et al. Qualitative and quantitative MALDI imaging of the positron emission tomography ligands raclopride (a D2 dopamine antagonist) and SCH 23390 (a D1 dopamine antagonist) in rat brain tissue sections using a solvent-free dry matrix application method. *Anal Chem.* 2011;83:9694–9701.
41. Zaima N, Hayasaka T, Goto-Inoue N, Setou M. Imaging of metabolites by MALDI mass spectrometry. *J Oleo Sci.* 2009;58:415–419.
42. Korte A R, Yandeau-Nelson M D, Nikolau B J, Lee Y J. Subcellular-level resolution MALDI-MS imaging of maize leaf metabolites by MALDI-linear ion trap-Orbitrap mass spectrometer. *Anal Bioanal Chem.* 2015;407:2301–2309.
43. Minakata K, Yamagishi I, Nozawa H, et al. Diphenidine and its metabolites in blood and urine analyzed by MALDI-Q-TOF mass spectrometry. *Forensic Toxicol.* 2015;33:402–408.
44. Nozawa H, Minakata K, Yamagishi I, et al. Simultaneous determination of cyclic antidepressants and their related drugs and the estimation of new metabolites in human whole blood and urine by MALDI-QTOF-mass spectrometry. *Forensic Toxicol.* 2016;34:244–255.
45. Yamada Y, Hidefumi K, Shion H, Oshikata M, Haramaki Y. Distribution of chloroquine in ocular tissue of pigmented rat using matrix-assisted laser desorption/ionization imaging quadrupole time-of-flight tandem mass spectrometry. *Rapid Commun Mass Spectrom.* 2011;25:1600–1608.
46. Signor L, Varesio E, Staack R F, Starke V, Richter W F, Hopfgartner G. Analysis of erlotinib and its metabolites in rat tissue sections by MALDI quadrupole time-of-flight mass spectrometry. *J Mass Spectrom.* 2007;42:900–909.
47. Trim P J, Henson C M, Avery J L, et al. Matrix-assisted laser desorption/ionization-ion mobility separation-mass spectrometry imaging of vinblastine in whole body tissue sections. *Anal Chem.* 2008;80:8628–8634.
48. Hatsis P, Brombacher S, Corr J, Kovarik P, Volmer D A. Quantitative analysis of small pharmaceutical drugs using a high repetition rate laser matrix-assisted laser/desorption ionization source. *Rapid Commun Mass Spectrom.* 2003;17:2303–2309.
49. Kovarik P, Grivet C, Bourgogne E, Hopfgartner G. Method development aspects for the quantitation of pharmaceutical compounds in human plasma with a matrix-assisted laser desorption/ionization source in the multiple reaction monitoring mode. *Rapid Commun Mass Spectrom.* 2007;21:911–919.

50. Poetzsch M, Steuer A E, Hysek C M, Liechti M E, Kraemer T. Development of a high-speed MALDI-triple quadrupole mass spectrometric method for the determination of 3,4-methylenedioxymethamphetamine (MDMA) in oral fluid. *Drug Test Anal.* 2016;8:235–240.
51. Prideaux B, Dartois V, Staab D, et al. High-sensitivity MALDI-MRM-MS imaging of moxifloxacin distribution in tuberculosis-infected rabbit lungs and granulomatous lesions. *Anal Chem.* 2011;83:2112–2118.
52. Flinders B, Beasley E, Verlaan R M, et al. Optimization of sample preparation and instrumental parameters for the rapid analysis of drugs of abuse in hair samples by MALDI-MS/MS imaging. *J Am Soc Mass Spectrom.* 2017;28:2462–2468.
53. Lamont L, Eijkel G B, Jones E A, et al. Targeted drug and metabolite imaging: desorption electrospray ionization combined with triple quadrupole mass spectrometry. *Anal Chem.* 2018;90:13229–13235.
54. Bodnar W M, Blackburn R K, Krise J M, Moseley M A. Exploiting the complementary nature of LC/MALDI/MS/MS and LC/ESI/MS/MS for increased proteome coverage. *J Am Soc Mass Spectrom.* 2003;14:971–979.
55. Tomlinson L, Fuchser J, Fütterer A, et al. Using a single, high mass resolution mass spectrometry platform to investigate ion suppression effects observed during tissue imaging. *Rapid Commun Mass Spectrom.* 2014;28:995–1003.
56. Swales J G, Tucker J W, Strittmatter N, et al. Mass spectrometry imaging of cassette-dosed drugs for higher throughput pharmacokinetic and biodistribution analysis. *Anal Chem.* 2014;86:8473–8480.
57. Vaikkinen A, Shrestha B, Kauppila T J, Vertes A, Kostiainen R. Infrared laser ablation atmospheric pressure photoionization mass spectrometry. *Anal Chem.* 2012;84:1630–1636.
58. Stopka S A, Rong C, Korte A R, et al. Molecular imaging of biological samples on nanophotonic laser desorption ionization platforms. *Angew Chem Int Ed Engl.* 2016;55:4482–4486.
59. Palermo A. Charting metabolism heterogeneity by nanostructure imaging mass spectrometry: from biological systems to subcellular functions. *J Am Soc Mass Spectrom.* 2020.

Chapter 19

Imaging mass spectrometry: gangliosides in brain tissue

Chapter Outline

Gangliosides are glycosphingolipids that contain a ceramide base and a carbohydrate chain with at least one sialic acid.[1–2] The oligosaccharide chains containing sialic acid are linked via a β-glycosidic to the hydrophobic ceramide backbone. Gangliosides accumulate in a wide range of lysosomal storage disorders, such as Tay–Sachs and Sandhoff disease characterized by massive storage of GM2 ganglioside.[3] Gangliosides are found in the central nervous system and believed to play roles in many neural functions and dysregulation of many neurodegenerative conditions such as aging, Alzheimer's disease, Parkinson's disease, Huntington's disease, and stroke.[4–6]

Imaging mass spectrometry (MS) can be used to map ganglioside spatial distribution within the tissue section to study neurodegenerative lipid storage diseases or injuries. For example, matrix-assisted laser desorption/ionization (MALDI) imaging MS was utilized to find GM2 and GM3 gangliosides are elevated in Niemann–Pick disease type C1, and localized in the posterior lobules of the cerebellum.[7] In comparison to immunohistochemistry (IHC) and in situ hybridization, imaging MS can directly perform gangliosides without any label, multiplex image of several gangliosides simultaneously, and image an individual molecular entity with a high degree of specificity.[8] For example, the differential spatial distribution of GM1 gangliosides with different fatty acid chains can be simultaneously imaged by MS. The role of imaging MS for spatial mapping distribution of gangliosides in the healthy and diseased or injured brain's neural anatomy is reviewed here.[6] A detailed general protocol for imaging major brain gangliosides using MALDI is given in the reference.[8]

Introduction to Spatial Mapping of Biomolecules by Imaging Mass Spectrometry. DOI: 10.1016/C2018-0-03962-6

Since its adoption in the early 1990s, MALDI has been used to ionize glycosphingolipid for detection by time-of-flight (ToF) mass spectrometers.[9–10] Several gangliosides were analyzed with MALDI using α-cyano-4-hydroxycinnamic acid or 2,5-dihydroxybenzoic acid (DHB) matrix. DHB in the negative ion mode was found to be a better matrix for gangliosides due to lower loss of sialic acid in gangliosides that ToF/ToF or quadrupole ion trap-ToF.[5,11] Chemical derivatization using esterification of a methyl group of the carboxyl group in sialic acid was proposed to increase the stability of the ganglioside species in high vacuum MALDI ion source.[12] Glycoblotting method utilizing selective ozonolysis of the C–C double bond in the ceramide moiety and subsequent derivatization of glycosphingolipids aldehydes by aminooxy-functionalized gold nanoparticle (AuNP) allowed MALDI analysis of gangliosides, such as GM1, GD1a/GD1b, GT1b, and GD3.[13] However, on-tissue chemical derivatization using methyl esterification for imaging ganglioside by MALDI might result in loss of the low concentration species in tissue due to sample perpetration.

Matrix selection for ganglioside for MALDI imaging

For MALDI analysis and imaging, the choice matrix plays a vital role in the ability to analyze a group of molecules. Many common matrices readily detected and imaged abundant phospholipids but struggled to detect larger lipids like gangliosides. Comprehensive imaging of several ganglioside species without chemical derivatization or labeling can be done by using 2,6-dihydroxyacetophenone (DHA) matrix with ammonium sulfate and heptafluorobutyric acid (HFBA). Ammonium sulfate limits the formation of salt adducts, and HFBA improves the structural stability of DHA in a vacuum.[14]

MALDI using a standard DHB matrix (50 mg/mL, 70% methanol, and 0.1% TFA) was able to image gangliosides in negative ion mode between *m/z* 1500 and 2300. Fig. 19.1 shows the MALDI image of GM1, GD1, and GT1 with either C18- or C20-sphingosine.[5] A commercially available IR-780 was proposed as a novel matrix for MALDI imaging of higher molecular weight lipids, such as gangliosides, cardiolipins, and (poly-)phosphoinositides.[15] Fig. 19.2 shows mass spectra and MALDI images of high molecular weight lipids detected in the coronal mouse brain section.

5-Chloro-2-mercaptobenzothiazole (CMBT) matrix was able to image a series of ganglioside molecules, such as GM1, GD1, and GT1b (d18:1, d20:1) in mouse brain sections with middle cerebral artery occlusion reperfusion injury.[16] CMBT matrix was used to measure changes in A-series ganglioside species (GD1a, GM1, GM2, and GM3) to study Alzheimer's disease and stroke. CMBT matrix (15 mg/mL in a 4:4:1 mixture of chloroform:ethanol:water solvent) was sprayed dried, quickly minimizing the delocalization of molecules within the tissue.[17] Stroke, Aβ toxicity, or both were induced in rats by injection of endothelin-1, or toxic 25-35 fragment of the Aβ peptide, or both. A significant increase in GM2 was observed in the ischemic brain region for the stroke

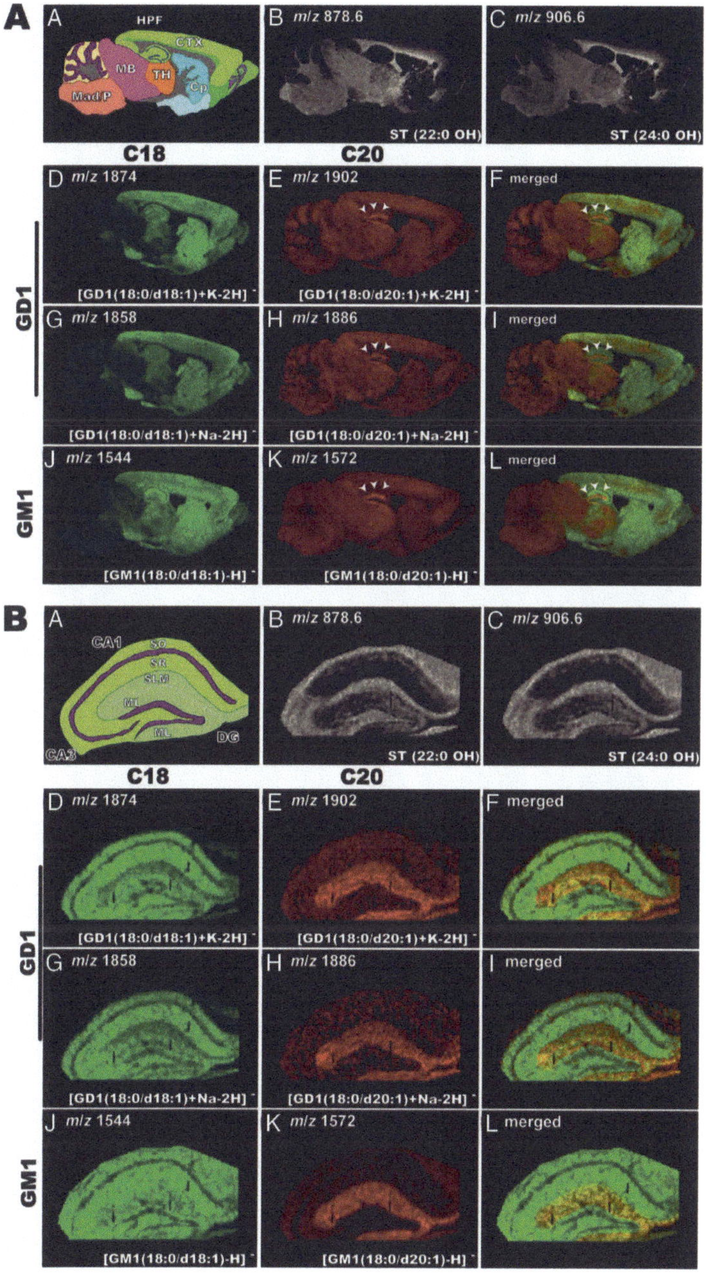

FIG. 19.1 MALDI images of gangliosides in mouse brain section acquired at (A) 50-μm pixel size to gain an overview of ganglioside distribution in different brain regions and (B) 15-μm pixel size for more detail the distribution in the hippocampus. Reprinted from Sugiura Y, Shimma S, Konishi Y, Yamada MK, Setou M. Imaging mass spectrometry technology and application on ganglioside study, visualization of age-dependent accumulation of C20-ganglioside molecular species in the mouse hippocampus. PLOS One 2008;3:e3232.[5] Creative commons.

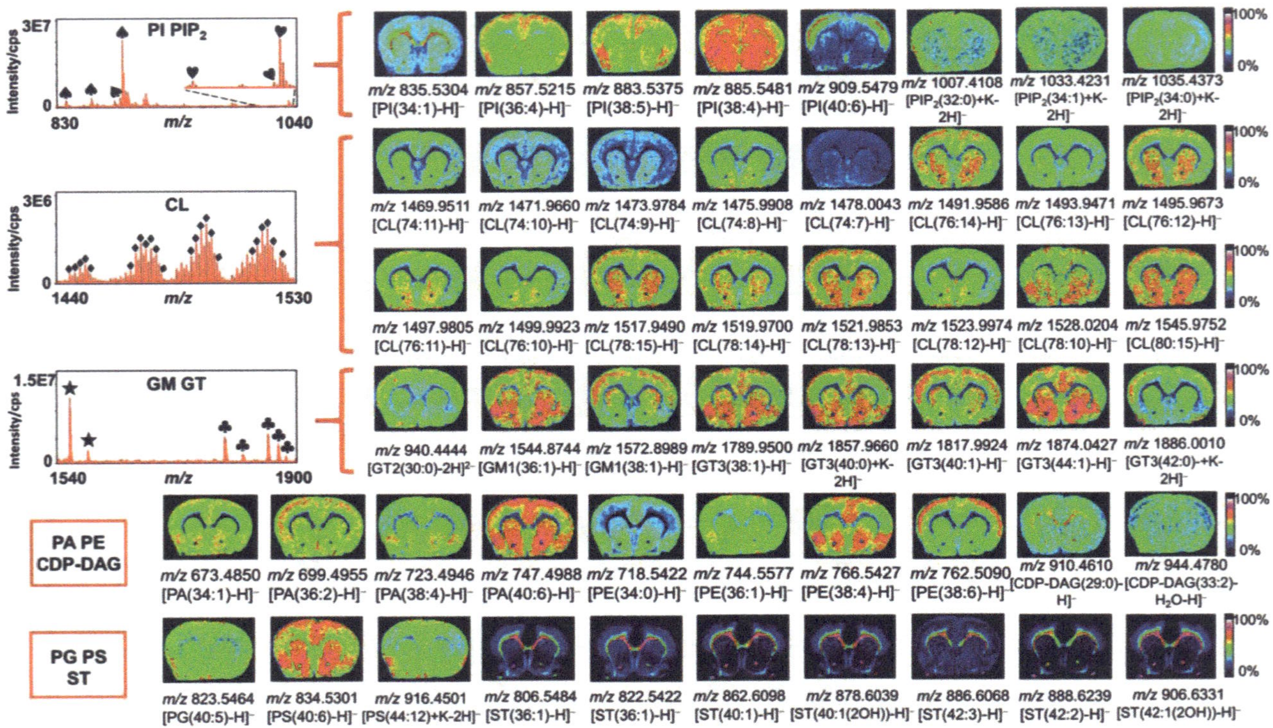

FIG. 19.2 MALDI images of larger lipids, such as gangliosides, cardiolipins, (poly-)phosphoinositides, in the coronal section of mouse brain tissue using IR-780 as matrix (2 mg/mL in methanol). Pixel size is 140 μm. Symbols: spades, PI; hearts, PIP2; diamonds, CL; stars, GM1; and clover, GT3. Reprinted with permission from Li N, Wang P, Liu X, et al. Developing IR-780 as a novel matrix for enhanced MALDI MS imaging of endogenous high-molecular-weight lipids in brain tissues. Anal Chem. 2019;91:15873–15882. Copyright 2019 American Chemical Society.[15]

group with or without Aβ by day 3. GM2 levels only remained elevated in the combined group by day 21. GM3 was elevated in the ischemic brain region only in the combined group on day 3. By day 21, GM3 was also elevated in the stroke group.

AuNPs modified with alkylamine has been proposed as a new matrix to enhance the detection of glycosphingolipids, such as minor gangliosides and sulfatides, in mouse brain sections.[18] The AuNP matrix showed approximately 20 times more sensitivity for the detection of glycosphingolipids compared to DHB.

Atmospheric pressure MALDI for ganglioside imaging

MALDI matrix, DHA or 2,5-dihydroxyacetophenone, can detect and image the distribution of many ganglioside species in rat brain.[4,19] However, a matrix such as DHA sublimes in a low-pressure environment under vacuum and can be used only for a quick imaging experiment in a TOF mass spectrometer.[20] DHA can be used with atmospheric pressure (AP) MALDI ion source without noticeable sublimation.[21] The combination of the AP-MALDI with the DHA matrix also reduced the fragmentation observed for gangliosides. DHA detected six ganglioside species (GM_1 d36:1, GM_1 d38:1, GD_1 d36:1, GD_1 d38:1, GT_1 d36:1, and GT_1 d38:1) and 1,5-diaminonaphthalene (DAN) detected four ganglioside species (GM_1 d36:1, GM_1 d38:1, GD_1 d36:1, and GD_1 d38:1). DAN showed a potential higher degree of gangliosides fragmentation than DHA using AP-MALDI. For example, the mass spectrum acquired with DAN is dominated by GM_1 potentially due to fragmentation of GD_1 and GT_1. In contrast, the mass spectrum produced using DHA has the three major brain gangliosides in the order, GD_1, GM_1, and GT_1—confirming with previous high-performance thin-layer chromatography study showing mouse brain gangliosides composition as GD_1 (49%), GM_1 (23%), and GT_1 (21%).[12] Signal enhancement observed in the AP-MALDI mass spectrum for the whole tissue section and selected area of the brain with DHA compared to the DAN matrix is shown in Fig. 19.3.

Sample preparation for enhanced ganglioside imaging by MS

The addition of ammonium sulfate in the DHA matrix with 0.05% HFBA lead to the detection of acetylated ganglioside species and improvement in the detection of other gangliosides by decreasing sodium adduct formation for GD1 and GT1.[19]

Aqueous washes, with ammonium formate (pH 6.4) or ammonium acetate (pH 6.7), showed up to fivefold increase sensitivity for gangliosides without notable delocalization up to 10-μm pixel size.[22] MALDI imaging MS with sublimed DHB coating after washing was able to detect other lipid species, including glycerophosphoinositols, glycerophosphates, glycerolphosphoglycerols, glycerophosphoethanolamines, glycerophosphoserines, and sulfatide (Fig. 19.4).

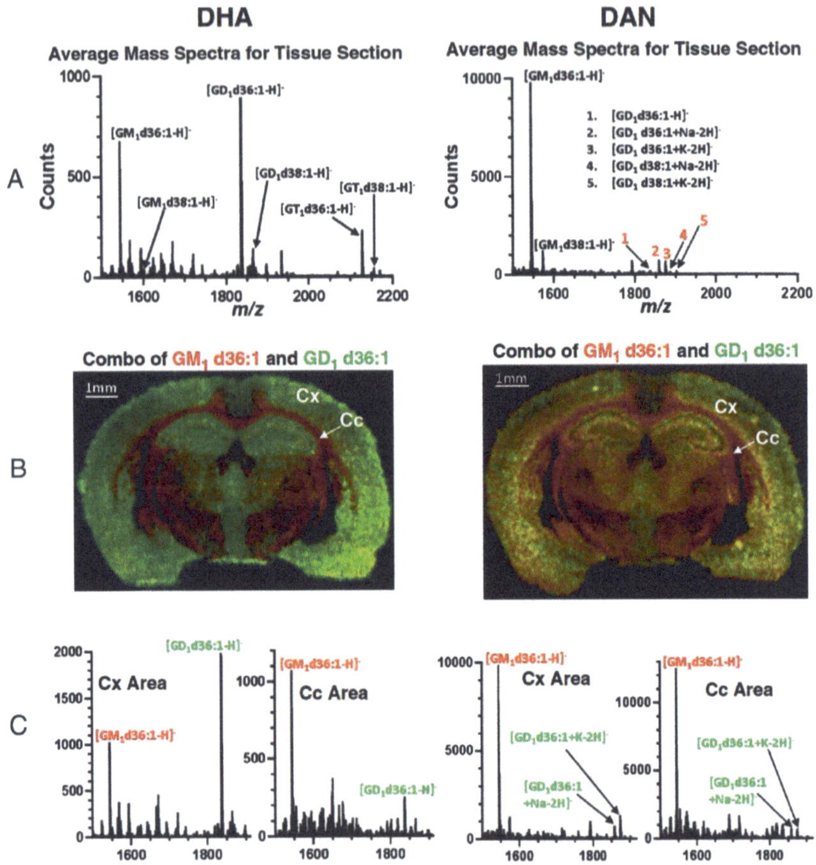

FIG. 19.3 (A) Mass spectra for brain tissue sections using DHA and DAN matrix in atmospheric pressure (AP) MALDI, (B) overlaid images of GM1 (d36:1) in red and GD1 (d36:1) in green with DHA and DAN matrix, (C) average mass spectra of gangliosides in the cortex (Cx) and corpus callosum (Cc) region with DHA and DAN matrix. Reprinted with permission from Jackson SN, Muller L, Roux A, et al. AP-MALDI mass spectrometry imaging of gangliosides using 2,6-dihydroxyacetophenone. *J Am Soc Mass Spectrom.* 2018;29:1463–1472. Copyright 2018 American Chemical Society.[21]

The distribution of acetylated ganglioside species, such as O-acetylGD1 and O-acetylGT1, in tissue depended on the sphingoid base (d18:1 sphingosine or d20:1 eïcosasphingosine). Both fresh-frozen tissue sections and formalin-fixed paraffin-embedded tissue washed with ammonium formate before matrix application showed an increase in sensitivity for gangliosides, including GM1 molecules without any delocalization.[23]

Novel developments in ganglioside imaging by MS

Multimodal imaging of gangliosides by MALDI and IHC can be performed on the same tissue sections. For example, GM3 and GM2 were overexpressed in

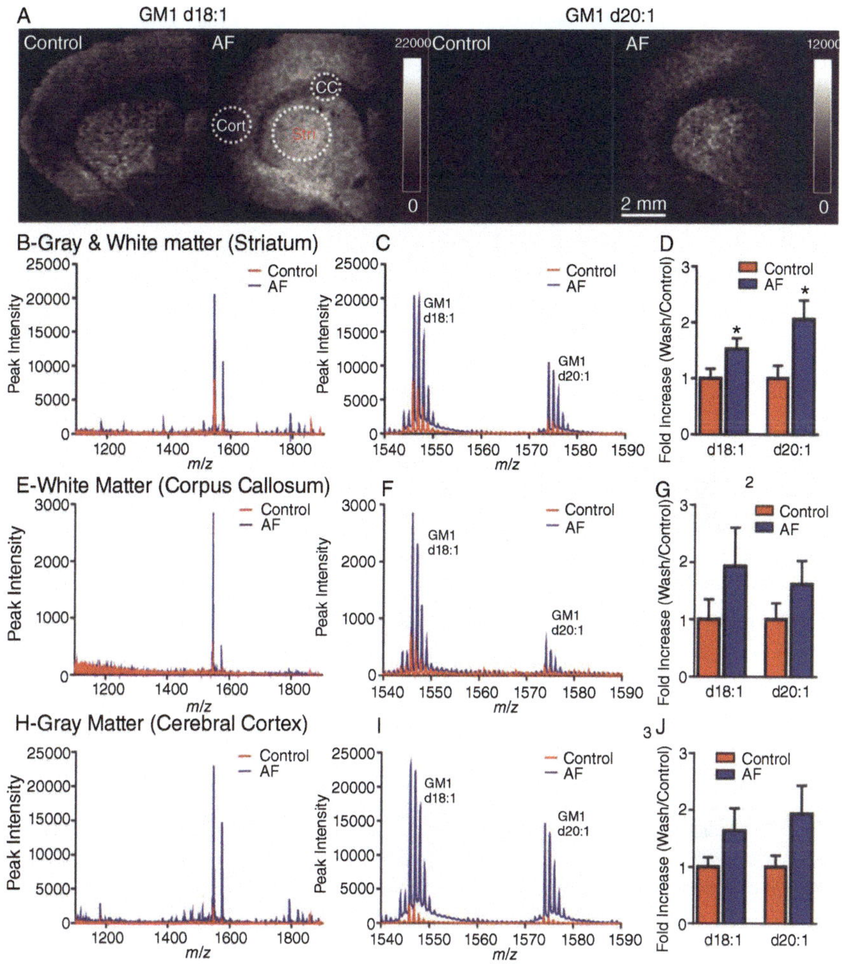

FIG. 19.4 The effect of ammonium formate (AF) wash on of gangliosides MALDI signal in the white and gray matter of a formalin-fixed rat brain tissue. A brain tissue sections were coated with a thin layer of 1,5-diaminonapthalene (DAN) matrix using sublimation. Reprinted with permission from Harris A, Roseborough A, Mor R, Yeung KKC, Whitehead SN. Ganglioside detection from formalin-fixed human brain tissue utilizing MALDI imaging mass spectrometry. *J Am Soc Mass Spectrom.* 2020;31:479–487. Copyright 2020 American Chemical Society.[23]

Hunter's disease mouse model imaged by MALDI was confirmed by anti-GM3 IHC on the same sections.[24] Besides MALDI, other ionization techniques such as ambient desorption electrospray ionization (DESI) have shown the capability to image multiple ganglioside species without any sample preparation.[25] DESI with ion mobility separation enabled detection and imaging doubly charged cardiolipins and gangliosides in biological tissues.[26] Gas cluster ion beam secondary ion mass spectrometry can be used to image intact gangliosides and cardiolipin in the brain section with better than 10-μm pixel size.[27] The low-abundance gangliosides and cardiolipin were imaged by suppressing the signals from the abundant phospholipids, such as phosphatidylcholine and phosphatidylethanolamine, using enzymatic and chemical treatments in normal brain and cortical impact model of traumatic brain injury.

Nanosecondary ion mass spectrometry has been utilized to image the ganglioside GM1 after orthogonal isotopic labeling of every lipid bilayer component and monofluorination of GM1. The ultrahigh spatial resolution image showed evidence for the colocalization of cholesterol and GM1 in lipid bilayers and indicated the presence of three compositionally distinct phases.[28]

Outlook

Uncovering the spatial distribution of ganglioside species is critical for investigating the biochemical mechanism in the central nervous system, neurodegenerative diseases, inherited lysosomal storage metabolic diseases, and brain injury. MALDI dominates the gangliosides' imaging by the mass spectrometer. Some novel MALDI matrices have been developed to improve the detection of not only gangliosides but also other larger molecular weight lipids, such as cardiolipins and (poly-)phosphoinositides. Development in sample preparation, such as buffer washing of tissue, is also helping to improve the ganglioside imaging. In addition, there is research and development in imaging gangliosides in higher spatial resolution and other accessible sources such as DESI and AP-MALDI. With mature workflow and accessible imaging systems, the broader adoption of gangliosides imaging by mass spectrometer will be beneficial to study neurological disease and condition.

References

1. Van Echten G, Sandhoff K. Ganglioside metabolism. Enzymology, topology, and regulation. *J Biol Chem*. 1993;268:5341 5341.
2. Mauri L, Sonnino S, Prinetti A. *Chemical and Physicochemical Properties of Gangliosides: Gangliosides*, Springer; 2018:1–17.
3. Walkley S U. Secondary accumulation of gangliosides in lysosomal storage disorders. *Semin Cell Dev Biol*. 2004;15:433–444.
4. Hirano-Sakamaki W, Sugiyama E, Hayasaka T, Ravid R, Setou M, Taki T. Alzheimer's disease is associated with disordered localization of ganglioside GM1 molecular species in the human dentate gyrus. *FEBS Lett*. 2015;589:3611–3616.
5. Sugiura Y, Shimma S, Konishi Y, Yamada M K, Setou M. Imaging mass spectrometry technology and application on ganglioside study, visualization of age-dependent accumulation of c20-ganglioside molecular species in the mouse hippocampus. *PLOS One*. 2008;3:e3232.
6. Wang W X, Whitehead S N. Imaging mass spectrometry allows for neuroanatomic-specific detection of gangliosides in the healthy and diseased brain. *Analyst*. 2020;145:2473–2481.
7. Tobias F, Pathmasiri K C, Cologna S M. Mass spectrometry imaging reveals ganglioside and ceramide localization patterns during cerebellar degeneration in the Npc1−/− mouse model. *Anal Bioanal Chem*. 2019;411:5659–5668.

8. Sugiyama E, Setou M. Visualization of brain gangliosides using MALDI imaging mass spectrometry. In: Sonnino S, Prinetti A, eds. *Gangliosides: Methods and Protocols*. New York, NY: Springer New York; 2018:223–229.
9. Egge H, Peter-Katalinic J, Karas M, Stahl B. The use of fast atom bombardment and laser desorption mass spectrometry in the analysis of complex carbohydrates. *Pure Appl Chem*. 1991;63:491–498.
10. Harvey D J. Matrix-assisted laser desorption/ionization mass spectrometry of sphingo- and glycosphingo-lipids. *J Mass Spectrom*. 1995;30:1311–1324.
11. Sugiyama E, Hara A, Uemura K-i, Taketomi T. Application of matrix-assisted laser desorption ionization time-of-flight mass spectrometry with delayed ion extraction to ganglioside analyses. *Glycobiology*. 1997;7:719–724.
12. Zarei M, Bindila L, Souady J, et al. A sialylation study of mouse brain gangliosides by MALDI a-TOF and o-TOF mass spectrometry. *J Mass Spectrom*. 2008;43:716–725.
13. Nagahori N, Abe M, Nishimura S - I. Structural and functional glycosphingolipidomics by glycoblotting with an aminooxy-functionalized gold nanoparticle. *Biochemistry*. 2009;48:583–594.
14. Colsch B, Woods A S. Localization and imaging of sialylated glycosphingolipids in brain tissue sections by MALDI mass spectrometry. *Glycobiology*. 2010;20:661–667.
15. Li N, Wang P, Liu X, et al. Developing IR-780 as a novel matrix for enhanced MALDI MS imaging of endogenous high-molecular-weightlipids in brain tissues. *Anal Chem*. 2019;91:15873–15882.
16. Whitehead S N, Chan K H N, Gangaraju S, Slinn J, Li J, Hou S T. Imaging mass spectrometry detection of gangliosides species in the mouse brain following transient focal cerebral ischemia and long-term recovery. *PLOS One*. 2011;6:e20808.
17. Caughlin S, Hepburn J D, Park D H, et al. Increased expression of simple ganglioside species GM2 and GM3 detected by MALDI imaging mass spectrometry in a combined rat model of Aβ toxicity and stroke. *PLOS One*. 2015;10:e0130364.
18. Goto-Inoue N, Hayasaka T, Zaima N, et al. The detection of glycosphingolipids in brain tissue sections by imaging mass spectrometry using gold nanoparticles. *J Am Soc Mass Spectrom*. 2010;21:1940–1943.
19. Colsch B, Jackson S N, Dutta S, Woods A S. Molecular microscopy of brain gangliosides: illustrating their distribution in hippocampal cell layers. *ACS Chem Neurosci*. 2011;2:213–222.
20. Ogrinc Potočnik N, Porta T, Becker M, Heeren R M, Ellis S R. Use of advantageous, volatile matrices enabled by next-generation high-speed matrix-assisted laser desorption/ionization time-of-flight imaging employing a scanning laser beam. *Rapid Commun Mass Spectrom*. 2015;29:2195–2203.
21. Jackson S N, Muller L, Roux A, et al. AP-MALDI mass spectrometry imaging of gangliosides using 2,6-dihydroxyacetophenone. *J Am Soc Mass Spectrom*. 2018;29:1463–1472.
22. Angel P M, Spraggins J M, Baldwin H S, Caprioli R. Enhanced sensitivity for high spatial resolution lipid analysis by negative ion mode matrix assisted laser desorption ionization imaging mass spectrometry. *Anal Chem*. 2012;84:1557–1564.
23. Harris A, Roseborough A, Mor R, Yeung K K C, Whitehead S N. Ganglioside detection from formalin-fixed human brain tissue utilizing maldi imaging mass spectrometry. *J Am Soc Mass Spectrom*. 2020;31:479–487.
24. Dufresne M, Guneysu D, Patterson N H, et al. Multimodal detection of GM2 and GM3 lipid species in the brain of mucopolysaccharidosis type II mouse by serial imaging mass spectrometry and immunohistochemistry. *Anal Bioanal Chem*. 2017;409:1425–1433.

25. Škrášková K, Claude E, Jones E A, Towers M, Ellis S R, Heeren R M A. Enhanced capabilities for imaging gangliosides in murine brain with matrix-assisted laser desorption/ionization and desorption electrospray ionization mass spectrometry coupled to ion mobility separation. *Methods*. 2016;104:69–78.
26. Feider C L, Elizondo N, Eberlin L S. Ambient ionization and FAIMS mass spectrometry for enhanced imaging of multiply charged molecular ions in biological tissues. *Anal Chem*. 2016;88:11533–11541.
27. Tian H, Sparvero L J, Amoscato A A, et al. Gas cluster ion beam time-of-flight secondary ion mass spectrometry high-resolution imaging of cardiolipin speciation in the brain: identification of molecular losses after traumatic injury. *Anal Chem*. 2017;89:4611–4619.
28. Lozano M M, Liu Z, Sunnick E, Janshoff A, Kumar K, Boxer S G. Colocalization of the ganglioside GM1 and cholesterol detected by secondary ion mass spectrometry. *J Am Chem Soc*. 2013;135:5620–5630.

Index

Page numbers followed by "*f*" and "*t*" indicate, figures and tables respectively.

S

T

U

V

X

Kolibri 360 GmbH
Ferdinand-Jühlke-Straße 7
99095 Erfurt, Deutschland
produktsicherheit@kolibri360.de